SPACECRAFT THERMAL CONTROL, DESIGN, AND OPERATION

Edited by
Howard E. Collicott
The Boeing Company
Seattle, Washington

Paul E. Bauer
McDonnell Douglas Astronautics Company
St. Louis, Missouri

Volume 86
PROGRESS IN
ASTRONAUTICS AND AERONAUTICS

Martin Summerfield, Series Editor-in-Chief
Princeton Combustion Research Laboratories, Inc.
Princeton, New Jersey

Technical papers selected from the AIAA 20th Aerospace Sciences Meeting, January 1982, and the 3rd AIAA/ASME Joint Thermophysics, Fluids, Plasma, and Heat Transfer Conference, June 1982, and subsequently revised for this volume.

Published by the American Institute of Aeronautics and Astronautics, Inc.
1633 Broadway, New York, NY 10019

American Institute of Aeronautics and Astronautics, Inc.
New York, New York

Library of Congress Cataloging in Publication Data
Main entry under title:

Spacecraft thermal control.

(Progress in astronautics and aeronautics; v. 86)
Papers presented at the thermophysics sessions of the AIAA 20th Aerospace Sciences Meeting in Orlando, Fla., in Jan. 1982, and the 3rd Joint AIAA/ASME Thermophysics, Fluids, Plasma, and Heat Transfer Conference in St. Louis, Mo., in June 1982.
Includes index.
1. Space vehicles—Design and construction—Addresses, essays, lectures. 2. Heat engineering—Addresses, essays, lectures. I. Collicott, Howard E. II. Bauer, Paul E. III. American Institute of Aeronautics and Astronautics. IV. AIAA Aerospace Sciences Meeting (20th : 1982 : Orlando, Fla.) V. AIAA/ASME Joint Fluids, Plasma, Thermophysics, and Heat Transfer Conference (1982 : St. Louis, Mo.) VI. Series.
TL507.P75 vol. 86 [TL900] 629.1s [629.47′4] 83-15474
ISBN 0.915928-75-2

Table of Contents

Preface xi

Editorial Committee xvi

List of Series Volumes xvii

Chapter I. Vehicle Thermal Control 1

Military Spacecraft Thermal Management: The Evolving Requirements and Challenges 3
E.T. Mahefkey, *U.S. Air Force Wright Aeronautical Laboratories, Wright-Patterson Air Force Base, Ohio*

Shuttle Orbiter Thermal Control Postflight Evaluation 17
A.A. Decrisantis and J.R. Nason, *United Technologies Corporation, Windsor Locks, Conn.*

The Orbital Test Satellite Thermal Experience after 3½ Years in Orbit 46
J.P. Bouchez and D.H. Howle, *ESA-ESTEC, Noordwijk, the Netherlands*

Thermal Design and Experiment Thermal Integration of the Long Duration Exposure Facility 71
R.F. Greene Jr., *NASA Langley Research Center, Hampton, Va.*

Satellite Thermal Design and Analyses for Expendable and Shuttle Launch Environments 89
H. Factor and E.A. Stipandic, *General Electric Space Division, Valley Forge, Pa.*

Chapter II. Subsystem and Components 107

Thermo-Mechanical Design and Analysis System for the Hughes 76-in. Parabolic Antenna Reflector 109
S.C. Clark, *Therma-Tech, Inc., Pasadena, Calif.*, and G.E. Allen, *Hughes Aircraft Company, El Segundo, Calif.*

Thermodynamic Optimization of a Cryogenic Storage System for Minimum Boiloff 130
G.R. Cunnington, *Martin Marietta Denver Aerospace, Denver, Colo.*

Molecular Absorption Crygenic Cooler for Liquid Hydrogen Propulsion Systems 147
G.A. Klein and J.A. Jones, *Jet Propulsion Laboratory, California Institute of Technology, Pasadena, Calif.*

Chapter III. Heat Pipes 163

Design, Fabrication, and Test of Liquid-Metal Heat-Pipe Sandwich Panels 165
A. Basiulis, *Hughes Aircraft Company, Torrance, Calif.*, and C.J. Camarda, *NASA Langley Research Center, Hampton, Va.*

Osmotic Pumped Heat Pipes for Large Space Platforms 182
H.J. Tanzer and G.L. Fleischman, *Hughes Aircraft Company, Torrance, Calif.*, and D.D. Stalmach, *Vought Corporation, Dallas, Texas*

Development of a Double-Wall Artery High-Capacity Heat Pipe 202
R. Ponnappan, *Universal Energy Systems, Inc., Dayton, Ohio*, and E.T. Mahefkey, *Wright-Patterson Air Force Base, Dayton, Ohio*

The Marangoni Effect in Axially Grooved Variable-Conductance Heat Pipes 222
R.L. Kosson and W. Harwell, *Grumman Aerospace Corporation, Bethpage, NY*

Chapter IV. Material Properties and Interfaces 239

Some Thermophysical Properties of Paraffin Wax as a Thermal Storage Medium 241
A. Haji-Sheikh, J. Eftekhar, and D.Y.S. Lou, *The University of Texas at Arlington, Arlington, Texas*

Ultraviolet and Electron Irradiation of DC-704 Siloxane Oil on Zinc Orthotitanate Paint 254
D.L. Mossman, M.K. Barsh, and S.A. Greenberg, *Aerojet Electrosystems Company, Azusa, Calif.*

The Thermal Contact Conductance of Dissimilar Metals 266
D.L. Padgett, *Bell Telephone Laboratories, Whippany, N.J.*, and L.S. Fletcher, *Texas A&M University, College Station, Texas*

Chapter V. Finite-Element Analysis Techniques 279

Finite-Element Analysis of Structures with Reusable Surface Insulation 281
E.A. Thorton and K.K. Tamma, *Old Dominion University, Norfolk, Va.*

Control-Volume-Based Finite-Element Formulation of the Heat Conduction Equation 305
G.E. Schneider and M. Zedan, *University of Waterloo, Waterloo, Canada*

Finite-Element Analysis of Planar Conductive and Radiative Transfer with Flux Boundary 328
R. Fernandes and J. Francis, *University of Oklahoma, Norman, Okla.*

Author Index for Volume 86 345

Table of Contents for Companion Volume 85

Chapter I. Space Shuttle Aeroheating 1

Aerothermodynamic Entry Environment of the Space Shuttle Orbiter 3
D.B. Lee, *NASA Johnson Space Center, Houston, Texas*, and M.H. Harthun, *Rockwell International, Downey, Calif.*

Approximate Method of Predicting Heating on the Windward Side of Space Shuttle Orbiter and Comparisons with Flight Data 21
H.H. Hamilton II, *NASA Langley Research Center, Hampton, Va.*

Viscous Shock-Layer Predictions for Hypersonic Laminar or Turbulent Flows in Chemical Equilibrium over the Windward Surface of a Shuttle-Like Vehicle 54
R.R. Thareja, K.Y. Szema, and C.H. Lewis, *Virginia Polytechnic Institute and State University, Blacksburg, Va.*

Influence of Radiant Energy Exchange on the Determination of Convective Heat-Transfer Rates on Orbiter Leeside Surfaces during Entry 78
D.A. Throckmorton, *NASA Langley Research Center, Hampton, Va.*

Catalytic Efficiency of the Space Shuttle Heat Shield 97
J.V. Rakich and D.A. Stewart, *NASA Ames Research Center, Moffett Field, Calif.*, and M.J. Lanfranco, *Informatics, Inc., Palo Alto, Calif.*

Catalytic Recombination/Space Shuttle Heating 123
C.D. Scott and S.M. Derry, *NASA Lyndon B. Johnson Space Center, Houston, Texas*

Viscous Shock-Layer Heating Analysis for the Shuttle Windward Symmetry Plane with Surface Finite Catalytic Recombination Rates 149
J.L. Shinn, J.N. Moss, and A.L. Simmonds, *NASA Langley Research Center, Hampton, Va.*

Chapter II. Space Shuttle Thermal Protection 181

Shuttle Orbiter Reusable Surface Insulation Thermal Performance 183
R.L. Dotts, *NASA Johnson Space Center, Houston, Texas*, H.H. Battley, *Rockwell International, Houston, Texas*, and J.T. Hughes and W.E. Neuenschwander, *Rockwell International, Downey, Calif.*

Space Shuttle Orbiter Leading Edge Thermal Performance 206
D.M. Curry, *NASA Johnson Space Center, Houston, Texas*, J.A Cunningham, *Rockwell International, Downey, Calif.*, and J.R. Frahm, *Rockwell International, Houston, Texas*

Comparison of Orbiter STS-2 Development Flight Instrumentation Data with Thermal Math Model Predictions 234
I. Norman, W.C. Rochelle, B.S. Kimbrough, C.A. Ritrivi, and P.C. Ting, *Rockwell International, Houston, Texas*, and R.L. Dotts, *NASA Johnson Space Center, Houston, Texas*

Flight Measurements of Tile Gap Heating on the Space Shuttle 255
W.C. Pitts, *NASA Ames Research Center, Moffett Field, Calif.*, and M.S. Murbach, *Informatics General Corp., Moffett Field, Calif.*

Tile-Gap Flow in the Shuttle Orbiter Thermal Protection System 271
D.L. Dwoyer, P.A. Newman, F.C. Thames, and N.D. Melson, *NASA Langley Research Center, Hampton, Va.*

Assessment of Alternate Thermal Protection Systems for the Space Shuttle Orbiter 300
H.N. Kelly and G.L. Webb, *NASA Langley Research Center, Hampton, Va.*

Initial Shuttle External Tank Aerothermal Flight Evaluation 325
S.C. Praharaj and C.D. Engel, *REMTECH, Inc., Huntsville, Ala.*

Chapter III. Solar Starprobe Heat Shielding 349

STARPROBE Thermal Shield System Design Concepts 351
J.M. Millard, C.R. Maag, and R.N. Miyake, *Jet Propulsion Laboratory, California Institute of Technology, Pasadena, Calif.*

Preliminary Design of the Thermal Protection System for Solar Probe 385
R.B. Dirling Jr., W.C. Loomis, and C.N. Heightland, *Science Applications, Inc., Irvine, Calif.*

Chapter IV. Jupiter Galileo Probe Entry Heating and Protection 417

Galileo Probe Forebody Flowfield Predictions 419
J.N. Moss and A.L. Simmonds, *NASA Langley Research Center, Hampton, Va.*

An Evaluation of Turbulence Models for Massively Blown Surfaces 446
R.N. Gupta, J.N. Moss, E.V. Zoby, and A.L. Simmonds, *NASA Langley Research Center, Hampton, Va.*

Trajectories of Solid Particles Spalled from a Carbonaceous Heat Shield 472
C.B. Davies, *Informatics General Corp., Palo Alto, Calif.*, and C. Park, *NASA Ames Research Center, Moffett Field, Calif.*

Spallation of the Galileo Probe Heat Shield 496
J.H. Lundell, *NASA Ames Research Center, Moffett Field, Calif.*

Response of Galileo Aft Cover Components to Laser Radiation 518
J.W. Metzger, *General Electric Company, Philadelphia, Pa.*

Preface

Increased emphasis on thermal control requirements for spacecraft and the experiments they support has sustained a continued growth in the importance of thermophysics in the future exploration of space. This unique field combines science and engineering disciplines into a technology that supports the development and application of the principles of heat and mass transfer, thermodynamics, fluid mechanics, mathematics, and materials into a singular system to meet the environmental control needs of all types of vehicles in the space exploration program. Spacecraft mission complexity continues to increase, bringing new requirements for thermal systems. Space probes require increased mission life and operation in planetary environments. Satellites and other devices carried aboard transport vehicles such as the STS Orbiter face several distinctly different environments, handling interfaces, and operational modes during a single mission. Traditional thermal control techniques will naturally continue to have an important role in future designs, but continued development of new techniques, materials, and computational capability must proceed. High heat flux, complex electronics, laser interactions, fiber optics, and increased cryogenic needs are but a few of the factors to be considered in future designs.

This volume brings together a collection of papers concerned with the development and application of thermophysics to spacecraft, specifically vehicle thermal control systems, component design, material properties and interfaces, and numerical techniques. The papers were selected from those presented at the thermophysics sessions of the AIAA 20th Aerospace Sciences Meeting in Orlando, Fla., in January 1982, and from the 3rd AIAA/ASME Joint Thermophysics, Fluids, Plasma, and Heat Transfer Conference in St. Louis, Mo., in June 1982. The technical scope of this volume is limited to spacecraft thermal control and is presented in five chapters: vehicle thermal control, component design and development, heat pipe design and application, material properties and interface characteristics, and the application of the finite element method in computational requirements. A companion volume covers the entry heating and thermal protection system aspects of these same two AIAA conferences.

Chapter I addresses the overall vehicle thermal control system, leading with a review of near- and far-term requirements for thermal control of military spacecraft by *Mahefky*. While these missions require considerations of hostile external forces and survivability, the major portion of the design criteria—such as power, heat flux, and weight parameters—are valid for nonmilitary applications. Of the four remaining papers in this chapter, the first two deal with vehicle flight results and the last two with design experience. While much has been written about the STS thermal protection system, little data have been available on other thermal control aspects. *Decrisantis* and *Nason* provide design data and flight results for the Shuttle Orbiter primary life support system. In the third paper, nearly four years of orbit operation for the Orbital Test Satellite (OTS) are described by *Bouchez* and *Howle*. Computed vehicle temperatures and flight data are discussed and then combined to postulate on thermal coating characteristics.

The last two papers in Chap. I describe thermal control system designs for other missions. The design and integration of many experiments into the long-duration exposure facility (LDEF) is described by *Greene*. LDEF is a reusable spacecraft that is transported into space by the STS Orbiter. Design criteria are identified for both the LDEF and the experiments it will carry, and a plan for integrating experiment thermal analysis models into the vehicle model is described. The fifth paper, by *Factor* and *Stipandic*, assesses the problems of designing a thermal control system for a satellite that must operate with two distinctly different launch systems. One operational mode is aboard a conventional booster and the second is being carried as an experiment on the Shuttle Orbiter. Design considerations and expected performance data are discussed for a three-axis stabilized satellite launched with a Titan 34D or STS Orbiter configuration.

Chapter II brings the level of interest from the vehicle down to the component or subsystem. The first paper by *Clark* and *Allen* provides a detailed insight into the thermal design and analysis of a 76 in. parabolic antenna reflector and the method of using thermal data as input to a structural analysis. Thermal-vacuum and other laboratory testing was used in correlating the analytical procedure. The remaining two papers in the chapter are concerned with cryogenic component and subsystem operation. *Cunnington* utilizes the second law of thermodynamics to develop a concept for optimizing a cryogenic system for minimum boiloff. The results, applied to the design of a typical liquid-hydrogen storage system, demonstrate the impact on the structure and insulation and show a significant reduction in boiloff rate. *Klein* and *Jones* describe a long-life molecular absorption cryogenic cooler and provide an analytical evaluation of cooler performance. The system is designed to use low-temperature waste heat to provide refrigeration capability suitable for maintaining a liquid-hydrogen storage system.

The design, performance, and application of heat pipes form the subject of Chap. III. *Basiulis* and *Camarda* describe the design and fabrication of a liquid-metal heat pipe panel. The panels are formed with wickable honeycomb core and wickable internal faces. Three different designs were fabricated and tested in a radiant facility. Next, *Tanzer*, *Fleischman*, and *Stalmach* define the concept of a thermal bus, or uniform thermal control source, for a future space platform. The predicted performance of an osmotically pumped heat pipe indicates the feasibility of such a system. The development of a double-wall artery heat pipe is described by *Ponnappan* and *Mahefkey* in the last paper. A 1.2 m pipe was designed and fabricated from copper. With water as the fluid, the pipe demonstrated a capacity of 1600 W·m and evaporator fluxes of 16 W/cm^2. Even higher capacities are proposed. *Kosson* and *Harwell* present a derivation of the momentum equation applied to the Marangoni effect in axially grooved variable conductance heat pipes. The results were used to modify an existing computer program and then used to calculate heat pipe performance. Test data were used to demonstrate the validity of the approach.

Three papers on materials and material interface effects are presented in Chap. IV: the first two discuss thermal storage and thermal coatings, and the last the thermal contact conductance effects related to dissimilar metals. *Haji-Sheikh*, *Eftekhar*, and *Lou* conducted an experimental evaluation of a specially selected paraffin wax for use as a thermal energy storage material. Measured properties included thermal conductivity, thermal expansion, viscosity, and latent heat. The effect of contamination on the response of zinc orthotitanate to ultraviolet and electron irradiation is described by *Mossman*, *Barsh*, and *Greenberg*. The degradation of the optical properties of this important spacecraft thermal control coating is shown for both clean and contaminated conditions. Thermal contact conductance data are presented first for dissimilar metals by *Padgett* and *Fletcher*. The data represent the output of a new experimental facility developed specifically for dissimilar metal testing, including the ability to make measurements with reversed heat flow direction and high contact pressure.

Chapter V includes three papers on the application of finite element analysis techniques to thermophysical problems. *Thornton* and *Tamma* develop new two-dimensional elements to model conduction in the thermal protection system and its support structure. The results presented show that these elements will predict structural temperatures accurately with a significant reduction in the model size and computer time required. Another approach to conduction analysis is given by *Schneider* and *Zedan* in a finite element formulation based on a control-volume approach. The formulation solution is compared to those based on a Galerkin method. The new method is shown to provide accurate results at a reduced computational cost. In the final paper, *Fernandes* and *Francis* apply the finite element method to a transient analysis with combined conduction and radiation in the unique case with heat flux boundary conditions. Results are given for representative analysis using both temperature and radiative flux profiles.

Many people have helped in the generation of this volume, and I wish to acknowledge their efforts and to thank them for their support. The Editorial Committee (listed separately) was responsible for the immense task of securing and evaluating the technical reviews of all papers presented at the two meetings and then again for consideration for this volume.

A debt of gratitude and sincere appreciation must go to Mrs. Camille S. Koorey, AIAA Managing Editor for the Progress Series, for her continuing support and endless patience during the preparation of the volume. Assistance came from other AIAA staff members, including: Mrs. Norma J. Brennan, Mrs. Joann Trongone, Miss Ruth F. Byrans, and certainly Mr. Walter J. Brunke in the preparation and conduct of the meetings.

The members of the AIAA Thermophysics Technical Committee for the past two years contributed to the successful outcome of the meetings that formed the basis for this volume. During that time, Mr. Jesse F. Keville and Dr. Thomas E. Horton served as Chairmen of the Committee and contributed from their experiences. Dr. David P. DeWitt organized the thermophysics sessions of the AIAA 20th Aerospace Sciences Meeting and provided inputs for the technical review of those papers. Mr. Paul E. Bauer served as General Chairman of the 3rd Joint AIAA/ASME Thermophysics and Heat Transfer Conference and provided welcome assistance throughout the preparation for the conference and the volume. Dr. James N. Moss also contributed valuable suggestions based on his experience as technical chairman of a previous thermophysics conference. Dr. Martin Summerfield continued in his service as Editor-in-Chief of the AIAA Progress Series.

The support of Boeing Computer Services Company, specifically Dr. Samuel L. Jacoby, Dr. James L. Tocher, and Dr. Ervin D. Herness is gratefully acknowledged. Mrs. Luella A. Collicott provided the extensive secretarial and administrative support for the thermophysics conference and the preparation of the volume.

Finally, I would thank the authors who have contributed to the meetings and subsequently to this volume. Their participation, cooperation, and effort in the submittal of their papers is appreciated.

Howard E. Collicott
April 1983

Editorial Committee for Volume 86

Progress in Astronautics and Aeronautics

Volume Titles	Volume Editors
*1. **Solid Propellant Rocket Research.** 1960	Martin Summerfield *Princeton University*
*2. **Liquid Rockets and Propellants.** 1960	Loren E. Bollinger *The Ohio State University* Martin Goldsmith *The Rand Corporation* Alexis W. Lemmon Jr. *Battelle Memorial Institute*
*3. **Energy Conversion for Space Power.** 1961	Nathan W. Snyder *Institute for Defense Analyses*
*4. **Space Power Systems.** 1961	Nathan W. Snyder *Institute for Defense Analyses*
*5. **Electrostatic Propulsion.** 1961	David B. Langmuir *Space Technology Laboratories, Inc.* Ernst Stuhlinger *NASA George C. Marshall Space Flight Center* J. M. Sellen Jr. *Space Technology Laboratories, Inc.*
*6. **Detonation and Two-Phase Flow.** 1962	S. S. Penner *California Institute of Technology* F. A. Williams *Harvard University*
*7. **Hypersonic Flow Research.** 1962	Frederick R. Riddell *AVCO Corporation*
*8. **Guidance and Control.** 1962	Robert E. Roberson *Consultant* James S. Farrior *Lockheed Missiles and Space Company*
*9. **Electric Propulsion Development.** 1963	Ernst Stuhlinger *NASA George C. Marshall Space Flight Center*

*Now out of print.

***10. Technology of Lunar Exploration.** 1963	Clifford I. Cummings and Harold R. Lawrence *Jet Propulsion Laboratory*
***11. Power Systems for Space Flight.** 1963	Morris A. Zipkin and Russell N. Edwards *General Electric Company*
***12. Ionization in High-Temperature Gases.** 1963	Kurt E. Shuler, Editor *National Bureau of Standards* John B. Fenn, Associate Editor *Princeton University*
***13. Guidance and Control—II.** 1964	Robert C. Langford *General Precision Inc.* Charles J. Mundo *Institute of Naval Studies*
***14. Celestial Mechanics and Astrodynamics.** 1964	Victor G. Szebehely *Yale University Observatory*
***15. Heterogeneous Combustion.** 1964	Hans G. Wolfhard *Institute for Defense Analyses* Irvin Glassman *Princeton University* Leon Green Jr. *Air Force Systems Command*
***16. Space Power Systems Engineering.** 1966	George C. Szego *Institute for Defense Analyses* J. Edward Taylor *TRW Inc.*
***17. Methods in Astrodynamics and Celestial Mechanics.** 1966	Raynor L. Duncombe *U. S. Naval Observatory* Victor G. Szebehely *Yale University Observatory*
***18. Thermophysics and Temperature Control of Spacecraft and Entry Vehicles.** 1966	Gerhard B. Heller *NASA George C. Marshall Space Flight Center*
***19. Communication Satellite Systems Technology.** 1966	Richard B. Marsten *Radio Corporation of America*

*20. **Thermophysics of Spacecraft and Planetary Bodies: Radiation Properties of Solids and the Electromagnetic Radiation Environment in Space.** 1967 — Gerhard B. Heller, *NASA George C. Marshall Space Flight Center*

*21. **Thermal Design Principles of Spacecraft and Entry Bodies.** 1969 — Jerry T. Bevans, *TRW Systems*

*22. **Stratospheric Circulation.** 1969 — Willis L. Webb, *Atmospheric Sciences Laboratory, White Sands, and University of Texas at El Paso*

*23. **Thermophysics: Applications to Thermal Design of Spacecraft.** 1970 — Jerry T. Bevans, *TRW Systems*

24. **Heat Transfer and Spacecraft Thermal Control.** 1971 — John W. Lucas, *Jet Propulsion Laboratory*

25. **Communications Satellites for the 70's: Technology.** 1971 — Nathaniel E. Feldman, *The Rand Corporation*; Charles M. Kelly, *The Aerospace Corporation*

26. **Communications Satellites for the 70's: Systems.** 1971 — Nathaniel E. Feldman, *The Rand Corporation*; Charles M. Kelly, *The Aerospace Corporation*

27. **Thermospheric Circulation.** 1972 — Willis L. Webb, *Atmospheric Sciences Laboratory, White Sands, and University of Texas at El Paso*

28. **Thermal Characteristics of the Moon.** 1972 — John W. Lucas, *Jet Propulsion Laboratory*

29. **Fundamentals of Spacecraft Thermal Design.** 1972 — John W. Lucas, *Jet Propulsion Laboratory*

30. **Solar Activity Observations and Predictions.** 1972 — Patrick S. McIntosh and Murray Dryer, *Environmental Research Laboratories, National Oceanic and Atmospheric Administration*

31. Thermal Control and Radiation. 1973	Chang-Lin Tien *University of California, Berkeley*
32. Communications Satellite Systems. 1974	P. L. Bargellini *COMSAT Laboratories*
33. Communications Satellite Technology. 1974	P. L. Bargellini *COMSAT Laboratories*
34. Instrumentation for Airbreathing Propulsion. 1974	Allen E. Fuhs *Naval Postgraduate School* Marshall Kingery *Arnold Engineering Development Center*
35. Thermophysics and Spacecraft Thermal Control. 1974	Robert G. Hering *University of Iowa*
36. Thermal Pollution Analysis. 1975	Joseph A. Schetz *Virginia Polytechnic Institute*
37. Aeroacoustics: Jet and Combustion Noise; Duct Acoustics. 1975	Henry T. Nagamatsu, Editor *General Electric Research and Development Center* Jack V. O'Keefe, Associate Editor *The Boeing Company* Ira R. Schwartz, Associate Editor *NASA Ames Research Center*
38. Aeroacoustics: Fan, STOL, and Boundary Layer Noise; Sonic Boom; Aeroacoustics Instrumentation. 1975	Henry T. Nagamatsu, Editor *General Electric Research and Development Center* Jack V. O'Keefe, Associate Editor *The Boeing Company* Ira R. Schwartz, Associate Editor *NASA Ames Research Center*
39. Heat Transfer with Thermal Control Applications. 1975	M. Michael Yovanovich *University of Waterloo*
40. Aerodynamics of Base Combustion. 1976	S. N. B. Murthy, Editor *Purdue University* J. R. Osborn, Associate Editor *Purdue University* A. W. Barrows and J. R. Ward, Associate Editors *Ballistics Research Laboratories*

41. Communication Satellite Developments: Systems. 1976
Gilbert E. LaVean
Defense Communications Engineering Center
William G. Schmidt
CML Satellite Corporation

42. Communication Satellite Developments: Technology. 1976
William G. Schmidt
CML Satellite Corporation
Gilbert E. LaVean
Defense Communications Engineering Center

43. Aeroacoustics: Jet Noise, Combustion and Core Engine Noise. 1976
Ira R. Schwartz, Editor
NASA Ames Research Center
Henry T. Nagamatsu, Associate Editor
General Electric Research and Development Center
Warren C. Strahle, Associate Editor
Georgia Institute of Technology

44. Aeroacoustics: Fan Noise and Control; Duct Acoustics; Rotor Noise. 1976
Ira R. Schwartz, Editor
NASA Ames Research Center
Henry T. Nagamatsu, Associate Editor
General Electric Research and Development Center
Warren C. Strahle, Associate Editor
Georgia Institute of Technology

45. Aeroacoustics: STOL Noise; Airframe and Airfoil Noise. 1976
Ira R. Schwartz, Editor
NASA Ames Research Center
Henry T. Nagamatsu, Associate Editor
General Electric Research and Development Center
Warren C. Strahle, Associate Editor
Georgia Institute of Technology

46. Aeroacoustics: Acoustic Wave Propagation; Aircraft Noise Prediction; Aeroacoustic Instrumentation. 1976	Ira R. Schwartz, Editor *NASA Ames Research Center* Henry T. Nagamatsu, Associate Editor *General Electric Research and Development Center* Warren C. Strahle, Associate Editor *Georgia Institute of Technology*
47. Spacecraft Charging by Magnetospheric Plasmas. 1976	Alan Rosen *TRW Inc.*
48. Scientific Investigations on the Skylab Satellite. 1976	Marion I. Kent and Ernst Stuhlinger *NASA George C. Marshall Space Flight Center* Shi-Tsan Wu *The University of Alabama*
49. Radiative Transfer and Thermal Control. 1976	Allie M. Smith *ARO Inc.*
50. Exploration of the Outer Solar System. 1977	Eugene W. Greenstadt *TRW Inc.* Murray Dryer *National Oceanic and Atmospheric Administration* Devrie S. Intriligator *University of Southern California*
51. Rarefied Gas Dynamics, Parts I and II (two volumes). 1977	J. Leith Potter *ARO Inc.*
52. Materials Sciences in Space with Application to Space Processing. 1977	Leo Steg *General Electric Company*

53. Experimental Diagnostics in Gas Phase Combustion Systems. 1977	Ben T. Zinn, Editor *Georgia Institute of Technology* Craig T. Bowman, Associate Editor *Stanford University* Daniel L. Hartley, Associate Editor *Sandia Laboratories* Edward W. Price, Associate Editor *Georgia Institute of Technology* James G. Skifstad, Associate Editor *Purdue University*
54. Satellite Communications: Future Systems. 1977	David Jarett *TRW Inc.*
55. Satellite Communications: Advanced Technologies. 1977	David Jarett *TRW Inc.*
56. Thermophysics of Spacecraft and Outer Planet Entry Probes. 1977	Allie M. Smith *ARO Inc.*
57. Space-Based Manufacturing from Nonterrestrial Materials. 1977	Gerard K. O'Neill, Editor *Princeton University* Brian O'Leary, Assistant Editor *Princeton University*
58. Turbulent Combustion. 1978	Lawrence A. Kennedy *State University of New York at Buffalo*
59. Aerodynamic Heating and Thermal Protection Systems. 1978	Leroy S. Fletcher *University of Virginia*
60. Heat Transfer and Thermal Control Systems. 1978	Leroy S. Fletcher *University of Virginia*
61. Radiation Energy Conversion in Space. 1978	Kenneth W. Billman *NASA Ames Research Center*
62. Alternative Hydrocarbon Fuels: Combustion and Chemical Kinetics. 1978	Craig T. Bowman *Stanford University* Jorgen Birkeland *Department of Energy*

63. Experimental Diagnostics in Combustion of Solids. 1978	Thomas L. Boggs *Naval Weapons Center* Ben T. Zinn *Georgia Institute of Technology*
64. Outer Planet Entry Heating and Thermal Protection. 1979	Raymond Viskanta *Purdue University*
65. Thermophysics and Thermal Control. 1979	Raymond Viskanta *Purdue University*
66. Interior Ballistics of Guns. 1979	Herman Krier *University of Illinois at Urbana-Champaign* Martin Summerfield *New York University*
67. Remote Sensing of Earth from Space: Role of "Smart Sensors." 1979	Roger A. Breckenridge *NASA Langley Research Center*
68. Injection and Mixing in Turbulent Flow. 1980	Joseph A. Schetz *Virginia Polytechnic Institute and State University*
69. Entry Heating and Thermal Protection. 1980	Walter B. Olstad *NASA Headquarters*
70. Heat Transfer, Thermal Control, and Heat Pipes. 1980	Walter B. Olstad *NASA Headquarters*
71. Space Systems and Their Interactions with Earth's Space Environment. 1980	Henry B. Garrett and Charles P. Pike *Hanscom Air Force Base*
72. Viscous Flow Drag Reduction. 1980	Gary R. Hough *Vought Advanced Technology Center*
73. Combustion Experiments in a Zero-Gravity Laboratory. 1981	Thomas H. Cochran *NASA Lewis Research Center*
74. Rarefied Gas Dynamics, Parts I and II (two volumes). 1981	Sam S. Fisher *University of Virginia at Charlottesville*

75. Gasdynamics of Detonations and Explosions. 1981	J. R. Bowen *University of Wisconsin at Madison* N. Manson *Université de Poitiers* A. K. Oppenheim *University of California at Berkeley* R. I. Soloukhin *Institute of Heat and Mass Transfer, BSSR Academy of Sciences*
76. Combustion in Reactive Systems. 1981	J. R. Bowen *University of Wisconsin at Madison* N. Manson *Université de Poitiers* A. K. Oppenheim *University of California at Berkeley* R. I. Soloukhin *Institute of Heat and Mass Transfer, BSSR Academy of Sciences*
77. Aerothermodynamics and Planetary Entry. 1981	A. L. Crosbie *University of Missouri-Rolla*
78. Heat Transfer and Thermal Control. 1981	A. L. Crosbie *University of Missouri-Rolla*
79. Electric Propulsion and Its Applications to Space Missions. 1981	Robert C. Finke *NASA Lewis Research Center*
80. Aero-Optical Phenomena. 1982	Keith G. Gilbert and Leonard J. Otten *Air Force Weapons Laboratory*
81. Transonic Aerodynamics. 1982	David Nixon *Nielsen Engineering & Research, Inc.*
82. Thermophysics of Atmospheric Entry. 1982	T. E. Horton *The University of Mississippi*

83. Spacecraft Radiative Transfer and Temperature Control. 1982	T. E. Horton *The University of Mississippi*
84. Liquid-Metal Flows and Magnetohydrodynamics. 1983	H. Branover *Ben-Gurion University of the Negev* P. S. Lykoudis *Purdue University* A. Yakhot *Ben-Gurion University of the Negev*
85. Entry Vehicle Heating and Thermal Protection Systems: Space Shuttle, Solar Starprobe, Jupiter Galileo Probe. 1983	Paul E. Bauer *McDonnell Douglas Astronautics Company* Howard E. Collicott *The Boeing Company*
86. Spacecraft Thermal Control, Design, and Operation. 1983	Howard E. Collicott *The Boeing Company* Paul E. Bauer *McDonnell Douglas Astronautics Company*

(Other volumes are planned.)

Chapter I. Vehicle Thermal Control

Military Spacecraft Thermal Management: The Evolving Requirements and Challenges

E. T. Mahefkey*
U.S. Air Force Wright Aeronautical Laboratories, Wright-Patterson Air Force Base, Ohio

Abstract

A summary is presented of near-term and future requirements for thermal management of military spacecraft, as well as perceived technology limitations and future needs. Military space missions for the post-1985 period are summarized. Evolutionary communications, data processing, and other systems will give rise to 10kW heat rejection missions, while revolutionary missions will require average heat rejection power of the order of 100kW, with pulse power heat rejection rates several orders of magnitude greater. The technology challenges include scale-up to higher power, management of pulsed power and high heat loads, increased survivability, and STS motivated weight and volume improvements.

Nomenclature

COP_{hp} = spacecraft heat pump coefficient of performance

COP_r = spacecraft refrigerator(s) coefficient of performance

f_e = fraction of bus power to spacecraft electrical loads

f_r = fraction of bus power to spacecraft refrigerator loads

P_s = source power

Presented as Paper 82-0827 at the 3rd AIAA/ASME Joint Thermophysics, Fluids, Plasma & Heat Transfer Conference, St. Louis, Mo., June 7-11, 1982. This paper is declared a work of the U.S. Government and therefore is in the public domain.

*Aerospace Engineer, Thermal Systems Group, Energy Conversion Branch.

P_{cs} = conditioned source power
P_b = spacecraft bus power
$\dot{Q}_{gen}$ = power train waste heat generation rate
$\dot{Q}_s$ = source waste heat generation rate
$\dot{Q}_{spc}$ = source power conditioning waste heat generation rate
$\dot{Q}_{bpc}$ = spacecraft bus power conditioning waste heat generation rate
$\dot{Q}_{ss}$ = spacecraft subsystems waste heat generation rate
η_{ps} = source conversion efficiency
η_{spc} = source power conditioning efficiency
η_{bpc} = spacecraft bus power conditioning efficiency
η_{ss} = spacecraft subsystems efficiency

Introduction

Military space missions under consideration for the post-1985 period introduce significant new technology needs for the spacecraft energy system (SCES) and the attendant spacecraft thermal management system (SCTMS). Thermal control, heat storage, heat transport, and heat rejection for evolutionary and revolutionary types of missions represent difficult problems when coupled to the envisioned payload weight and volume limitations and operational environment scenarios for that period.

The reliance on military space systems as part of the force structure is expected to expand in the next several decades. Traditional military space missions such as surveillance; communications, command, and control (C^3); navigation; and global weather monitoring will continue to evolve in sophistication and capabilities. Use of in-space data processing and data relay techniques will be expanded, resulting in decreasing spacecraft dependence on ground control. These evolutionary missions point toward increased spacecraft electrical power from the now prevalent 1-2kW class. This power trend will be accompanied by a similar evolutionary trend in onboard waste heat generation with extended design lifetimes in a potentially more hostile operational environment.[1,2]

Revolutionary military missions under study point toward average spacecraft power levels in the 100kW regime with pulsed peak power requirements greater than an order of magnitude more.[2]

In a continuing planning study, the Air Force has identified 34 military spacecraft system concepts which are evaluated in terms of technical feasibility and technology status.[1] The envisioned systems perform a variety of communication, surveillance, and support missions; employ a variety of functional payload subsystems; and utilize orbits ranging from 600n.mi. to five times geosynchronous.

Survivability to hostile force radiation threats will be a necessity for some missions. The composite effect of the envisioned mission scenarios greatly expands the set of design requirements for the SCTMS.

This paper primarily addresses thermal management of spacecraft power systems and related components in the nominal 0-100°C operating range. Other efforts are underway in defining thermal management issues associated with launch and orbital insertion and passive and active cryogenic cooling. Reference 3 describes some of these additional considerations.

Thermal Management Design Drivers

The technology needs for these future missions can be characterized in terms of several design issues concerning the nature of the spacecraft payload equipment, the geometrical scale of the future systems, the mission power profile and corresponding instantaneous and average heat generation rates, and the envisioned waste heat fluxes. Some of these design issues can be readily illustrated by

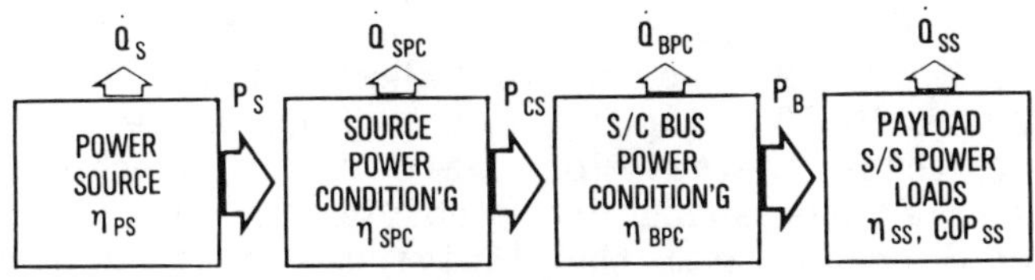

η_{PS} - SOURCE CONVERSION EFFICIENCY

η_{SPC} - SOURCE POWER COND. EFFICIENCY

η_{BPC} - S/C BUS POWER COND. EFFICIENCY

η_{SS} - S/C SUBSYSTEMS EFFICIENCY (IES)

COP_{SS} - S/C REFRIGERATOR(S) COEFF. OF PERF.

P_S - SOURCE POWER

P_{CS} - CONDITIONED SOURCE POWER

P_B - S/C BUS POWER

$\dot{Q}_S$ - SOURCE WASTE HEAT GENERATION RATE

$\dot{Q}_{SPC}$ - SOURCE POWER COND. WASTE HEAT GEN. RATE

$\dot{Q}_{SS}$ - S/C SUBSYSTEMS WASTE HEAT GEN. RATE

$\dot{Q}_{BPC}$ - S/C BUS POWER COND. WASTE HEAT GEN. RATE

Fig. 1 Spacecraft power/converter/load train.

considering a simplified spacecraft power train, as shown schematically in Fig. 1. This power train is composed of a power source, such as a solar array/battery or reactor/direct energy conversion system, with source and spacecraft bus power conditioning, and, finally, electrical or refrigeration spacecraft functional subsystem loads. The use of refrigerators for cryogenic cooling of sensors and overcoming heat leaks into large cryogenic reactant tanks for long duration missions represents a major new thermal design problem.

Quantitatively, the ratio of spacecraft waste heat rate to instantaneous bus power is given by

$$\frac{\dot{Q}_{gen}}{P_b} = \frac{1-\eta_{ps}}{\eta_{ps}\eta_{bpc}\eta_{spc}} + \frac{1-\eta_{spc}}{\eta_{bpc}\eta_{spc}} + \frac{1-\eta_{bpc}}{\eta_{bpc}} + f_e(1-\eta_{ss}) + f_r(1+COP_r) \tag{1}$$

This equation is useful in illustrating the general nature of the spacecraft generated thermal management problem. The electrical efficiencies are generally a function of the instantaneous power and component temperature, while the refrigerator coefficient of performance (COP) is a function of both the refrigerator heat load and refrigerator temperature.

Representative "future" subsystem efficiencies include power conditioning in the range of 90-98%, energy storage at 40% (high rate discharge) - 80% (normal rate discharge), payload subsystems at 5-98%, and cryogenic refrigerators with COP's in the 0.01 - 0.02 range. Representative average bus powers for near-term future missions are in the range of 10-100kW; "burst" and/or peak powers may be 10-1000 times these average values.

Since these thermal loads are distributed spatially within the various spacecraft subsystems, the waste heat must be collected from the individual heat sources and ultimately rejected. As with current spacecraft the heat generation rate for a given component and subsystem can approach or exceed the input power for future systems; however, this problem is further magnified due to the increased average and peak bus powers envisioned, the increased number of spacecraft subsystems, and the larger transport distances characteristic of these future larger spacecraft. Additionally, volume and weight limit considerations force the design of compact payloads, and survivability dictates compact radiation-shielded electronics. Hence, volumetric heat generation rates in payload systems will increase, necessitating higher heat flux handling capabilities. The scale, peak power, and heat flux issues are illustrated further in the following subsections.

Scale and Capacity

Figure 2 shows the trend in SCTMS system specific weight and weight fraction as functions of heat rejection capability. Three time domains are shown. The current time domain is characterized by relatively low requirements of 1-5kW. In the near-future domain, that is, within the next 5-10 years, average heat rejection requirements from the payload will grow to the 5-100kW regime. The future domain, 10-20 years, could require average payload heat rejection rates much in excess of 100kW.

Current low-power near-Earth spacecraft thermal design practice commonly employs semipassive-cold bias design practice. Heat transport from the heat loads to the radiator is accomplished by direct component to radiator coupling and by conduction through structural components. Low-temperature protection of temperature sensitive components is accomplished by electrical

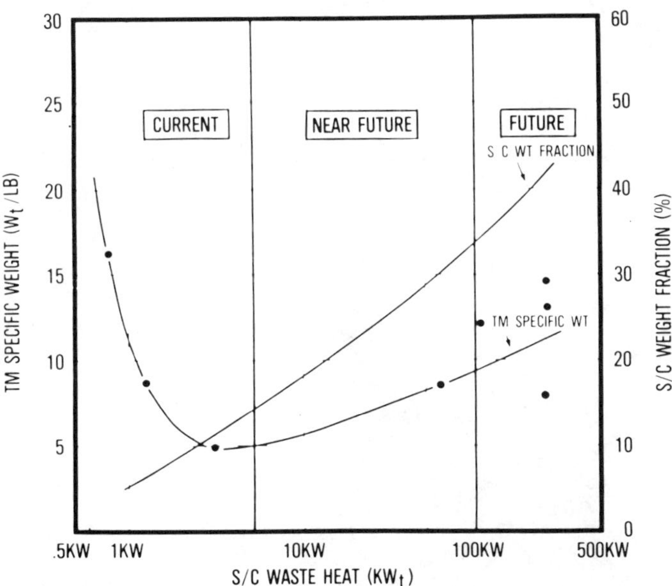

Fig. 2 Spacecraft thermal management specific weight and weight fraction trends.

heaters. The specific weight of the thermal management system for these low-power satellites is relatively high due to their predominately passive nature. As powers approach the multi-kilowatt regime, passive TM techniques become inadequate and active techniques (heat pipes, pumped loops) must be employed, and attainable specific weights decrease markedly.

Capacity (W·m) limits of current heat pipes also limit performance of large, passive heat transport and radiator components by necessitating hybrid pumped-loop/heat pipe concepts, as exemplified in the Air Force/Lockheed SIRE spacecraft design. Design interface complexities, radiator structural scale-up, increased transport distances, and parasitic power penalties sharply reduce specific weights achievable with current technology in the 10kW regime.

Trends in specific weight at higher powers are based largely on recent NASA 25-250kW_t space platform studies. The corresponding payload fraction shown assumes a 60,000 lbm low Earth orbit (LEO) spacecraft payload weight limitation. While a single Shuttle limit is not an implicit constraint of the large platform studies, the figure points out the nonlinear growth in required payload fraction in the 100kW regime. This consideration is even more serious for the midaltitude (600-5600n.mi.) and geosynchronous orbits of interest for most future military spacecraft.

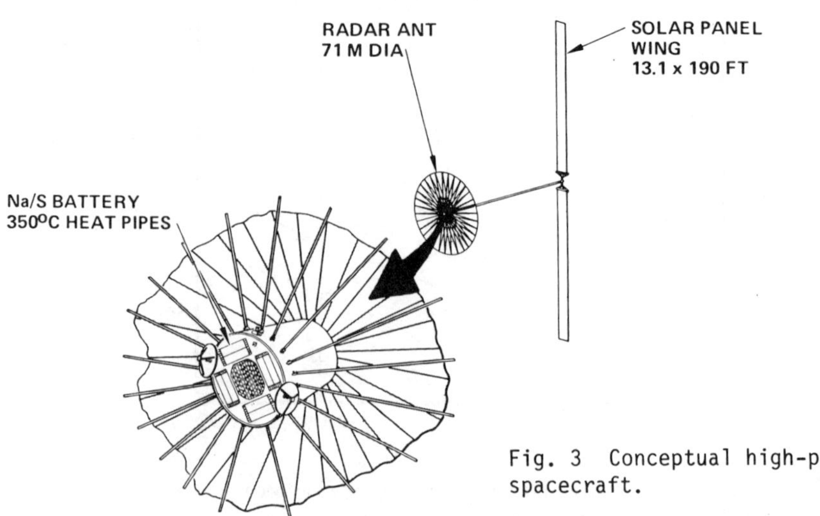

Fig. 3 Conceptual high-power spacecraft.

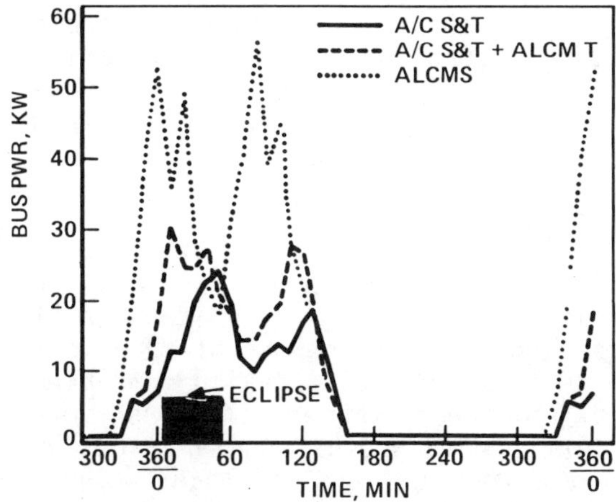

Fig. 4 Representative pulsed power profiles.

For ultrahigh powers, such as those considered in the solar power satellite concept, it is estimated that the technology limit for pumped-loop/heat pipe radiators is 5 kg/m^2 (or approximately 30-60 w_t/lbm for a 50^{0}C radiator). Generally, the radiator represents at least half of the entire SCTMS weight.

Pulsed Power Thermal Loads

Figure 3 shows a conceptual design of a large solar-powered spacecraft.[4] Note the scale of the payload antenna (71 m diam) and solar array (500 m^2). Figure 4 illustrates the spacecraft payload bus power profile over an orbital period.

The solid, dashed, and dotted lines represent different mission operations. In one scenario, the bus power peak exceeds 50kW and one of the peaks occurs in the orbital eclipse period where the power is provided solely by the energy storage system. During the noneclipse peaks, the battery energy storage system and solar array both provide power to the load. A similar peak power generation scheme could be used with a nuclear reactor sized to provide baseload and battery charging power. The reactor could also be designed without energy storage, which would require dissipation of both nonpeak electrical energy as heat and average waste heat proportional to peak power.

Other mission concepts introduce similar transient thermal management/peak power dissipation design problems. While current spacecraft can handle peak-to-average heat generation profiles in the range of 5-10/1 by a combination of sensible heat storage, louvers, or variable conductance radiators, these techniques cannot necessarily be employed for higher power or higher peak-to-average missions due to the component temperature limits and weight penalties.

High Heat Flux Large Area Loads

Currently communications transponders employing traveling wave tube amplifiers (TWTA's) with waste heat thermal footprints exceeding 1 W/cm^2 represent the most common high flux thermal control design problem encountered in present low-power spacecraft. As more combined frequency functions, data processing, and antijam features are built into communications payloads, input heat fluxes and total baseplate heat inputs could grow by a factor of 10. High-rate primary and secondary batteries will require baseplate heat removal fluxes in the 0.1-5 W/cm^2 range over baseplate areas of 1-10 m^2.[5] Heat dissipation from ultracompact very-large-scale integral circuits may exceed 20 W/cm^2.[6] Large externally generated heat flux loads on spacecraft radiators represent difficult design problem which can be solved by either hardening, tolerance, or avoidance. Current low thermal resistance, high flux input limits are typified by the heat flux surface area limits of heat pipe evaporators. Figure 5 shows the conceptual arrangement of a hardened concentrating photovoltaic array under study by TRW for the Air Force.[7] The design concentration ratio

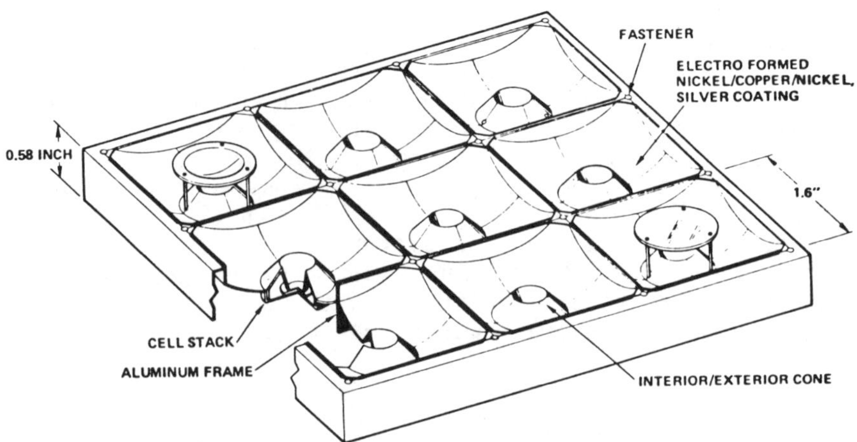

Fig. 5 Hardened solar cell concentrator configuration.

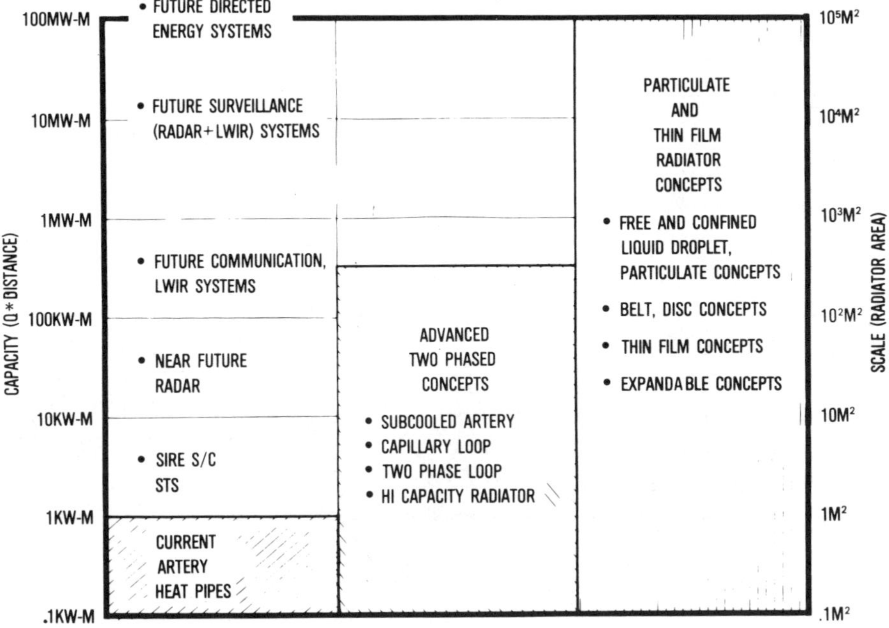

Fig. 6 Transport capacity and radiator scale requirements and technology.

is low (CR ~ 100), based on cell efficiency/operating temperature trades with conductive, radiative back surface cooling of the device. Major conversion efficiency and survivability gains could be made if active, lightweight substrate heat rejection techniques were available.

Future Technology Directions

In the previous section four future problem categories were identified concerning large distributed loads generated by the spacecraft subsystems, specific weight, pulsed power, and high heat fluxes. Some apparent current technology limits and potential future technology capabilities are summarized here.

Capacity and Scale Technology Limits and Potential

NASA is currently studying advanced concepts for a LEO science platform thermal utility bus in the 25-250 kW_t range. Conventional heat pipe capacity limits for these long multiple load-to-radiator configurations are

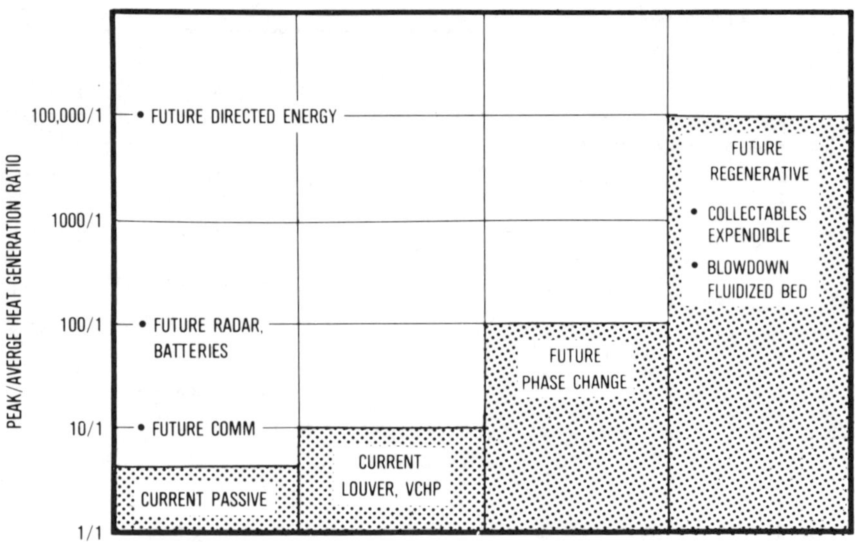

Fig. 7 Pulsed power heat rejection requirements and technology.

being extended by exploration of subcooled artery heat pipes and two-phase pumped-loop concepts. This work may extend the capability for low thermal resistance, distributed load transport capacities from the current 1-2 kW·m range to a 20-200 kW·m regime.[8-10] How these or other advanced transport concepts can be applied to distributed high-power, high-flux, and/or high-peak loads remains to be explored. Figure 6 presents an estimate of the current, near-future, and future SCTMS capacity and scale requirements. Current payload waste heat transport and radiator needs can be met with present artery heat pipe technology. Required capacity and scale increases for higher power near-future missions will require evolutionary changes of the transport technology, such as those suggested in the middle block of the figure. Extremely high capacity and scale future requirements will necessitate revolutionary developments in lightweight radiator concepts such as those listed in the right-hand block of Fig. 6.

The current 10-15 W_t/lb specific weight and 10-20% parasite power penalties associated with current hybrid pumped-loop/heat pipe radiator concepts represent a serious weight problem for near-future high-power spacecraft, half of which is associated with the radiator. Evolutionary engineering improvements may extend radiator

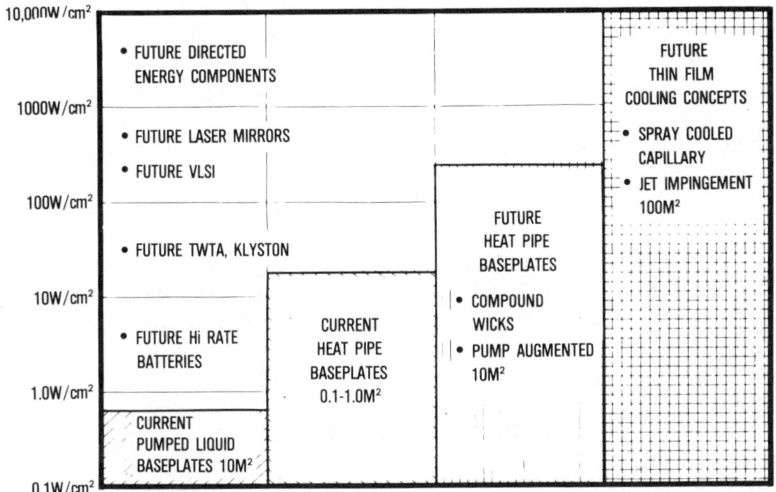

Fig. 8 High heat flux/large area requirements and technology.

specific weights to 60 W_t/lb (50°C). Novel approaches such as the moving belt, disk, and particulate radiator concepts, which were explored in the early 1960's, offer some promise and some fundamental compatibility problems.[11-12]

Use of heat pumps to increase radiator temperatures may yield overall system weight, deployed area, and packaged volume advantages if the heat pump power and weight penalties are offset by radiator weight reductions or compactness advantages.

For spacecraft employing expendables for cooling, the expendable's enthalpy of vaporization ($\sim$ 1.0 MJ_t/lbm) and density (1000 kg/m^3) impose fundamental limits on spacecraft weight, volume, and operating time (without re-supply) for both peak and average power dissipation applications. This limit prompts interest in "collectible" expendable concepts, which offer inherent contamination advantages over particle and droplet radiator concepts presently being explored.

Pulsed Power Heat Rejection

Current spacecraft louver and variable conductance heat pipe (VCHP) radiators are sized to reject peak powers. They provide turndown radiator areas ratios of

5-10/1 to accommodate off-peak heat rejection at near-constant thermal resistance and hence near-constant load temperature. For high-power, weight-constrained applications it appears the inverse approach must be utilized, that is, the radiator must be sized for the average heat rejection and the peak energy fluxes stored and dissipated during off-peak generation periods.[13,14] Figure 7 presents an estimate of the mission requirements for pulsed power heat rejection, and the technologies applicable to these requirements are listed. It appears the very high peak/average requirements can only be met with compact regenerative techniques sized to meet peak power input and average power heat rejection. A technique to collect and reuse the working fluid would greatly aid the system weight penalty.

High Heat Flux/Large Areas

Currently demonstrated heat pipe evaporators (15-30 W/cm^2 flux, 0.1 m^2 area) represent the approximate limit of low thermal resistance baseplate technology. Capillary augmented heat-transfer surfaces have demonstrated capabilities in excess of 300 W/cm^2 for small areas (1-2 cm^2).[15] Spray-augmented capillary surfaces and jet impingement techniques may raise this limit to 3000 W/cm^2 with no area limit and minimal pump power penalties.[6,16]

The envisioned requirements for future high heat flux/large area thermal loads and technologies applicable to these needs are shown in Fig. 8.

Summary and Conclusions

It is evident that major spacecraft thermal management system technology gains are necessary to enable conduct of the envisioned near- and far-term military mission scenario. It is also evident that major technology gains are attainable.

The trend in military spacecraft thermal management requirements is clear. The technology will be driven by performance needs (weight, deployed area, packaged volume), extended life, autonomy, and, in some cases, survivability requirements. These performance requirements will be superimposed on requirements to manage higher average and peak waste heat generation rates and/or higher volumetric and surface heat generation rates.

The potential for improvement exists and such

improvement must be realized if the future candidate missions are to become realities.

Acknowledgments

Many of the ideas and concepts described in this paper have been contributed freely by referenced and unreferenced members of the spacecraft thermal design and thermodynamics community. These contributions are sincerely appreciated.

References

[1]Hord, R. and Cohen, M., "Military Space Systems Technology Model," Vols. I-IV, Air Force Space Division, Rept. SD-TR 82-01, Jan. 1982 (available from DDC).

[2]Cohen, M., Fornoles, E., and Mahefkey, T., "Requirements and Technology Trends for Future Military Space Power Systems," Proceedings of 16th Intersociety Energy Conversion Engineering Conference, ASME, New York, 1981, pp. 2122-2125.

[3]Johnson, A., "Thermal Control," Proceedings of AIAA/NSIA Space Systems and Technology Workshop, Alburquerque, N. Mex., Sept. 1982.

[4]Marcoux, L. S., "High Energy Density Rechargeable Battery for Satellite Applications," Quarterly Progress Report under U.S. Air Force Contract F33615-79-C-2044, July 1981.

[5]Mahefkey, T. and Barthelemy, R., "Heat Pipe Applications for Future Air Force Spacecraft," AIAA Paper 80-1463, 15th Thermophyics Conference, Snowmass, Colo., July 1980.

[6]Tuckerman, D. and Pease, R., "High-Performance Heat Sinking for VLSI," IEEE Electron Device Letter, Vol. 2, May 1981.

[7]Patterson, R. E., "Cassegrainian Concentrator Solar Array Exploratory Development Module", Proceedings of 17th Intersociety Energy Conversion Engineering Conference, IEEE, New York 1982.

[8]Alario, J. et al., "The Monogroove High Performance Heat Pipe," AIAA Paper 81-1156, 16th Thermophysics Conference, Palo Alto, Calif., June 1981.

[9]Bacilius, A., Hughes Aircraft Co., private communication, Feb. 1982.

[10]Fowle, A., "A Pumped, Two-Phase Flow Heat Transport System for Orbiting Instrument Payload," AIAA Paper 81-1075, 16th Thermophysics Conference, Palo Alto, Calif., June 1981.

[11]Speeds, J. A. and Fuller, L. E., "Feasibility Investigation of a Moving Belt Radiator," Rocketdyne Corp., ASD TDR 63-551, Aug. 1963, prepared under Contract AF33(657)-8127.

[12]Knapp, K., "Lightweight Moving Radiators for Heat Rejection in Space," AIAA Paper 81-1076, 16th Thermophysics Conference, Palo Alto, Calif., June 1981.

[13]Mahefkey, T., "Thermal Management for Future Air Force High Power Spacecraft," Proceedings of 14th Intersociety Energy Conversion Engineering Conference, ACS, Washington, D.C., 1979, pp. 1374-1376.

[14]Haslett, R., "Thermal Management of Large Pulsed Power Systems," Proceedings of AFOSR Special Conference on Prime Power for High Energy Space Systems, Bolling AFB, Washington, D.C., Feb. 1982, pp 941-972.

[15]Jacobson, D., Rabe, D., and Barthelemy, R., "Heat Pipe Mirrors for High Power Lasers," AIAA Paper 78-391, Third International Heat Pipe Conference, Palo Alto, Calif., May 1978.

[16]Ernst, D. and Wren, P, Thermocore Corp. and Sundstrand Corp., private communication, Feb. 1982.

Shuttle Orbiter Thermal Control Postflight Evaluation

A. A. Decrisantis* and J. R. Mason†
United Technologies Corporation, Windsor Locks, Conn.

Abstract

Space Shuttle Orbiter primary life support is provided by the atmospheric revitalization subsystem (ARS), which controls cabin temperature, humidity, carbon dioxide level, and odor. ARS waste heat is transferred to the freon collant loop (FCL) by way of an intermediate water loop. In addition, the water loop removes heat from avionics equipment, cabin window frames, and access hatch. Heat rejection from the FCL is accomplished by boiling water in the flash evaporator subsystem (FES), boiling ammonia, thermal radiation to space or to ground equipment by way of the ground support equipment heat exchanger. Cooling is provided to the hydraulic and auxiliary power unit (APU) lubricating oil systems on demand by the water spray boiler (WSB), which utilizes boiling water as a heat sink. This report reviews the design requirements and actual performance of these systems during the first Space Shuttle flight. In addition, a brief comparison of system performance will be made between Space Transportation System-1 and -2 (STS-2 and STS-2).

Introduction

The first Space Shuttle flight (designated STS-1) initiated a series of five development test flights which will prove the space worthiness of this vehicle. This development flight series is extraordinary

---

Presented as Paper 82-0006 at the AIAA 20th Aerospace Sciences Meeting, Orlando, Fla., Jan. 11-14, 1982.

*Analytical Engineer, Hamilton Standard Division.
+Senior Analytical Engineer, Hamilton Standard Division.

in that all flights are manned. Previous launch vehicles have been thoroughly flight tested in an unmanned configuration.

A significant part of the early flight tests is the evaluation of all onboard systems including the environmental control and heat rejection systems. This is the first manned test of this hardware under flight conditions.

STS-1 and STS-2 represent the first flight tests of some gravity sensitive components such as the condensing heat exchanger slurper, centrifugal water separators, FES and WSB. These systems have been thoroughly tested under laboratory conditions in an Earth gravity environment.

Orbiter OV-102 (Columbia) is unique in that it contains an extensive amount of test instrumentation. This information is recorded on magnetic tapes contained in the development flight instrumentation (DFI) package, which is mounted in the Shuttle payload bay. As the Shuttle passes over ground stations, data frin these instruments are made available to ground stations.

Rockwell International has supplied Hamilton Standard with flight test data for STS-1 and STS-2. These data reflect the instrumentation readouts at selected time periods during the missions when the vehicle is in contact with a ground station. From these data, a basic understanding of the performance of the environmental control and heat rejection systems was obtained.

Mission Overview

STS-1 was approximately 54 h in length, resulting in some 36 orbits of the Earth at an inclination of 40.3 deg to the equator and at an altitude of 150 nm. Columbia was launched into orbit on April 12, 1981. The mission is considered a total success since all primary objectives were accomplished.

From liftoff to postrollout, the imposed dynamic loads and thermal loads on the Shuttle Orbiter during STS-1 did not stress the design limits of the vehicle. Flight planning dictated checkout of all vehicle systems without jeopardizing safety by stressing the vehicle at maximum design conditions. In addition, this nominal profile applies to cooling requirements for the Orbiter thermal control systems including the hydraulic and APU lubricating oil systems. The combined heat load for the cabin ventilation system and thermal control water loop on STS-1 was 20% less than design point conditions.

STS-2 was also approximately 54 h in length. Following a successful launch on Nov. 12, 1981, the mission was terminated early due to the failed fuel cell. A majority of mission objectives was accomplished within the reduced timeline. Greater thermal and dynamic loads were imposed on the vehicle during STS-2.

The ARS was sized to provide a comfortable environment for four to seven crew members under normal operating conditions and up to ten crew members for rescue purposes. On STS-1 and STS-2 only two crew members were present, which significantly reduces cabin metabolic load. The reduced heat load combined with a cabin temperature sensor biased by avionics heat generation resulted in a cooler than normal cabin temperature during sleep periods. In addition, fewer crew members results in less activity and minimal usage of onboard systems reducing cabin heat generation.

Heat input to the cabin water loop from heat exchangers and cold plates located in the three cabin avionics bays was as expected. In these areas, heat generation (avionics activity) is primarily a function of mission phase and not crew members or crew activity.

In the FCL, the payload heat exchanger was performing at 15% of capacity because of the small heat load generated by the DFI package. Consequently, the flow proportioning module in each loop remained in the low heat load position, which proportions approximately 220 lb/h of Freon-21 to the payload heat exchanger and 2100 lb/h to the interchanger to meet cabin cooling requirements. The fuel cells were operating at near predicted capacity during ascent; however, orbital and re-entry electrical needs were approximately 20% less. Therefore, the heat load to the FCL was slightly less than expected. Heat load on the aft and midbody cold plates was as expected. Consequently, on STS-1 the FCL heat load was approximately 20% lower on ascent and return and 15% lower for orbital operations. Heat loads on STS-2 were 4-5% greater than STS-1 during ascent and return.

Total on-orbit heat rejection was much greater for the FES on STS-2 as a result of longer operating time. Large variations in orbital heat load extremes were encountered on STS-2.

During the the ascent phase of STS-1 and STS-2, the No. 3 WSB system experienced a transient freeze-up. It was determined that as a result of low APU heat load, the 5 lb of water preload in the WSB core overcooled the APU lubricating oil. The steam saturation pressure within the core dropped below freezing long enough to allow the lube section spray bar to freeze. However, after approximately 2 min. into the STS-1 ascent, the bar thawed as the boiler increased in temperature, allowing the boiler to return to normal operation.

On STS-2, the No. 3 WSB experienced a similar freeze-up that was more severe than the STS-1 outage. No recovery occurred after the freeze-up, necessitating premature shut-down of the associated APU. Like STS-1, each boiler was preloaded with 5 lb of water. Anomalous operation on both flights prompted the call for an investigation into the problem. Hydrazine contamination of the lube oil by the APU was considered a contributing factor.

System Description and Requirements

Four major subsystems of the Shuttle Orbiter environmental control and active cooling systems are discussed in this paper. Figure 1 depicts the arrangement of these systems within the Orbiter. The ARS conditions the cabin environment and cools electronic equipment in the cabin. It interfaces with the FCL, which cools equipment located outside of the cabin. The FES provide total FCL cooling during ascent and return and partial cooling on-orbit. The WSB subsystem cools the vehicle hydraulics systems and APUs during ascent and return.

Atmospheric Revitalization Subsystem

The ARS provides ventilation, temperature control, humidity control, carbon dioxide removal and trace gas contaminant removal within the cabin. In addition, secondary air loops provide cooling to various electronics and avionics packages. ARS waste

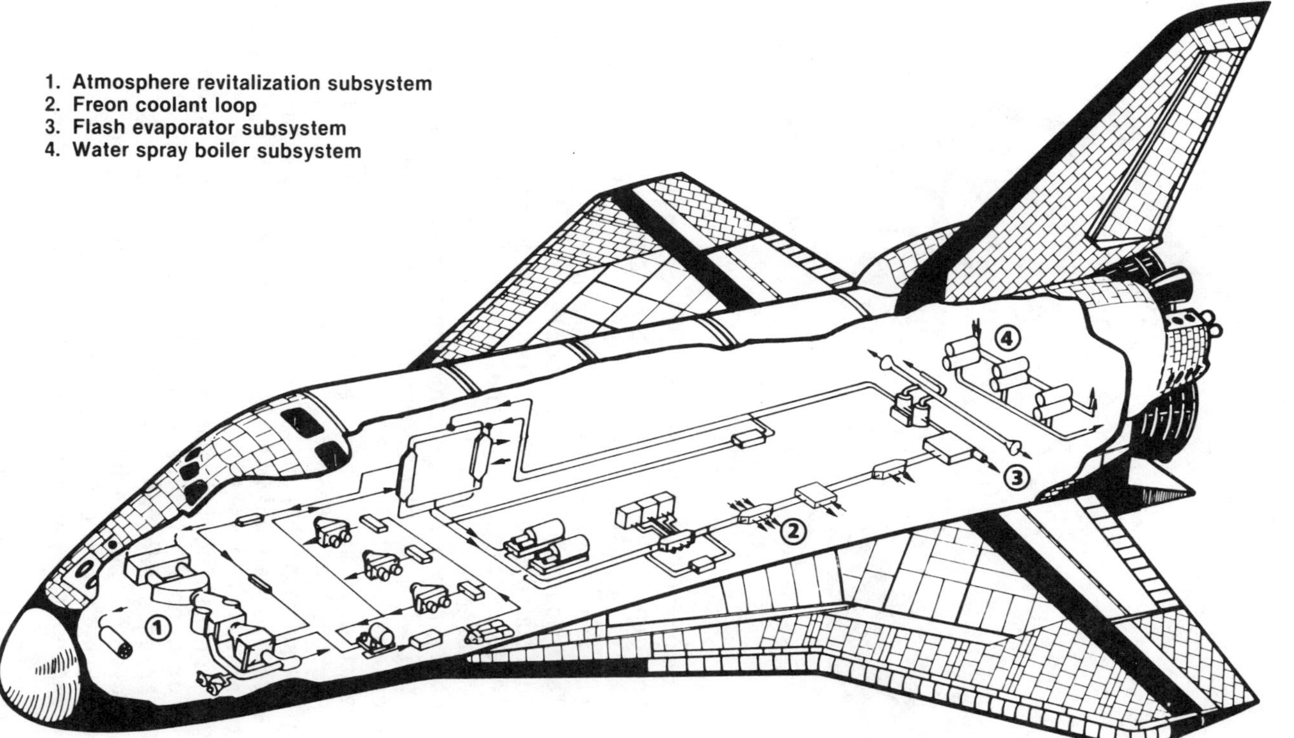

Fig. 1 Shuttle Orbiter environmental control equipment.

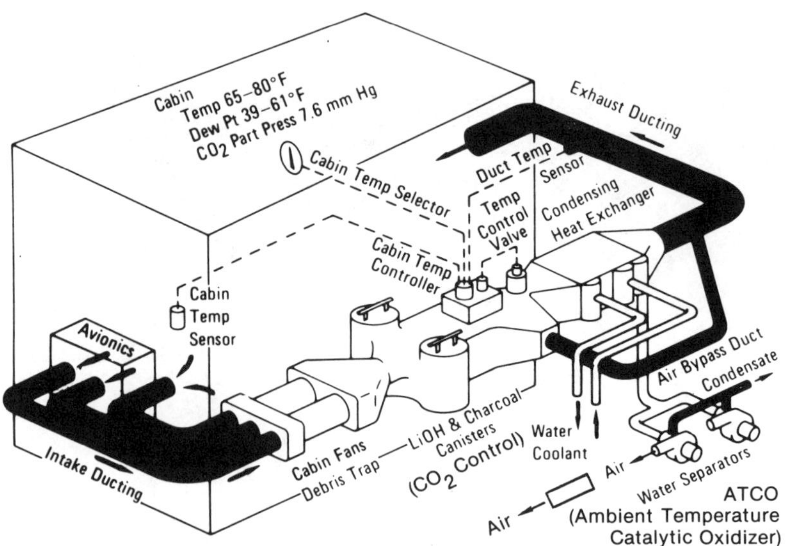

Fig. 2 ARS cabin air circuit.

heat is transported to the FCL by way of an intermediate water loop. A portion of the electronics equipment located in the cabin is air cooled by the cabin ventilation flow while the remaining equipment is cold plate mounted and water cooled. Additionally, air and liquid cooling of avionics equipment located in three separate bays and air cooling of three inertial measurement units (IMUs) are provided. The ARS also cools potable water for crew consumption and two space suit liquid cooled undergarments when worn inside the cabin.

A schematic of the cabin ventilation air loop is shown in Fig. 2. Air from the cabin is drawn through electronic equipment by one of two redundant cabin fans at a rate of approximately 320 ft^3/min. A portion of the total airflow is then directed through two cartridges containing lithium hydroxide and activated charcoal for carbon dioxide removal and odor control. Cartridges are replaced alternately every 24 h for the two-man crews on both flights in order to assure that the cabin carbon dioxide partial pressure will not exceed the maximum allowable level of 7.6 mm Hg.

Air then flows through or bypasses a condensing heat exchanger[1] that rejects heat to the water cooling loop and removes sensible heat and humidity from the cabin atmosphere. The cabin air bypass temperature control valve is modulated by an electronic controller to maintain the cabin at any selected temperature between 65° and 80°F. The controller monitors both cabin temperature and return air temperature and varies the return air temperature by modulating the condenser bypass valve which maintains cabin temperature to the selected level. With this temperature control arrangement, the cabin dew point temperature is maintained within the allowable band or 39° to 61°F.

Coolant water is supplied to the heat exchanger at approximately 45°F. The air that passes through the heat exchanger is cooled below the dew point and condensation occurs. The condensed moisture is removed from the heat exchanger at the heat exchanger air outlet face by a vacuum-cleaner-type device called a "slurper." Suction is generated by one of two redundant, electric motor driven, centrifugal fan/separators. All of the condensate and about 8 ft^3/min of ventilation airflow is processed by the fan/separator. The condensate is pumped to the waste water storage tanks, and the air is returned to the cabin.

Cooling throughout the cabin is provided by one of two redundant water coolant loops. Heat is then

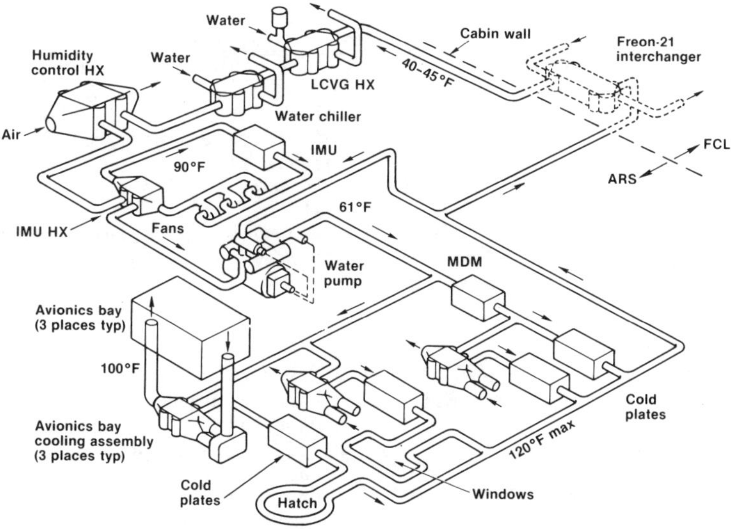

Fig. 3 ARS cooling water circuit.

transferred from the water loop to two active Freon loops in a four-fluid interface heat exchanger (called an interchanger) which is part of the FCL and located outside of the pressurized cabin. A schematic of the water loops is shown in Fig. 3. Cold water from the interchanger at about 45°F flows through two four-fluid heat exchangers before entering the condenser. One cools two space suit cooling loops to 50°F, and the other provides chilled water at 55°F for crew consumption. After leaving the condenser, the water loop cools three IMUs through an intermediate air loop. Fans draw 35 ft^3/min of cabin air through the IMU and return it to the cabin at 70°F through a heat exchanger. Air leaving the IMU is limited to 150°F.

Circulation pump packages are located downstream of the IMU heat exchanger. The packages contain redundant, canned electric motor driven, centrifugal pumps that provide 950 lb/h of water flow; accumulators for water thermal expansion; filters; and bypass valves. The bypass valves are used for water loop temperature control and to reduce water flow through the interchanger to approximately 600 lb/h during periods of low cabin heat loads on-orbit. In addition, the avionics cold plates can be maintained above the cabin dew point to prevent condensation.

Table 1 ARS performance requirements

Cabin conditions			Water chiller		
Maximum temperature	4 men	70°F	Inlet temperature	100°F	max
	7 men	77°F	Outlet temperature	55°F	max
	10 men	80°F	Potable flow rate	225	lb/h
Temperature range		65-80°F	Liquid cooling garment Hx		
Dewpoint range		39-61°F	Heat load	4000 Btu/h	
Carbon dioxide level			LCG outlet temperature	50°F	max
Nominal		5 mm Hg	LCG flow (2 loops)	500	lb/h
Range		0-7.6 mm Hg			
LiOH replacement time			**Interchanger Hx**		
Based on 2.11 lb	4 men	12 h	Water flow		
per man-day	7 men	5.5 h	Launch, entry	950	lb/h
	10 men	3.2 h	On-orbit	600	lb/h
Avionics bays			Water outlet temperature		
Cold plate temperature	120°F	max	Launch, entry		40°F
Air inlet temperature	100°F	max	On-orbit		45°F
IMU platform			Freon inlet temperature		37°F
Air inlet temperature	90°F	max	Maximum heat load	50,900 Btu/h	

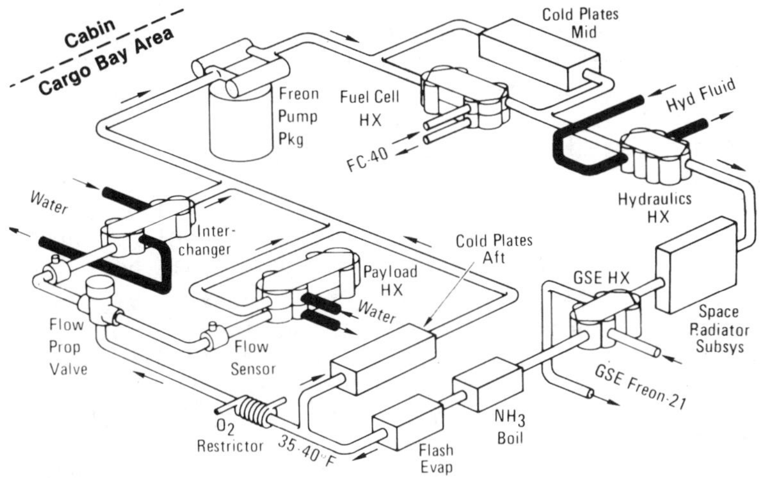

Fig. 4 Freon coolant loop.

Downstream of the pumps, the water flow splits into three branches, with each branch cooling equipment in one of three avionics bays. One of two redundant fans in each bay circulates 225 ft^3/min of air from a heat exchanger at 100°F maximum through the avionics equipment. Water also flows through cold plates and areas around cabin windows and hatches at a maximum temperature of 120°F before returning to the interchanger. Table 1 summarizes the performance requirements for the ARS cabin air and water loops.

Freon Coolant Loop

The FCL cools the ARS and other equipment located outside of the Orbiter cabin. Two identical cooling loops circulate a total of approximately 5500 lb/h of Freon-21 through a series of heat sources and heat sinks. A schematic of the FCL is shown in Fig. 4. The heat sinks maintain a maximum Freon outlet temperature of 40°F. Cold Freon is then supplied to three parallel branches. One branch contains the interchanger that cools the ARS water loops. A second branch contains a four-fluid payload heat exchanger that will cool up to two payload loops. During the STS-1 flight, only one payload loop containing development flight instrumentation was connected to the payload heat exchanger. On STS-2, the payload was increased by adding the Office of Space and Terrestrial Applications (OSTA) package, which also increased cooling requirements for the payload heat exchanger. An electrically driven flow porportioning valve splits Freon flow between the interchanger and payload heat exchanger in two proportions depending on mission phase. During ascent and entry, most of the flow is diverted to the interchanger because cabin electronics heat load is high and payload heat load is low. Flow is split more evenly on-orbit when cabin heat load is reduced and payload heat load increases. Since the development flight instrumentation package heat load was low, the flow proportioning valves were maintained in the ascent/entry position (220 lb/h each loop to the payload heat exchanger) throughout both flights. In the on-orbit position, the system is capable of cooling a Freon-21 payload loop flow of 2045 lb/h to 45°F with a maximum payload heat load of 29,000 Btu/h. The third branch cools cold plate mounted electronic equipment in the aft avionics bay.

The three branches merge and enter circulation pump packages which contain canned electric motor driven, centrifugal pumps; accumulators for Freon thermal expansion; and filters. Normally, one pump in each of the Freon loops is operating. If one of the two loops is lost, sufficient heat rejection capacity exists with single loop operation to effect a safe abort under reduced power conditions.

From the pump packages, Freon flow splits into two parallel paths. A small portion of the flow cools cold plate mounted electronic equipment in the midbody area of the Orbiter. The majority of flow cools three fuel cells which generate all of the Orbiter electrical power during flight. Three independent fuel cell FC-40 coolant loops are cooled to 140°F by the two Freon loops in a single five-fluid heat exchanger. Cooling can be provided with any one, two, or three fuel cells in operation; and with either or both Freon loops in operation. In the event of a Freon loop failure, the fuel cells would be powered down in order that temperature control could be maintained. Downstream of the fuel cells the Freon flow merges and passes through a five-fluid hydraulics heat exchanger. Three independent vehicle hydraulic loops are warmed by the two Freon loops. The hydraulic fluid must be maintained above 0°F on orbit when the hydraulic systems are not operating.

A small flow of hydraulic fluid in each of the three loops is circulated in a cyclic manner through the heat exchanger to maintain temperature. Freon then flows through a series of four heat sinks which are operated according to mission phase. Space radiators[2] located on the payload by doors provide a major portion of the heat rejection during orbital operations. Radiator panels are deployed after the Shuttle attains orbit and the payload bay doors are opened. In some vehicle attitudes, the radiator may not be able to provide total heat rejection, and supplementary heat rejection is provided by the FES.

At altitudes above 100,000 ft during ascent and entry when the radiator is not deployed, heat rejection is provided only by the FES. During the period from launch following disconnection of GSE cooling until the FES is operational, there is no active cooling. Therefore successful startup of the FES during launch is critical to mission survival. During the final stages of return, landing, and post rollout, heat rejection is provided to the FCL by an ammonia boiler[3] and by accessing subcooled Freon stored in the radiators.

During prelaunch and postrollout mission phases, the Freon loops are cooled by GSE. GSE coolant fluid is circulated through a GSE heat exchanger located in the vehicle Freon loops. Connection is through an umbilical and fly-away disconnect on the outside skin of the vehicle.

Table 2 summarizes the requirements for all FCL components.

Table 2 FCL performance requirements

	Space radiator	GSE heat exchanger	Ammonia boiler	Flash evaporator
Heat load, Btu/h	7000-75,000	0-111,640	0-120,000	625-131,000
Temperature, °F				
Inlet	49-97	35.5-120	32-135	32-133
Outlet	35.5-39.5	32-37	32-37	38.5-40.0

Freon flow per loop, lb/h	
System total	
Launch and return	2750-3000
On-orbit	2250-2500
Payload Hx	195-1110
Aft cold plates	210-275
Interchanger Hx	1360-2480

Inlet temperature, °F	
Interchanger Hx	32-40
Payload Hx	32-40
Aft cold plates	32-40
Fuel cell Hx	53-105
Midbody cold plates	53-105
Hydraulics Hx	60-133

Flash Evaporator Subsystem

The FES[4] provides heat rejection through evaporation of expendable water in one or both of two identical evaporator chambers, maintained at or below the water triple point pressure, that cools liquid Freon circulated in series through the walls of both chambers. Steam is vented overboard through a heated duct system. An electronic controller modulates the frequency of pulsing feedwater spray nozzles in the chamber in order to maintain a constant Freon outlet temperature under varying heat load conditions. A separate electronic circuit is provided to shut down the unit if performance is abnormal.

The dual evaporator concept provides a wider range of heat load while attaining the required outlet temperature of 39°F. Also, the two cores and three electronic controllers are used to satisfy system reliability requirements. Two of the controllers and two sets of solenoid spray valves on each evaporator are provided for normal operation. The first evaporator in series is called the "high load" unit since it operates during periods of high heat load during ascent and entry. The second evaporator is called the "topping" unit since it operates alone on-orbit to "top off" the radiator when on-orbit heat loads exceed radiator capacity. Both evaporators operate during ascent and entry. A schematic of the FES is shown in Fig. 5.

There are two separate overboard steam ducts, one for the high load and one for the topping evapora-

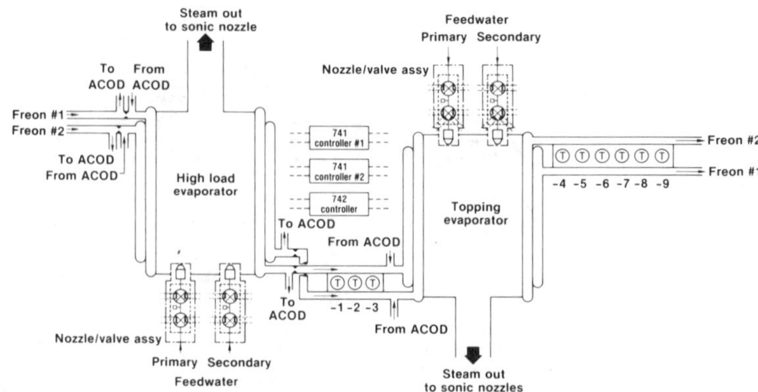

Fig. 5 Flash evaporator subsystem schematic.

Table 3 FES performance requirements

Requirements or interface	Ascent and entry/both evaporators	On-orbit topper only	Abort either evaporator
Heat load, Btu/h	131,000	0-30,000	67,500 max.
Freon outlet temperature, °F	38.5-40	38.5-40	62±2
Freon flow rate, lb/h/loop	2750	2500	2750
Maximum inlet temperature, °F	133	60	111
Feedwater pressure, psia	20-120	30-40	20-40
Feedwater temperature, °F	65-110	65-110	65-110
Carryover, lb/h	Topping core all modes		1.1 max.
	High load core all modes		2.9 max.

tor. The high load duct is relatively short with few bends. Electrical heaters attached to the outside of the duct maintain the wall temperature at a level of 250°F. Because the wall is maintained at a high temperature, unevaporated ice particles or water droplets that enter the duct from the evaporator and approach the duct walls will be repelled by their own vapor pressure. It is theorized that these particles will not have time to be effectively evaporated in the duct because the vapor cushion between the particles and the wall will act as thermal insulation and prevent heat transfer to the particles. It is anticipated that the particles will leave the high load duct with only a small reduction in mass by evaporation.

The topping duct exhausts overboard through two thrust balancing nozzles, one on each side of the vehicle. Because of the duct length and complexity, a different thermal concept was selected. By running the initial sections at high temperature, the energy delivered to the steam in these sections is enough to evaporate all of the expected carryover. Subsequently, the remaining duct sections are maintained at a temperature only slightly above the steam temperature to prevent condensation on the duct wall.

Table 3 summarizes the performance requirements of the FES.

Water Spray Boiler

The WSB[5] is a completely self-contained system that provides an expendable heat sink for the Orbiter hydraulic system and APU lubricating oil, both on the ground and in space. There are three separate boilers, one for each hydraulic and APU system, mounted in the aft end of the Orbiter. Designed for high fluid pressures, the boiler is adapted from an aircraft-type tube bundle oil cooler. At sea level, the tube bundle cools the hydraulic loops to 228°F and the lube oil to 252°F using 212°F boiling water as a heat sink.

In space, weightlessness negates pool boiling and an alternate evaporation mechanism must be utilized. The low steam back pressure greatly increases the temperature difference between the boiler water heat sink and the cooled fluids requiring evaporation from only 25% of the tube surface area. This area is wetted by injecting water into the core in thin streams at intervals along the core length from five peripheral spray bars. Water impinging upon the tubes generates a spray that wets the required area as the droplets ricochet through the tortuous tube bundle. Water is metered into the core in short pulses with the pulse frequency matched to heat load.

The WSB also contains a water supply, a regulated gas pressurization system for water expulsion, separate feedwater metering valves for independent control of hydralic fluid, and APU lube oil cooling and hydraulic fluid control valves. Electronic controllers provide completely automatic operation. A system schematic is shown in Fig. 6.

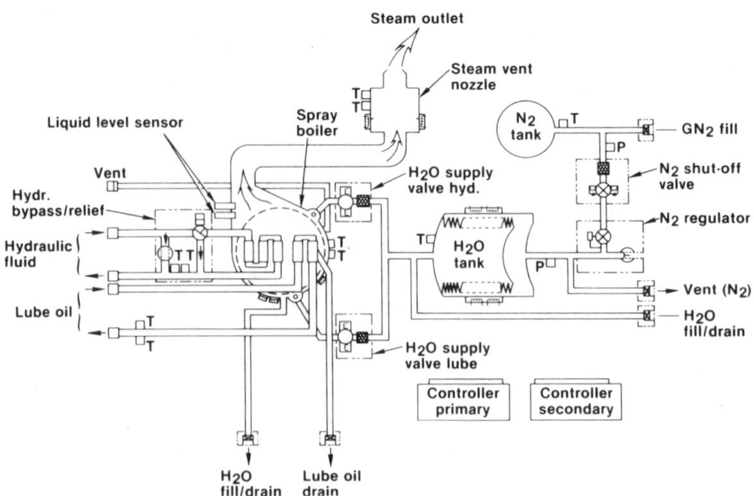

Fig. 6 Water spary boiler subsystem schematic.

Since the hydraulic systems provide power to actuate the main rocket engine gimbals and valves, the aerodynamic flight control surfaces, and the landing gear and brakes, cooling is required primarily during the launch, entry, and landing phases of the mission. Except for on-orbit APU checkouts, the WSB is dormant while the Shuttle is in orbit.

When the Orbiter is in the vertical launch attitude, cooling is required only in the lube oil section. Therefore, the lube oil section of the boiler is located toward the aft end of the Orbiter so that the pool of boiling water will cover just the lube oil tubes. During entry and landing, with the Orbiter in the horizontal attitude, cooling is necessary in both sections, thereby requiring that all the tubes be covered.

In the pool boiling mode, a liquid level sensor located in the core provides supervisory override control to prevent excessive amounts of feedwater from entering the core and spilling overboard. This reduces the need to carry excessive expendables, saving launch weight.

In a typical mission, the APUs are activated 5 min prior to launch. APU lube oil cooling is needed within 3-5 min depending upon heat load. The boiler core is preloaded with 5 lb of water to provide pool boiling for the lube oil section upon demand. During launch, the back pressure decreases, lowering the sink temperature and reducing the level of water required to maintain temperature control. Lube oil section feedwater flow is initiated when the pool is consumed down to the level where adequate cooling cannot be maintained. This mode continues until MECO, when the pool of water is lost due to weightlessness. Then spray boiling commences as needed until APU shutdown.

Simultaneously, the hydraulic fluid section starts out cold on the ground with no spraying and with the hydraulic fluid diverter valve in the bypass position. When the fluid temperature reaches 210°F during launch, the diverter valve is activated to direct hydraulic fluid through the boiler and spray operation commences. Spraying ceases at APU shutdown.

Two hours prior to APU startup on-orbit, the WSB is activated and heaters warm the overboard steam vent nozzle. After APU startup, spraying is initiated in each boiler section as required. During entry, as the back pressure rapidly increases, fluid temperatures rise and the spray valves are pulsed at maximum frequency until the core is filled. Then during pool boiling operation, liquid level control prevents spillage of water overboard.

Table 4 summarizes the performance requirements of the WSB.

System Performance

The following discussion presents the detailed analysis of the STS-1 flight test data. In addition, brief comparisons will be made with the limited amount of data available from STS-2 at the time of this publication. Because of the extensive amount of information that was recorded, only major system performance parameters will be treated. Problems encountered with cabin overcooling and a transient freeze-up in the WSB will be reviewed, and proposed solutions to these problems will be presented.

Atmospheric Revitalization Subsystem

The expected cabin temperature range is 65-80°F depending on the number of crew members. Cabin temperature is maintained by an automatic system that varies the amount of condenser bypass air. The con-

Table 4 WSB performance requirements

	Hydraulic fluid		Lubricating oil	
Outlet temperature, °F	208±2 nominal 228 max. roll-out		250±2 nominal 265 max.	
Flow rate, gpm	1.25-21.0 nominal 63.0 transients		2.0-4.0 nominal 0-25% entrained N_2	
Heat load, Btu/h	75,600 approach	52,000 orbit	24,000 launch	24,000 orbit
Cooling mode	pool boiling	spray boiling	pool boiling	spray boiling
Steam temperature, °F	217	126 max.	217	126 max.
Function	Cooling during all phases of APU operation			
Carryover	10% max.			

trol system relies on the inputs from the cabin temperature sensor located behind, above, and to the right of the copilot's position. As shown in Table 5, the actual cabin temperature on STS-1 appeared to be some 5-10°F higher than was expected. On ascent, the crew experienced a maximum cabin temperature of 82°F as indicated by the sensor. This phase of flight is characterized by high metabolic loads, maximum solar load through the windows and natural convection of hot air to the top of the cabin. The maximum expected temperature range was approximately 65-70°F. As a result, it was determined by Rockwell International that the location of the temperature sensor does not adequately reflect the bulk temperature of the cabin. The sensor is influenced by heat generated in nearby avionics.

This problem was further amplified during the two sleep periods. The cabin sensor indicated 74°F minimum, while the computer simulation predicted 67°F. Actually, the temperature was rather low, as indicated by crew activity. During the STS-1 sleep periods, the commander selected the full hot position on the cabin temperature selector and noted the cool cabin conditions. The cabin had probably cooled down to 64-65°F, which is close to the prediction. This error is further demonstrated by the low relative humidity. With a 5-10°F higher dry bulb temperature, the relative humidity appeared to be very dry, with a range of 15-35%.

Heat generated by cabin avionics influences the cabin temperature sensor at the present location such that a setting of nearly 90-100% bypass is needed before the control valve will begin to move toward the full hot position. The cabin temperature was higher during the second sleep period after the crew manually pinned the valve in the full hot position. A contributing factor to the sensor problem is the fact that the crew elected to sleep in their flight deck couches rather than the middeck sleeping quarters. Therefore the cool cabin air supply vents located over the instrument panel created enough air circulation to make the crew feel uncomfortable.

Similar cooling problems were encountered on STS-2; however, based on experience from STS-1, the crew changed procedures to reduce cabin temperature excursions. During crew activity, the bypass valve was manually pinned in the full cool position. This resulted in the same comfortable working environment that existed previously, with cabin bypass valve settings of 90% or less. During the sleep periods, the valve was pinned in the full hot position, which improved sleeping conditions.

No immediate system changes are planned. In the near future, a new location for the cabin temperature sensor will be selected. On STS-2, a network of temperature sensors was added to the cabin that provided data on the cabin temperature distribution during the entire mission. With these data and subsequent data from future flights, a new senor location will be chosen that will reflect cabin bulk temperature and allow adequate automatic temperature control.

It should be pointed out that on STS-1 and STS-2, during normal work activity, the cabin temperature was at a comfortable 68-70°F. During re-entry, the cabin temperature did not exceed 80°F indicated by the sensor, which would imply that the actual temperature was closer to 70°F, taking into account the possible error.

Cabin heat exchanger cooling requirements for both STS-1 and STS-2 were running 25-30% of peak. Typical heat load range was 5500-7000 Btu/h for both latent and sensible loads. The maximum design capacity of the condensing heat exchanger is approximately 23,500 Btu/h, which applies to contingency operation with ten men. Instrumentation was not available to measure the slurper and rotary separator performance: however, a dew point range of 37-52°F indicates proper slurper and separator operation on both flights.

No problems were encountered in meeting the cooling requirements for all three avionics bays. The maximum allowable air inlet temperature to the avionics is 100°F. Table 5 shows that on STS-1 the heat exchanger outlet temperature varied from 93-123°F, which results in calculated inlet temperatures below 100°F. A maximum cold plate outlet temperature of 113°F was encountered during ascent, which is below the 120°F limit. All other cases were below 100°F. No temperature measurements are recorded on the IMU platform. Since the IMU draws upon cabin air for cooling, the 90°F maximum inlet temperature was easily met while maintaining a 70-80°F cabin. Similar results prevailed on STS-2.

Controlling cabin carbon dioxide level on these flights presented no problem for the lithium hydrox-

Table 5 ARS flight test results

Heat load and temperatures	Mission phase Ascent	Orbit	Return
Cabin			
Temperature, ^{0}F	82 max.	74 min.	80 max.
Relative humidity, %	15-35	15-35	15-35
Avionics bays			
Air, ^{0}F	98 max.	93 max.	123 max.
Cold plate, ^{0}F	113 max.	86 max.	98 max.
Interchanger			
Water flow, lb/h	1040	700-1250	1050
Outlet temperature, ^{0}F	41.0	38-46	41.0
Heat load, Btu/h	41,300	30,300	39,450

ide canisters. The design point conditions are actually for four crew members. With two crew members, the predicted change-out time for each canister was 24 h; i.e., the life of each canister was 48 h. Only four canisters were onboard for STS-1, three for basic needs and a fourth for contingency plans. On STS-2, several more canisters were onboard because of the planned 5-day mission. However, many canisters were not used when STS-2 was cut by 2 days. The Orbiter is launched without any canisters installed in the ARS. With the large cabin volume (2000 ft/^{3}) and small crew load, the carbon dioxide level would build rather slowly. It was desired to have as much cabin airflow at launch as possible in the event of an abort situation that would reduce cooling capability. Removal of the lithium hydroxide canisters reduces the air loop resistance which results in increased total airflow rate.

At approximately 6 h into both missions, two canisters, designated A and B, were installed in the carbon dioxide absorber and temperature control assembly. The A canister was replaced at 12 h as planned and the B canister replaced at 35 h. During re-entry, the lithium hydroxide canisters were removed from the assembly and stowed to provide maximum cabin ventilation.

Freon Coolant Loop

The primary components in the FCL are the heat sink devices which reject heat to maintain vehicle internal temperatures. Heat sinking requirements vary with mission phase. Table 6 depicts the actual STS-1 flight test results of the FCL heat sinks. The FES will be treated in detail in the following section.

During the prelaunch and postrollout phases of STS-1, the GSE heat exchanger is used to transfer heat from the Orbiter to a ground cart, Freon refrigeration system. During prelaunch, the GSE heat load was 82,500 Btu/h as compared to a peak capability of 111,600 Btu/h. Because of the location of temperature sensors in the system, the actual outlet temperature at the GSE heat exchanger is not available. As shown in Fig. 4, the heat exchanger is located between the space radiator and ammonia boiler. However, the closest Freon temperature measurement is the FES outlet. Prelaunch data indicate that the evaporator outlet was 39.3^{0}F as compared to a GSE outlet requirement of 35-37^{0}F. This discrepancy is a result of environmental effects at the ammonia boiler and evaporator. Similar results were obtained on STS-2. Prelaunch cooling operation required removal of approximately 85,000 Btu/h. Ascent and on-orbit phases for STS-2 were approximately 8-10% greater than STS-1. Return and postrollout phases were nominally the same as STS-1. Equally good Freon supply temperatures existed during STS-2 as in STS-1.

For approximately 2 min after launch, no cooling is provided from the FCL heat sinks. At approximately 140,000 ft of altitude or 2 min after launch, the flash evaporator is activated. As presented in Table 6, the maximum heat load for the evaporator during the launch phase of STS-1 was 106,000 Btu/h. The heat load increased slightly to 110,600 Btu/h for

Table 6 FCL heat sink flight test results

Heat load and temperatures	Pre-launch	Ascent	Mission phase Orbit	Return	Rollout
Heat sink	GSE Hx	Evaporator	Radiator/ Evaporator	Evaporator/ NH_3 Boiler	GSE Hx
Heat load, Btu/h	82,500	106,000	65,600/ 77,100	97,000/ 102,900	74,300
Temperature, ^{0}F					
Outlet	39.3	39.7	37.4/39.7	39.7/37.1	44.1
Inlet	97.2	108.4	87.2/94.4	107.5/109.7	96.6

STS-2. The maximum heat load that can be expected at launch is 131,000 Btu/h. This reduced heat load is reflected in the nominal nature of these missions as discussed earlier. The evaporator delivered an average of 39.7°F while in operation, meeting the required control band of 38.5-40.0°F. After achieving orbit, the Orbiter payload bay doors were opened and the space radiators were deployed. Just prior to radiator deployment, the radiator bypass control valve is closed, which initiates Freon flow through the radiators. In this mode, the radiators offer more resistance to the Freon pump, which reduces total Freon flow rate. Table 7 compares the changes in system flow for radiator and bypass operation. The minimum flow required per loop for radiator operation is 2500 lb/h and in the bypass mode, 2750 lb/h. Actual flows were 2635 and 2840 lb/h, respectively. The flow rates for STS-2 were nearly the same: however, there was a net decrease in flow of about 10-20 lb/h. This is not a significant decrease and is well within the ±5% instrumentation accuracy.

The peak heat load encountered by the radiator during STS-1 was 65,600 Btu/h, which occurred shortly after postorbit insertion deployment. Predicted peak heat load is 75,000 Btu/h. At various points in the orbit during radiator operation, the outlet temperature of the panels dropped as low as -30°F. Therefore the radiator temperature control valve could easily maintain 37.4°F by mixing cold panel outlet with hot bypass flow. This temperature is within the allowable band of 35.5-39.5°F. The availability of radiator cooling for STS-2 was even greater due to a change in vehicle attitude. Performance of the radiators was nominal for both missions.

At approximately 27 h into the STS-1 mission, the crew ran a payload bay door closing and opening exercise as part of a re-entry rehearsal. At that time, the FES began operating to reject heat buildup in the FCL. The evaporator functioned for a short time

Table 7 FCL Freon-21 flow rates

Flow rate, lb/h/loop	Radiator Bypassed	Radiator Operational
System	2840	2635
Payload Hx	235	221
Aft cold plates	330	305
Interchanger	2280	2130

until the radiators were deployed again. A peak on-orbit heat load for the evaporator of 77,100 Btu/h occurred just a re-entry with a delivery temperature of 39.7°F. As a result of the shortened STS-2 mission, a payload bay door closing test was also performed at approximately the same point in the mission as in STS-1. The heat rejection rate was slightly higher at 80,000 Btu/h.

During the return phase of the mission, the evaporator provides cooling from payload bay door closing to approximately 100,000 ft. At this point, the ammonia boiler becomes the heat sink. The ammonia boiler continues to operate 15 min after landing for a total of 30 min. Total heat rejection capability is 48,000 Btu as determined by the capacity of the ammonia tanks. During the return phase on STS-1, the FES heat load peaked at 97,000 Btu/h, and the ammonia boiler peaked at 102,900 Btu/h as depicted in Table 6. Both systems delivered Freon-21 within the required control band. On STS-2, a different approach was taken. Prior to closing, the space radiators were cold soaked to approximately -20°F to -30°F. During return at the point where FES operation ceases, flow was initiated through the radiator to take advantage of the cold soak conditions. The radiator provided FCL cooling to a point just before landing. At that time, ammonia boiler cooling was initiated.

Flight test data were also available on other FCL components. The interchanger provided adequate cooling to the cabin and water loop. The payload heat exchanger provided adequate cooling to the DFI package contained within the Shuttle payload bay. Table 7 shows that sufficient Freon flow was delivered to these heat exchangers.

Table 8 FCL component inlet temperatures

Inlet temperature, °F	Mission phase				
	GSE	Ascent	Orbit	Return	Rollout
Interchanger	41.0	39.7	38.7	37-40	44.9
Payload Hx	41.0	39.7	38.7	37-40	44.9
Aft cold plates	39.3	39.4	38.7	37-40	44.1
Fuel cell Hx	78.8	77.1	65.8	72-74	65.0
Midbody CP	78.8	77.1	65.8	72-74	65.0
Hydraulics Hx	96.2	106.5	89.5	103-105	76.1

Table 8 presents the inlet temperatures to the various FCL components. The 44.1°F temperature that occurred during rollout resulted from the ammonia boiler depleting tank A. Shortly thereafter, a switch occurred to tank B which dropped the FCL temperature down to 38°F. Comparable conditions existed on STS-2 except for a slight increase in heat load due to the OSTA payload. No pertubations existed on STS-2 during ammonia boiler operation.

The aft cold plates are located in parallel with the interchanger and payload heat exchanger as shown in Fig. 4. Because of the proximity to the FES in the aft section of the vehicle, the cold plates received properly conditioned Freon-21. The aft cold plate heat load for these missions varied from 4250 Btu/h on-orbit to 14,170 Btu/h at launch.

Of the remaining FCL components (fuel cell heat exchanger, hydraulics heat exchanger, and midbody cold plates), insufficient flight data prevent evaluation of each component. The collective heat load of these components is reflected in the Freon temperature rise from the pump outlet to the space radiator inlet. In Table 8, an estimate was made of the hydraulics heat exchanger inlet temperature based on fuel cell cooling requirements. The combined heat load of these three components varied over 26,000-44, 400 Btu/h, of which 90% is contributed by the fuel cells during launch and return and 65% on-orbit. Up to 15,000 Btu/h can be transferred to the hydraulics systems by the hydraulics heat exchanger on-orbit to maintain the fluid temperature above 0°F. These data are typical for both STS-1 and STS-2.

Flash Evaporator Subsystem

The FES is the most critical of the four heat sink subsystems because it is the only heat sink that provides cooling during ascent and the high altitude portion of re-entry, including the 0-g phase prior to entry interface at 400,000 ft. The performance of the evaporator is summarized in Table 6 for STS-1. A comparison of FES operation is shown in Table 9 for STS-1 and STS-2. During the STS-1 ascent, the evaporator experienced a maximum heat load of 106,000 Btu/h. As stated earlier, this is 20% short of peak capacity. Throughout both missions, the evaporator maintained 39.7°F, which is within the 38.5-40°F control band.

On orbit, the topping section of the evaporator is activated most of the time except during sleep periods when radiator performance is adequate to allow shutdown of the FES. Since the radiator delivers 35-37°F Freon, the evaporator will not provide any cooling since the control temperature is 39.7°F. Test data show that during STS-1 the evaporator was needed for short periods of time to "top" the radiator. On STS-2, the Shuttle on-orbit orientation was such that the FES was required to supplement the radiator heat rejection more often and at a higher heat load than STS-1 as shown in Table 9. The mission support personnel noted that the evaporator was activated in response to increased radiator outlet temperature when the sun angle was unfavorable to radiator operation.

An additional secondary function of the evaporator is to expend excess potable water produced by the fuel cells that has accumulated in the potable water tanks. The radiator temperature control is reset to 57°F, which activates the topping evaporator. In this mode, 30 lb/h of water can be evaporated. The evaporator was not used for dumping water on both flights. Potable water was dumped directly into space through heated nozzles.

During ascent, the evaporator achieved a successful startup at approximately 2 min 18 s. The evaporator was in control within approximately 1 min. During the startup transient, the maximum inlet temperature at the evaporator was approximately 115-120°F in both cases.

While the STS-1 mission went smoothly, a number of anomalies were experienced on STS-2. At main engine cutoff (MECO) during the ascent phase of

Table 9 FES performance comparison

	Design	STS-1	STS-2
Heat load, Btu/h			
Ascent	131,000	106,000	110,600
Topping	30,000	7,000	25,500
Entry	131,000	85,000	88,800
Control temperature, °F	39±1	39.0	39.0
Duration of operation, h	168	>9	>6
Carryover, % max.	3.0	None indicated	

STS-2, a perturbation in Freon outlet temperature appeared. This outage was in response to a sharp drop in feedwater pressure at MECO caused by the loss of vehicle acceleration, which creates pressure in the water system at the FES. The FES is designed to accommodate this type of transient; however, the transient did result in a FES shutdown and recovery.

Additional outages were experienced during orbital operations. The FES failed to come out of the standby mode so that a switch was made from the primary A controller to a primary B controller. A checkout was conducted prior to radiator stowage to gain confidence in dual evaporator operation. The FES functioned normally in this mode. Failure investigation revealed that a midpoint temperature sensor had drifted high on the primary B controller and an outlet sensor drifted low on the primary A controller. The out-of-specification sensors were replaced for STS-3.

The evaporator steam ducting system requires an extensive arrangement of heaters to prevent steam from freezing to the duct walls in the cold environment. The high load evaporator duct system was maintained at 237-300°F while in operation. After launch, the high load evaporator is deactivated or "inhibited," which shuts down the duct heaters. During dormant modes, the ducting dropped to 12-23°F. Prior to high load evaporator usage, the unit is activated to allow the ducting to heat up.

Since the topping evaporator is active continuously, the duct heaters remain active. Without steam flow, the ducting temperature cycled over 261-273°F in the hot section and 157-167°F in the cooler section. The overboard steam nozzles were held at 49-63°F. With steam flow, the ducting cools somewhat; however, the steam does not cool the ducting enough to even approach a freezing situation. The ducting was designed around this philosophy. On STS-2, similar duct temperatures existed. The only problem that existed was a section of high load evaporator duct that was overheating during prelaunch. It was determined that this problem should continue to be monitored on STS-3.

In order to prevent the feedwater lines and accumulator from freezing, a heater system is used. It is similar in design to the Orbiter water line heater system. At all points in the missions, the feedwater lines ranged in temperature from 66-95°F depending on the location. The present heater design on this Orbiter (OV-102) is provided by Rockwell International. Future Orbiter vehicles will have Hamilton Standard heater systems which are similar in design. The adequacy of the new heater design can be evaluated based on the successful operation of the heater system on STS-1 and STS-2.

Water Spray Boiler

The total heat load for the WSB systems was considerably less than design point conditions for both flights. Total water expended was approximately 19-20% of the available supply. Data from STS-1 indicate that WSB systems 2 and 3 consumed the most water at 31.4 and 29.9 lb, respectively. System 1 used only 17.9 lb. Data were not available for STS-2. Systems 2 and 3 operated longer than system 1 during the first flight prior to re-entry. Due to the lack of flow rate instrumentation, the actual hydraulic fluid and lubricating oil flow rates are not known. Therefore water utilization cannot be evaluated since total heat load cannot be determined. Table 10 summarizes boiler performance.

Water usage for each WSB can be monitored by ground control and the crew. By knowing the pressure and temperature of the nitrogen tank, a pressure-

Table 10 WSB STS-1 flight test results

Ascent profile	
Hydraulic outlet temperature	151°F max.
Lubricating oil temperatures	
Maximum inlet	295°F
Maximum outlet	283°F peak 252°F avg.
Return profile	
Hydraulic temperatures	
Maximum inlet	N/A
Maximum outlet	221°F
Lubricating oil temperatures	
Maximum inlet	278°F
Maximum outlet	252°F
Expendables	
Total water loaded	424.2 lb
Total water used	79.7 lb
Water usage	19% total

volume-temperature relationship is used to determine the change in volume of the water tank. The accuracy of this measurement technique was demonstrated on the first flight. Using the PVT method, the calculated water usage on STS-1 was 79.8 lb compared to an actual usage of 79.7 lb, which was determined by water-off-loading after landing. Therefore it was demonstrated that this method can reliably predict the quantity of water that remains for use in each system. Table 11 presents a comparison of PVT and off-loading data.

The ascent phase of the missions proved to be the most interesting part of the WSB operation, especially in the case of STS-2. Because of the very low heat load condition in the hydraulic system, the hydraulic fluid never reached the control temperature of 208°F. The peak temperature on launch was 151°F, which was well below limits. As a result, the hydraulic spray bars were not activated during the launch phase. In the lubricating oil section, the temperature rises at about 35°F/min. A combination of low thermal mass and a steady heat load of approximately 24,000 Btu/h raises the lube oil temperature quickly. Upon reaching 252°F, the electronic controller initiates water spray to begin cooling down the APU. During STS-1, WSB system No. 3 suffered a spray bar freeze-up so that the lube oil temperature finally peaked at 283°F before spray was activated. After activation, the WSB was in control at 252°F. The two remaining systems did not experience freeze-up. It was determined that an initial boiler load of 5 lb of water in the core resulted in

Table 11 Quantity gauging system data

	System 1	System 2	System 3
Water consumed, lb			
Amount loaded	141.4	141.4	141.4
Amount off-loaded	123.5	109.5	111.5
Amount used	17.9	31.9	29.9
QGS measurements, lb			
Ascent	2.6	3.6	0.4
Entry	15.3	28.3	29.6
Total	17.9	31.9	30.0
Error			
Pounds of water	0.0	0.0	0.1
Percent of total	0.0	0.0	0.3

overcooling of the APU, subsequently reducing the heat load on the boiler in absence of hydraulic section spray bar operation and allowing the back pressure to drop below the triple point. The lower back pressure allowed residual water in the core to freeze, which resulted in spray bar freezing in system No. 3. It should be noted that system No. 3 dropped below freezing for a longer time than the other systems, which probably contributed to the freezing.

The situation worsened on STS-2. WSB system No. 3 experienced a freeze-up during ascent to the point that premature shutdown of the associated APU became necessary approximately 30 s prior to MECO. The vehicle continued on a normal trajectory with the No. 2 engine fixed at 80-85% thrust requiring the remaining engines to throttle down to maintain a 3g acceleration limit.

This anomalous behavior did not affect operation during the on-orbit checkout and re-entry mission phases. For the return phase of this mission, no detailed test data were available to determine heat load and the operation of each section. The flight log indicates that all three boilers and APUs functioned normally during re-entry. Also, the heat load was probably sufficient to require full boiler operation, as indicated by water usage. STS-1 data indicate that only about 20 lb of water was used on launch so that on the order of 60 lb was used during return. Of that 60 lb, the lube oil section would have required approximately 30 lb. Therefore 30 lb must have been used by the boilers for hydraulic cooling. This includes the 5 lb of water required to reflood each boiler at the end of the return phase. Although data were not available for STS-2, a similar situation existed, since all three WSB's functioned normally under similar conditions during return.

As a result of consecutive launch anomalies on WSB system No. 3, an investigation was initiated to determine the cause of the problem and arrive at a potential solution. There was speculation that a leaking APU resulted in contamination of the lubricating oil and potential plugging of the WSB core. In addition, since the failure occurred on the same WSB on both flights, there was concern that position No. 3 had some peculiarities not present in the other locations. Also, there was concern that the configuration of WSB system No. 3 may be different than the others.

Conclusions

In general, the overall performance of the Orbiter thermal control systems was as predicted. Minor problems were encountered, such as the transient outage of the WSB and cabin overcooling. However, solutions to these problems will be incorporated on future flights.

In the ARS, the minimal heat loads did not stress the operation of the condenser and lithium hydroxide canisters. These components will be evaluated for their capacity during operational flights which require four to seven crew members. The overcooling of the cabin will be corrected through relocation of the cabin temperature sensor.

All FCL heat sinks operated properly at heat loads of 75-90% of the rated capacity, which gives confidence in the design of these systems. The FES has demonstrated that it can be successfully restarted and reach control temperature range within an adequate amount of time. The space radiators encountered no problems after repeated, successful payload bay door operation. Adequate cooling was provided on the ground by the GSE heat exchanger both before launch and during postrollout "safing" procedures, and the ammonia boiler provided necessary cooling during the interim from high altitude FES operation to postrollout GSE connect.

References

[1]Trusch, R. B. and Nason, J. R., "Component Heat Exchangers for the Space Shuttle," ASME 75-ENAs-54, April 1976.

[2]Stoll, O. T., Laubach, G. E., and Gibb, J. W., "Space Shuttle Orbiter ECLSS," ASME 73-ENAs-23, May 1974.

[3]Heinrich, S. R., "Shuttle Orbiter Ammonia Boiler Subsystem," ASME 81-ENAs-44, July 1981.

[4]Nason, J. R. and Decrisantis, A. A., "Shuttle Orbiter Flash Evaporator," ASME 79-ENAs-14, April 1979.

[5]O'Connor, E. W. and Rethke, D. W., "Shuttle Orbiter Water Spray Boiler," ASME 78-ENAs-17, April 1979.

The Orbital Test Satellite Thermal Experience after 3½ Years in Orbit

J.P. Bouchez and D.H. Howle
ESA-ESTEC, Noordwijk, the Netherlands

Abstract

The orbital test satellite is a geosynchronous orbit, three-axis stabilized, European experimental communications satellite which has been in orbit for nearly four years. Quasi-steady-state thermal tests carried out at all equinoxes and solstices allow observed spacecraft temperatures to be compared with temperature predictions using "beginning-of-life" thermal coating values. The gradual rise in general temperature due to surface degradation gives continually larger deviations from predictions. The data now permit analysis using exponential curves with a least-squares method. Extrapolation then allows prediction of long-term temperatures to be expected; by comparing equinox and solstice data, the degradation of the optical solar reflector surfaces on both radiators may be estimated.

Nomenclature

AE	=	autumnal equinox
BOE	=	beginning of eclipse
BOL	=	beginning of life
EOE	=	end of eclipse
MLI	=	multilayer insulation
OSR	=	optical solar reflector
SS	=	summer solstice
t	=	time

Presented as Paper 82-0831 at the 3rd AIAA/ASME Joint Thermophysics, Fluids, Plasma and Heat Transfer Conference, St. Louis, Mo., June 7-11, 1982.

t_o = initial time
VCM = volatile condensible material
VE = vernal equinox
WL = weight loss
WS = winter solstice
α = absorptance
ΔT = increase in temperature
ΔT_m = maximum increase in temperature (asymptotic value)
τ = time constant

Introduction

The orbital test satellite, OTS-2, is the first experimental telecommunication spacecraft of the European Space Agency (ESA). It was conceived to validate, in a three-year orbital test program, the telecommunication technologies and spacecraft hardware intended to be used in the European communication satellite (ECS) program. It is three-axis stabilized and was positioned in May 1978 in a geostationary orbit at 10 deg E longitude. In April 1982, it is planned to move it to 5 deg E, in order to make room for the ECS series of satellites, the launch of the first of these being scheduled for the second half of 1982.

Most of the tests performed to date within the orbital test program (OTP) have been dedicated to the telecommunication payload, but tests were also planned at regular intervals to evaluate the performance of other subsystems. In particular, the thermal subsystem performance has been observed in a dedicated series of tests which provides, every three months at solstice and equinox, for stable and well-known operational conditions. The present paper confines itself to a presentation and interpretation of these thermal tests, and is an extension of an earlier paper [Ref. 1] which reported on the first two years in orbit. The mathematical treatment of the data is elaborated in the present paper, and more statistical approach is now becoming possible with the ever-increasing amount of test data accumulated every three months.

System Impact on Thermal Evaluation

Bearing in mind the future development of ECS and other space communication systems, the OTS spacecraft was conceived as a "bus" (Fig. 1) which houses all service subsystems and carries a communication module dedicated

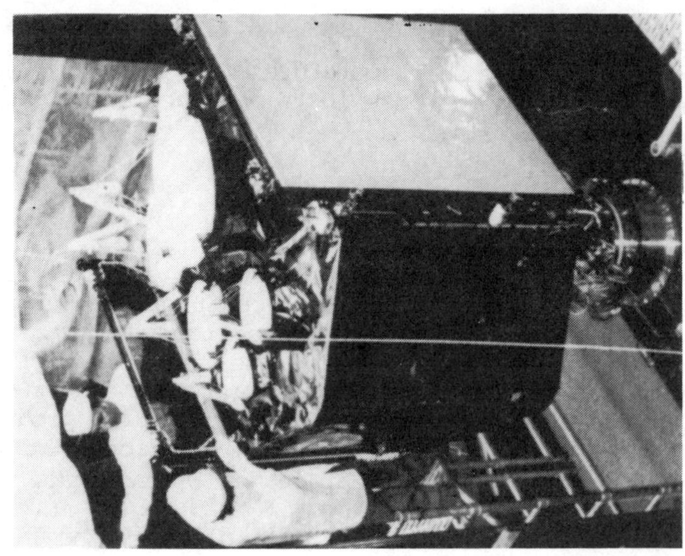

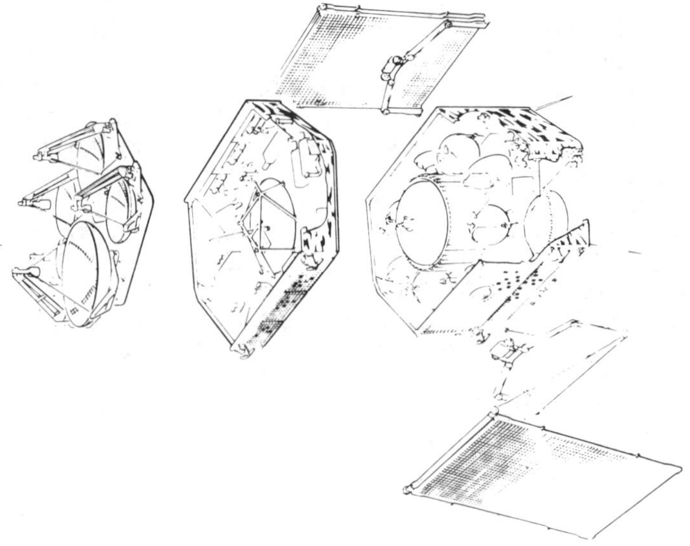

Fig. 1 OTS-2 spacecraft.

to the particular mission. With regard to thermal evaluation the following system characteristics are important: The power subsystem capability allows operation of a maximum of six traveling wave tubes (TWTs) during sun illumination and two channels during eclipse. As traffic demand varies, the power output of each TWT is variable, and so is its internal power dissipation. This variation is a few watts per tube and, at most, around 1 W for its associated electronic power conditioner (EPC). When a channel is switched off, a set of simulation heaters automatically replaces the power dissipation of the TWT and its electronic power conditioner (EPC). This was implemented to limit extreme variations in spacecraft power dissipation and to keep the internal temperature of the spacecraft relatively constant. In the service module most units are redundant, e.g., fixed momentum wheels, telemetry, tracking and command (TT&C) transponders, and the reaction control system (RCS), with two branches of ten thrusters each. During operation, a change from the normally operating equipment to the redundant units has very little thermal impact on the spacecraft, except for the momentum wheels, which for mass balance reasons are located on opposite areas of the spacecraft. A change there introduces a significant asymmetry in the power distribution but not in the overall dissipation value. A change from one thruster branch to the redundant one has only local influence on the flow control valve/thruster temperatures. The equipment with the most important power dissipation excursion is the shunt associated with the solar array output. Its dissipation varies between 40 and 145 W. Although the shunt electronics are mounted on a dedicated radiator area on the north face of the spacecraft, the unit is easily the largest contributor to changes in the spacecraft's temperature level. This fluctuation is of no real consequence to the spacecraft temperature limits because the design took this fact into account. It does, however, render direct orbit test comparision more difficult. This is particularly noticed in the case of tests which have identical operational status, but different shunt power dissipations, resulting from the slow degradation of the solar array output.

The thermal design of the OTS, which has been described in detail in previous papers,[1,2] is not reproduced here. It is, however, important to note that this present study quantifies the degradation of the different coatings used on the spacecraft, in particular the OSRs on the north and south radiators, and the white paint on the antenna dishes.

Tests Performed

During the course of the OTS program, thermal tests were performed on the ground, simulating space temperatures and solar irradiation at extreme equinox and solstice conditions under selected fixed solar input angles (steady-state tests). The purpose of these tests was to verify that, in a simulated space environment, the spacecraft performance was within the acceptable limits which had been specified. Another aspect was to verify that the mathematical model used for the extreme hot and cold design cases was capable of reasonably predicting the test results, thus giving confidence in the orbital predictions.

The OTS being a test satellite, it had been decided early in the program to include the thermal subsystem when planning the different in-orbit performance tests. The objectives of each an endeavour were somewhat different from the objective of the tests performed on the ground. They can be summarized as follows:

1) to assess the adequacy of the thermal control subsystem in providing the proper environment to the different equipment of the spacecraft;

2) to compare flight temperatures with analytical predictions for the same electrical configuration, and to ascertain the degree of accuracy of each orbital predictions;

3) to provide the possibility of assessing the thermal distortion of a large antenna dish; and

4) to detect and quantify, at specified intervals, any degradation of the thermal coatings under the influence of the space environment.

This paper is concerned only with that aspect of the tests that leads to achievement of the fourth of these objectives.

Since launch in May 1978, dedicated thermal tests have been performed at each equinox and solstice. To establish with good confidence thermally representative conditions, the dedicated tests were carried out for approximately two days with a well-defined and constant power dissipation and repeater status.

Available Test Data

To achieve the combined objectives stated in the preceding paragraph, the spacecraft was instrumented with

152 thermistors, thermocouples, and platinum resistance sensors to provide for temperature mapping. The number of these was constrained by the available telemetry channels, so that it was not always possible to have a sensor at every desired position. The summary of sensor apportionment is given in Table 1 for the different subsystems, including 36 sensors which are internal to equipment and required by the manufacturers. They do, however, provide the thermal subsystem with some additional temperature information. Of these 152 sensors, 60 are mounted on components outside the main body of the spacecraft, as, e.g., antennas, solar array.

All sensors have been calibrated, and the error associated with the temperature reading is partly due to the mathematical function introduced to represent the calibration curve and partly due to the quantization associated with a one bit of that telemetry channel. The first error is very small, but the change associated with a 1-bit variation in the telemetry signal corresponds to between 0.3 and 0.8°C, depending on the sensor. Thus the largest inaccuracy is essentially attributed to the platinum resistance sensors, which are all located on the antenna, the solar arrays, and the shunt radiator, i.e., equipment with a large daily temperature variation. This inaccuracy, which is basically random in nature, tends to be reduced when looking at all the sensors in a statistical manner. During the two days of each thermal test, the readings of all sensors are telemetered at 25.6-s intervals. The cyclic reproduction of the temperature time-

Table 1 Sensor Distribution

Subsystem	Sensors
RCS	38
Power	13
ABM	2
AOCS	14
TTC	9
Antenna	24
Repeater	40
Structure	12
Total	152

line is verified by comparing the temperatures at corresponding times of two consecutive days. Reasonably stable conditions have been achieved in all tests, although distinct differences are noticeable. Only the four hydrazine tanks, which have a high thermal capacity and are well insulated, systematically did not reach cyclic temperature reproducibility during the 48 h of test.

One of the major problems confronting the thermal analyst when making temperature predictions is determining the accuracy of the power dissipations to be introduced in the mathematical model. Individual power dissipations are usually known from previous testing at component or system level, but are very rarely determined for the exact flight configuration to be studied and analyzed. As an example, a TWT power dissipation (as well as that of its EPC) varies as a function of the power output, which is not as accurately known as the dissipation of a heater. In general, one can say that the equipment associated with the repeater subsystem (payload) and the batteries have dissipations which are functions of the operating status of the payload. For the OTS, this is the case for TWT, EPC, power conditioning unit (PCU), batteries and shunt, and, to a lesser extent, for filters and operating networks.

From the power subsystem telemetry of OTS, it is possible to estimate the power capability of the solar arrays and also to calculate the global value of the spacecraft consumption. The currents leaving the active and inactive sections of the solar arrays are measured, and from this data the power generated can be determined. On the spacecraft side, the power consumption is calculated at that point of the electrical bus where current and voltage are measured. The energy used before the current measuring point has to be evaluated separately for the power dis-

Table 2 Power Distribution (W) SS 1978

	Delivered array power	Distribution to S/C load					
		Shunt	Component	Battery	PCU	output	Total
Orbit data from telemetry	557.9	107.3	415.1	7	2	25.1	556.5
Thermal analysis input		107.4	400.8	7	0	External (not in S/C) 34.0	549

sipations of the shunt, battery, and PCU. This evaluation is performed accurately for the shunt dissipation and can be estimated quite precisely for the battery and the PCU. A comparison can be performed between the capability of the arrays and the spacecraft consumption. A typical example is given for summer solstice 1978 in Table 2, which illustrates the accuracy (to a few watts) within which the spacecraft power distribution is known. In this example, the different variables were constant during a 24-h period, but when there are variations, the assessment is performed for every time period for which the individual measurements can be considered constant.

Some of the electric energy produced in the solar array is not converted into heat within the spacecraft but radiated as power to the Earth, or converted into chemical energy during battery charging. It is on the power rejected outside of the spacecraft that the largest uncertainty in the evaluation may exist. Reduction of this uncertainty is possible by reducing the number of channels which are in operation or by using channels on which a direct measurement of the power radiated to space is available. In the OTS, a direct measurement is possible on three channels, and for our dedicated thermal tests we use these channels when possible.

Usually, for our tests the difference between the total power dissipation obtained by summing individual components (data used in the thermal analysis) and the telemetered data is of the order of a few watts, with the largest deviation noticed being at approximately 12 W. As the reference used is the telemetered data, this difference is used to correct the data knowing the sensitivity of the spacecraft to internal power dissipation change (1^0C/9 W).

Operating Condition of the S/C

The tests performed to date have always been very well defined beforehand in order to model accurately the dissipation of individual equipment. In particular, as thermal orbital tests require a constant well-known power dissipation, activity of the repeater is kept to a minimum and as few as possible channels are operated, which reduces the uncertainty as to TWT and EPC power dissipation. The remaining TWTAs are substituted by heaters of well-known power dissipation.

Nevertheless, owing to operational constraints the aim to have identical operating conditions at all comparable orbit tests cannot be fully achieved. When heaters have to be switched on and off during the 24-h orbit, the switching times are kept identical each day during the complete length of the test. (Switching of heaters occurs at certain periods of the year and especially at equinox, where boost heaters are used before entry to eclipse.) Any heater switching activity or other change in the power distribution is immediately registered in the power dissipation of the shunt, and its variations are well monitored during the complete test. This is very important information, because while the switching of a small heater may have a relatively minor influence on the shunt dissipation if it is operating in the middle of its range, a change of a few watts in internal power consumption near the upper or lower limit of the shunt capability may require one solar array section extra or one solar array section less, with a consequent flip of the shunt operating point to the other side of its range (a variation of more than 100 W).

Orbit Test Evaluation

In view of the bright future of space communications in general and of ESA's program in particular, some of the most important questions regarding thermal control concern the long-term stability of the employed thermal technologies in the geostationary orbit environment. It is known from ground tests performed during the last two decades, such as, for instance, those of DERTS,[3] but also from orbit observation, that the thermo-optical properties of materials degrade under the electromagnetic and particle environment in space. The extent of this degradation depends on the selected material, the technology of application, cleanliness of the surface and disturbances coming from the spacecraft (contamination either directly or by reflection at appendages). It is further suspected, and the possibility has been demonstrated in ground tests,[4] that the differential electrostatic charging and subsequent arc discharges which may occur on spacecraft surfaces in geostationary orbits, may damage thermal control materials like MLI and OSR.

The data available from the OTS thermal tests performed in orbit can help to provide some answers to these questions, valid not only for this particular spacecraft but also, with certain reservations, for other three-axis stabilized spacecraft in geostationary orbit which use thermal control materials and technologies similar to those of the OTS.

Methods of Evaluation

The initial thermal subsystem performance or, more precisely, the accuracy of the mathematical model, can be verified by a straightforward comparison of the actual orbit temperatures with those predicted by the mathematical model when fed with identical operating and environment conditions.

Degradation mainly affects the reflectivity of surfaces in the solar spectrum, and generally leads to a reduction in that reflectivity. (Changes in the infrared spectral region are, with few exceptions, of comparatively minor importance.) Thus degradation results in additional absorbed heat in the spacecraft, and consequently in a slow temperature rise during life. To assess the effect of degradation, the OTS orbit thermal test data can be used in three different manners:

1) Direct comparison of temperatures under identical external conditions. (This presumes that the operational configuration and in particular the power dissipation of the spacecraft is identical in the cases to be compared. When this is not the case, the knowledge of the sensitivity of the spacecraft temperatures to power variations is a prerequisite for this approach.)

2) Duplication of the orbit measurements by a mathematical model which takes into account the actual operational configuration of each test, and indicates the effect of degradation by increasing differences between measurement and analytical results. (A theoretical alternative to this would be to maintain the initial correlation of measurement and analysis constant, by allocating suitable degradation rates to the various surfaces.)

3) Direct comparison of individual sensor readings from areas which are mainly influenced by external heat input, and for which the possible internal coupling to the spacecraft is either negligible or quantifiable.

The first two approaches have an integrating character, in the sense that the increase of temperature results from the degradation of every surface of the spacecraft. The last method allows precise determination of local degradation and is often used for dedicated experiments with calorimeters. For this paper, only the second method is considered. In an earlier paper,[1] it was shown that good agreement could be obtained between results from the first two methods, but the second method has been favored by ESA for all subsequent work, because it is the more flexible, and gives good accuracy. The investigations into the initial thermal performance of OTS, and the correlation with, and updating of, the mathematical model, are also fully reported in Ref. 1 and will not be repeated here.

Computer-Aided Comparison

By performing thermal analysis for the exact operational condition of an in-orbit test, one can define a difference between the test and the beginning of life (BOL) analysis results. Interpretation of this difference as a function of time in orbit can then lead to the determination of the temperature effect of degradation of the thermal control materials. The results of the computer-aided comparison are interpreted statistically using histograms, constructed by plotting the temperature difference between the observed flight temperatures and the predicted temperatures for a set of (originally) 48 sensors, which are situated internally and on the radiators of the spacecraft. As a by-product of the selection of the set of sensors they have small temperature changes (of the order of 0.3 to 0.5°C) per telemetry bit. Now, after nearly four years, some sensors have been lost, and the population used is only 46.

The operating conditions implemented in the analysis are determined from orbit measurements as outlined in the section entitled "Available Test Data." For each equipment, the power dissipation known from ground testing is entered into the mathematical model. A typical power distribution in the analysis has been given in Table 2.

Presentation of Results. For a better understanding two typical histograms are reproduced from Ref. 1 for daily maximum and minimum of the first dedicated test of SS 1978 and also a new pair for SS 1981 (Figs. 2 and 3). The spread of each histogram, quantified by the standard deviation, is reasonable for a three-axis stabilized satellite, as explained in Ref. 1. [One has to remember that what is represented by the histogram is a difference with respect to a beginning-of-life (BOL) prediction, and does not indicate that equipments which are 1 or 2 standard deviation away from the mean temperature are outside acceptable operating limits.]

Results gathered to date are presented in Table 3 and in Figs. 4 and 5.

Accuracy of Results. The accuracy of each individual sensor is as stated earlier, i.e., of the order of 0.3 to 0.5°C for the chosen sample. However, due to the number

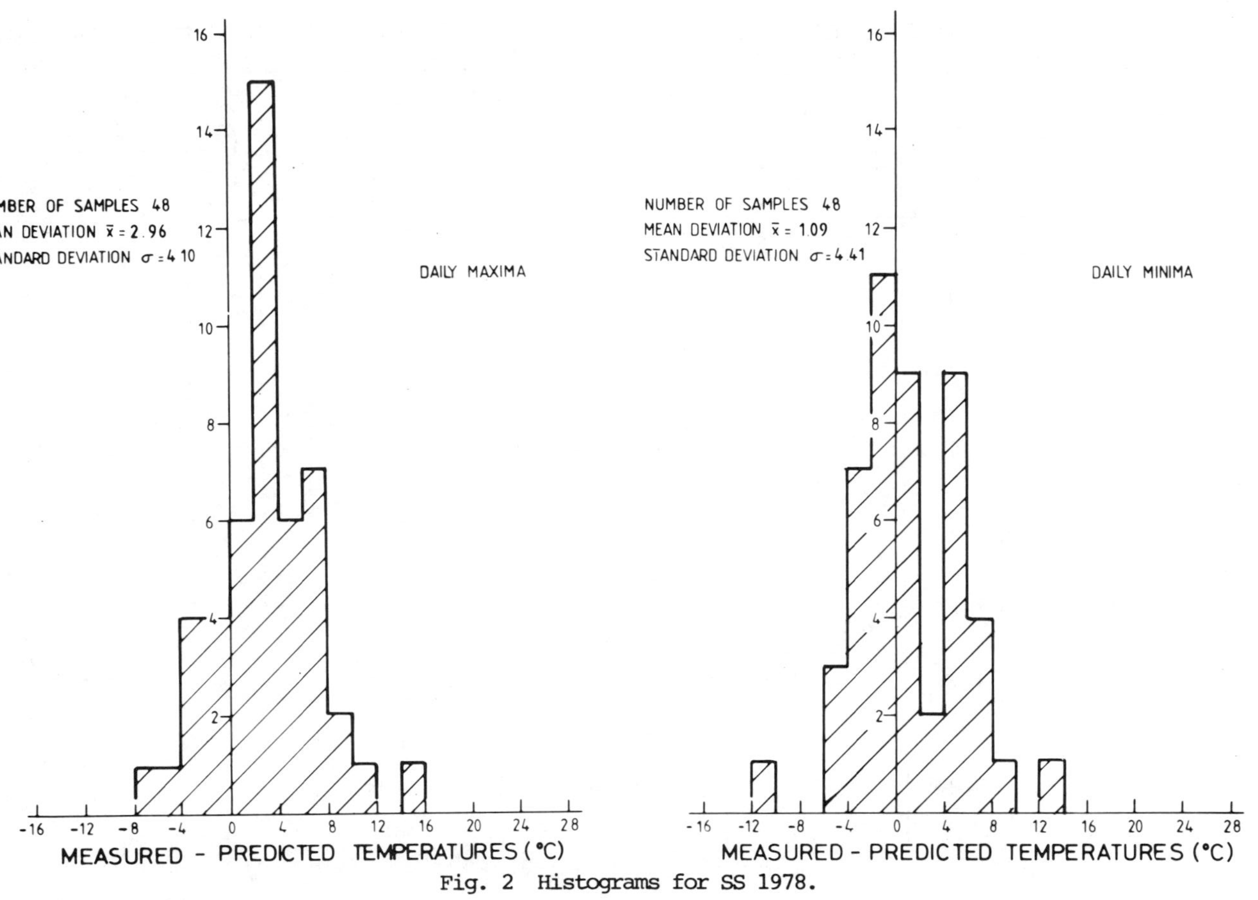

Fig. 2 Histograms for SS 1978.

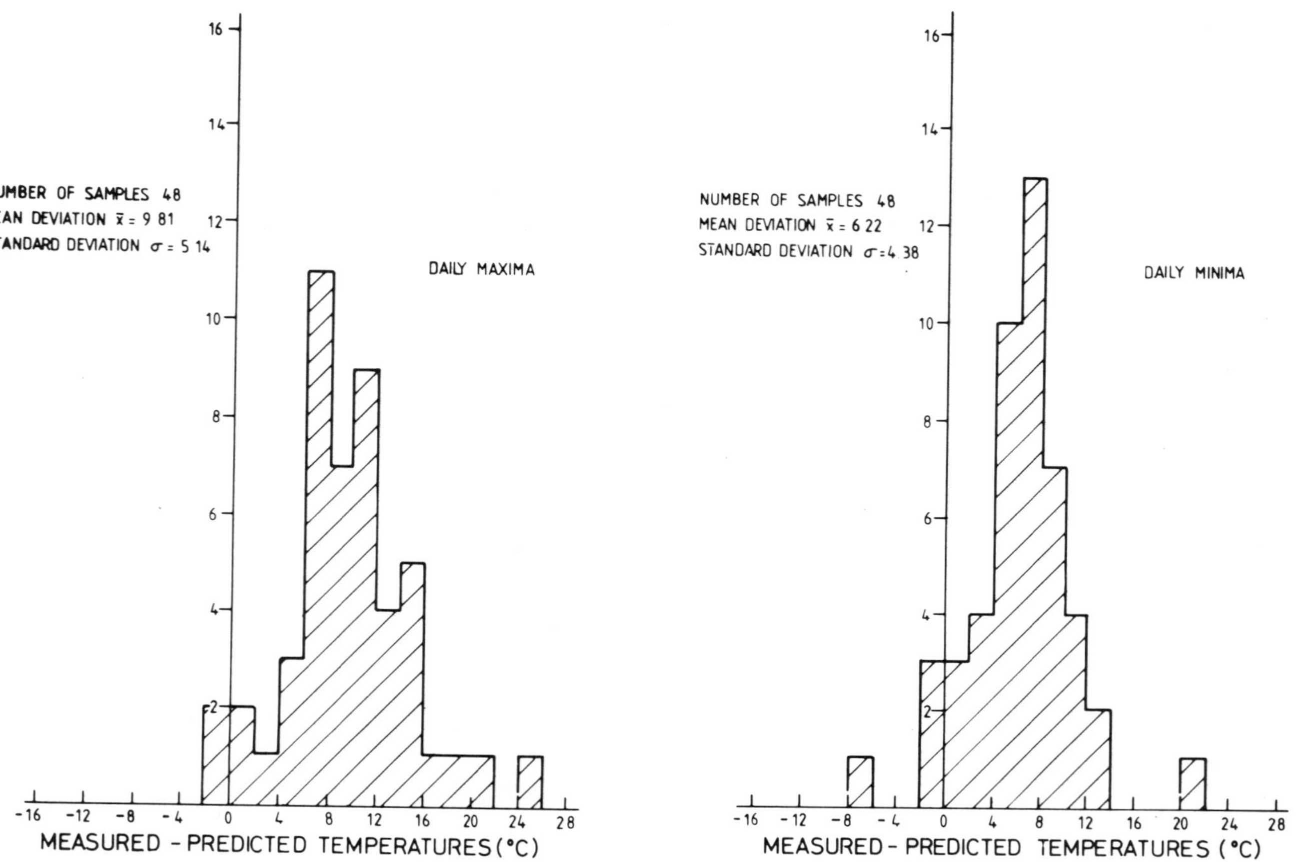

Fig. 3 Histograms for SS 1981.

of sensors selected (48 at beginning of life), the computation of the mean spacecraft temperature based on these sensors is known to an accuracy better than 0.1°C.

The most important contributor to the inaccuracy of this type of approach resides in the knowledge of the power dissipation, and, as shown in Table 2, two different values can be used as reference. The solar array output is known to 0.44% [Ref. 5], or approximately 3 W. However, the power distributed within the spacecraft is

Table 3 OTS-2 seasonal temperature increases. Reference BOL prediction, corresponding to actual flight power distribution. Days are from January 1, 1978.

		S.S.		W.S.		EQUINOX		
Day	Test Data	Daily maximum	Daily minimum	Daily maximum	Daily minimum	Non-Eclipse maximum	BOE	BOE
	May 25/26 78	1.81	-0.22					
171	S.S. 78	2.96	1.09					
264	A.E. 78					2.47	1.12	1.84
345	W.S. 78			3.84	1.52			
446	V.E. 79					3.90	2.55	3.14
540	S.S. 79	6.39	3.61					
631	A.E. 79					3.98	2.20	2.88
713	W.S. 79			7.63	4.68			
812	V.E. 80					4.53	3.14	3.83
904	S.S. 80	8.06	5.12					
994	A.E. 80					4.21	2.68	3.23
1080	W.S. 80			9.93	6.60			
1177	V.E. 81					5.19	3.55	4.04
1268	S.S. 81	9.81	6.22					
1359	A.E. 81					6.25	4.22	4.21
1443	W.S. 81			11.13	6.87			
1540	V.E. 82					6.76	4.62	5.56

taken as the reference for this analytical work because it not only gives the total power dissipation but also the general distribution within the spacecraft. Its accuracy can be determined by statistically combining the temperature influence of each individual power inaccuracy (shunt, component dissipation, etc.). As each measurement, or estimate based on previous ground measurements, is independent of the others, this statistical approach is valid, and leads to a value which is different from test to test but of the order of 0.75°C. It then follows that the overall inaccuracy for a solstice case can be assumed to be approximately 0.8°C. The power dissipation used in the analysis reproduces approximately (within a typical value of 10 W) the measurement received from the telemetry, and this systematic error is corrected using the results of a study of the spacecraft mean temperature sensitivity to internal power dissipation changes.[6]

Solstice tests usually involve few spacecraft switching activities, and the spacecraft operating mode is more or less constant. This is not true during an equinox test, which as a minimum requires, in addition to the operational mode changes, a certain extra switching of heaters during the pre-eclipse boost period, the recharging of the battery after eclipse in both full- and then trickle-charge mode, and a drastic change of external input as between the sun illuminated and the eclipse periods. The mathematical model of course takes into account these changes, but an additional inaccuracy occurs because relatively large time steps are used during the integration of the mathematical expression in order to limit the cost of the computations. This error as such has not been accurately evaluated, but it is estimated that an additional 0.2°C might be a reasonable assumption. In addition, true equinox (no sun on either radiator face) occurs only at a specific time in the orbit, and a parasitic heat input on one or other radiator is present during the test, although it has not been introduced in the mathematical analysis. This error can be even larger when the test period is not centered exactly on true equinox for operational reasons. This sun input on the radiators, which occurs at a grazing angle where the absorptance become large,[7] can be significant (possibly 6 W if 1 day out of true equinox), and is not a random but a systematic error. The temperature influence on the chosen example may even be different, depending on which radiator is illuminated. A small study has been started to investigate this influence. The effect of this systematic error, which at present is taken to be 0.5°C, would be to reduce the temperature increase in the equinox season. For all these reasons, in equinox, the inaccuracy is estimated to be of the order of 1.4°C.

Temperature Fit. As can be seen in both Figs. 4 and 5, the temperatures increase significantly and monotonically as a function of time. As this effect during the solstice season is essentially due to increase of solar absorptance on all the spacecraft thermal finishes, an attempt was made to fit exponential functions of the following type:

$$\Delta T = \Delta T_m \left(1 - e^{-\frac{t - t_0}{\tau}}\right)$$

to the data of Fig. 4 and Table 3 for solstices.

The results are shown in Tables 4 and 5 and do match rather well some experimental degradation data obtained in laboratory tests.[3] The difficulty of applying such a method is the time needed to build up a data base, and now, after nearly four years, we have enough such data to apply the method for the first time to the solstice results. We now have four data points for both winter and summer solstices, which is the minimum needed to determine a best fit curve which will not automatically go through each individual point. (The additional point of data obtained two weeks after launch has not been used in this mathematical fitting because it is not truly a solstice case, but is given in Table 3 for completeness of the data.) In equinox, however, one is more fortunate, because this condition occurs twice a year, and we have now eight data points on which to make our assessment (Table 3). It has been somewhat disconcerting to find that at present it is difficult to obtain a unique solution for this set, and in particular one may observe that the last two data points are significantly higher than anticipated. This may be a result of the measurement inaccuracy (estimated to be ± 1.4°C), or the simple fact that an exponential fitting of the type mentioned earlier is not appropriate for these conditions (no sun on radiators). In order to assess the absorptance degradation, a logarithmic least-squares curve has been generated (Fig.

TABLE 4

Coefficients for exponential fitting of data

	ΔT_m	to	τ
S.S. daily maximum	12.5	-76	898
S.S. daily minimum	8.1	48	850
W.S. daily maximum	12.9	109	673
W.S. daily minimum	7.6	244	462

TABLE 5

Comparisons of curve-generated ordinates with corresponding test values

Date	Daily maximum		Daily minimum		ΔT's	
	Test Value	Curve Value	Test Value	Curve Value	maximum	minimum
S.S. 78	2.96	3.01	1.09	1.10	-0.05	-0.01
W.S. 78	3.84	3.83	1.52	1.49	+0.01	+0.03
S.S. 79	6.39	6.21	3.61	3.58	+0.18	+0.03
W.S. 79	7.63	7.67	4.68	4.83	-0.04	-0.15
S.S. 80	8.06	8.31	5.12	5.17	-0.25	-0.05
W.S. 80	9.93	9.88	6.60	6.34	+0.05	+0.26
S.S. 81	9.81	9.70	6.22	6.20	+0.11	+0.02
W.S. 81	11.13	11.15	6.87	7.01	-0.02	-0.14

5), but only with further data shall we be in a position to give a better interpretation and understanding of this phenomena.

Interpretation of the Results. The equinox data show that the temperature is not yet stable and that additional heat input is progressively finding its way into the spacecraft. The surfaces which are liable to degradations during this period are the multilayer insulation blanket (MLI), the S 13 G/LO white point on the antenna dishes, the spacecraft adapter, and some small Earth and sun sensor apertures. The temperature sensitivity of OTS is known with respect to an increase in absorptance of the antenna white paint and of the aluminized Kapton forming the outer layer of the MLI. With the updated expected values for these two materials, it is possible from the results of Ref. 6 to deduce the increase of spacecraft temperature. Kapton absorptance changing from 0.45 to 0.60 would give an increase of temperature of 2.1°C, while an increase from 0.20 to 0.60 of the white paint would have an influence of 1.1°C.In total these expected degradations account for 3.3°C, and the results in Fig. 5, which show an increase of the order of 4°C since the first equinox, do not look unreasonable. This may point to a further degradation mechanism possibility, to the finish of the spacecraft adapter. The higher temperature increases observed in solstices are caused by the same phenomena, on which increase is superimposed the effect of the OSR degradation on the illuminated radiator as well as that of some extra blanketed areas on the appropriate radiator face. By subtraction, it is then possible to deduce the increase caused solely by the degradation of the quartz mirrors, and to use a sensitivity analysis to determine the actual absorptance increase. The two radiator curves seem at present to tend to the same end temperature, thus indicating that the degradation phenomena is nearly similar on both sides of the spacecraft. This at first is somewhat surprising, as one would expect that early contamination, if there is any, would most logically accumulate faster on the colder radiator, thus increasing its degradation potential. In our case, the south radiator illuminated during winter solstice was the colder in the postlaunch period and started to be illuminated only after autumnal equinox. A closer assessment of the situation with the correct sensitivities does reveal a slightly higher degradation of the north radiator, but the difference cannot be interpreted as significant. The time constants which are associated with the solstice rise are

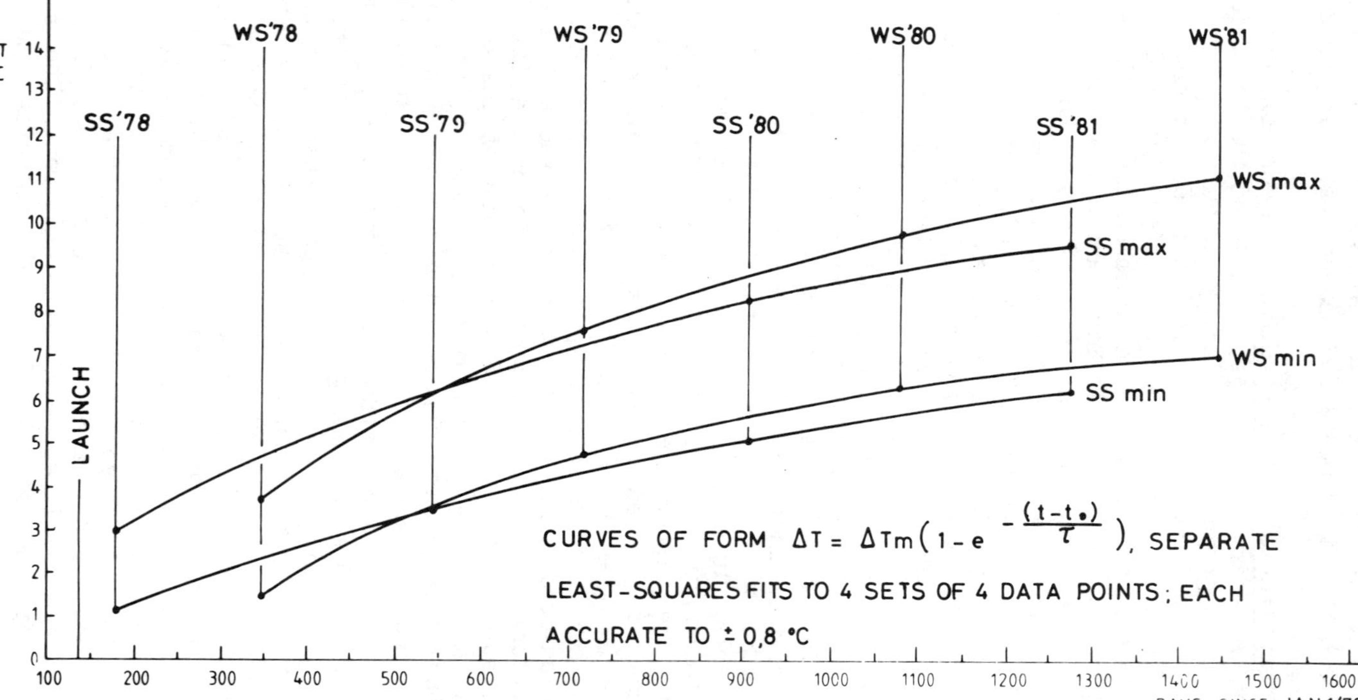

Fig. 4 Temperature increase in solstice.

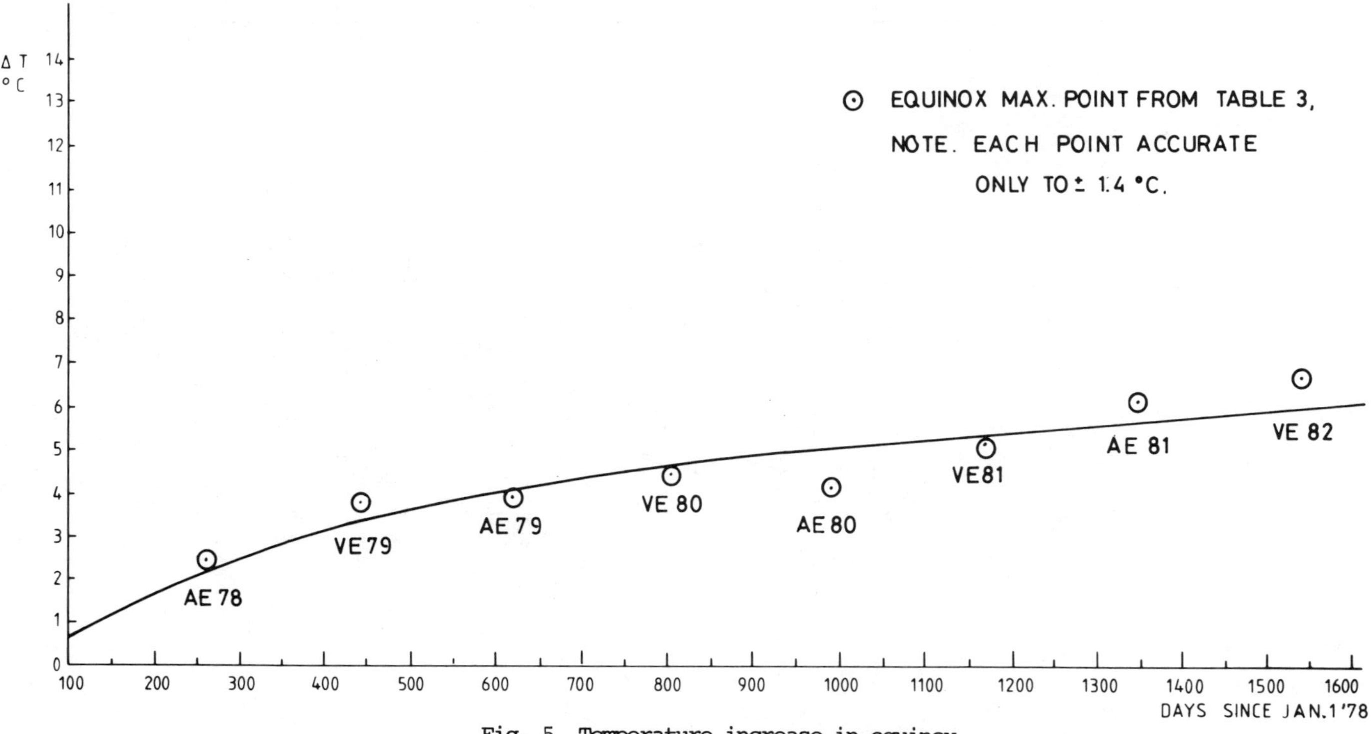

Fig. 5 Temperature increase in equinox.

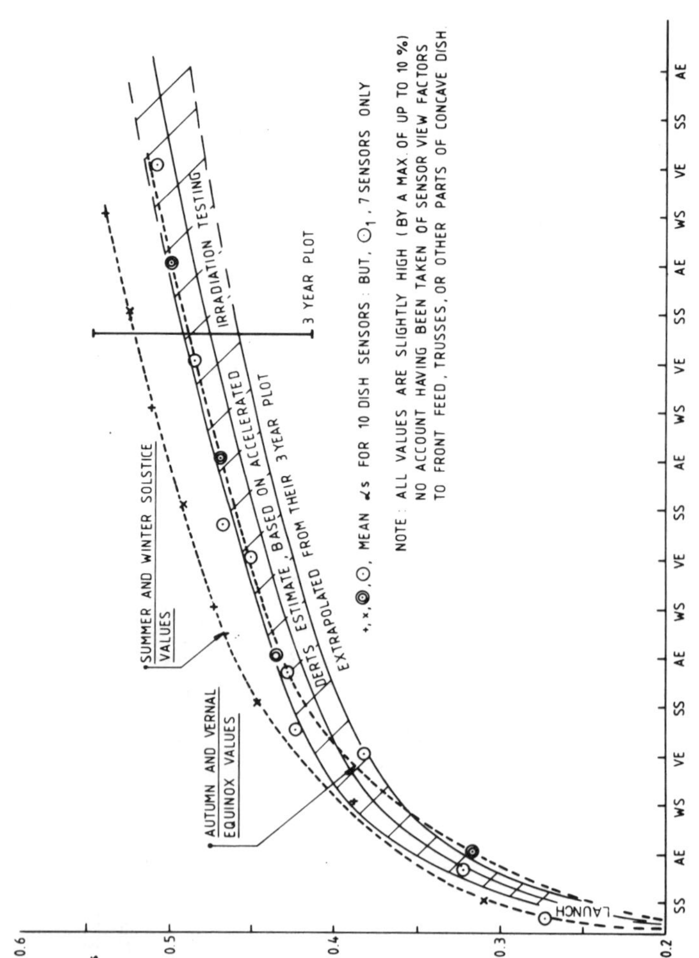

Fig. 6 OTS: estimated maximum α s on antenna dish white paint, S 13G/LO.

of the order of two or three years, which is also in good for agreement with that found OSR by DERTS,[3] while the level of degradation is much higher in orbit than in the laboratory. The possible absorptance degradations through the first three years in orbit are given in Table 6, and the asymptotic limit is, both north and south radiators, approximately 0.12. These values must be added to the beginning-of-life value of 0.09 to obtain the final value to be expected in orbit, i.e., 0.21. The exact cause of degradation is not very well known, but can be due to:

1) pure degradation of the OSR, which could be higher in the space environment, but is unlikely to be so much higher than in laboratory tests;

2) contamination of the OSRs, with subsequent degradation of the deposit; this contamination could come from the outgassing spacecraft materials, but also from expendables, i.e., the combustion products of the apogee kick motor (AKM) and/or attitude control thrusters;

3) damage caused by electrostatic discharges (in ground tests some destruction of both the quartz surface and the silver backing layer of OSR mirrors has been produced under electron charging and arc discharge[5]; or

4) a synergistic effect between any or all of the above.

While the exact cause of the degradation cannot be stated, the possibility of degradation by contamination may be supported by a rapid rise in temperature in the early days of the mission (May 26 and first summer solstice data), but uncertainties in the slopes of the equinox and solstice curves in this region render this judgment uncertain.

Table 6 OSR absorptance degradations derived from solstice data

Days from launch	409	582	773	949	1137	1312
SS	0.052	...	0.072	...	0.084	...
WS	...	0.069	...	0.097	...	0.110

(In connection with possible sources of contamination, it should be said that attention was paid to possible outgassing of both the OSR and the solar cell RTV 460-series adhesives, both via tight specification constraints on WL and VCM, and by outgassing the adhesives just prior to application. The mean 'view' of the radiators to the solar array wings was under 5%). The final section gives more information on the assessment of the degradation of αs for the white antenna dish paint.

Thermal Degradation of the Antenna White Paints S 13G/LO. The OTS antenna dishes are coated with a matt-finish white paint of relatively low initial solar absorptivity, although like all white paints it degrades significantly with time in the space environment. The spotbeam antenna dish was instrumented with a total of 14 temperature sensors (three platinum resistance type and 11 thermistors), mostly under the front skin of the dish but with a few at the rear skin to allow the investigation of thermal deformation effects. As the temperature gradients front to rear were observed to be of the order of only 2 to 6°C when the dish was hot, it was first decided to use all 14 sensor outputs, plus a few simplifying assumptions, to attempt to estimate the solar absorptivity of the front paint surface at different times in the spacecraft's history. On each occasion, the appropriate solar intensity and angle of the solar vector to the antenna platform were obtained, and the maximum incident solar flux density could then be calculated for each sensor. A maximum possible value of solar absorptivity was next calculated for each sensor, at that selected date, by assuming that an emissivity of 0.9 existed on the front face, and that no absorbed heat traveled through the dish from front to back (almost true) or laterally (more true still). Typically, the 14 values so obtained showed a fairly large percentage spread, even after normalization for position on the dish; much of this spread is due to the fact that these initial calculations took no account of view factor (either averaged or individual) of each sensor area element to the feed, the trusses, or the rest of the dish. However, as the various values changed their relative magnitude positions at different times of the year, again showing varying effects of individual view factors, it was decided to use the arithmetic mean of the 14 as a gross estimate of dish solar absorptivity. A sensitivity study was carried out to assess view-factor effects, and it was concluded that all gross solar-absorptivity values should be reduced by about 10%. However, only the gross values are used in Fig. 6, and these are calculated from soon after the launch of OTS-2 up to the recent 1982 vernal equinox. On the same graph is the nominal beginning-of-life absorptivity for the matt-finish white paint (0.18) and a predicted band of absorptivity values as a function of in-orbit time prepared from ground tests (irradiation with ultraviolet light, electrons, and protons, to simulate the space environment) carried out for ESA at DERTS in Toulouse. Owing to deterioration in performance of some sensors with time, a group of ten of the original 14 sensors has been used to calculate all the points in Fig. 6 (with the expection of VE 1982, where only seven are used). It is interesting to note that a good line can be drawn through the winter and the summer solstice points, although the angles that the sun vector makes with the planes of the sensors are quite different in the two cases. Similarly, a good line can be drawn through the autumn and vernal equinox points, although here it is to be expected, since the sun is at the same position relative to the antenna on both occasions. The reason for two separate lines is probably due to the fact that in equinox the antenna sensors never reach potential maximum temperatures because of the eclipse period, and thus the αs values calculated are also slightly low. The upper line therefore is more representative. It can be seen that while the first two values after launch are significantly greater than expected, subsequent values are generally only a little above the prediction band, with some in or below it; remembering that all gross values as plotted should be reduced by about 10%, the fit of orbital data and preflight estimates is seen to be good. White paint is notoriously susceptible to contamination and dirt in its original low-absorptivity state, and a degree of contamination may well have occurred before the first orbital results could be obtained, since telemetry from the sensors on the antenna dish was only available after ABM firing and solar-array deployment. This could explain the otherwise anomalous result for the immediate postlaunch period.

Conclusion

After almost four years of orbital data collection, valuable information has been gathered, and can now be interpreted with a good degree of confidence.

1) The equinox data show significant increase in temperature with time, this increase apparently still continuing. This was not expected to increase for so long,

and is related to degradation of the coatings on the surfaces rotating daily with respect to the sun vector (white paint on antenna dishes, outer Kapton layer of MLI, and possibly the spacecraft adapter).

2) In between successive solstice seasons, the increase in temperature has been larger than expected. This seems to be the result of an increase in absorptance of the OSRs, but the exact reason is not yet known. It is most likely that it is caused by some contamination generated at the beginning of the mission. For the whole spacecraft, good exponential curve fits occur for all four solstice plots.

In order to reduce the level of degradation observed, it is suggested that the radiators be protected from possible AKM/thruster contamination. A second suggestion would be to integrate heaters on the radiators for the sole purpose of warming up the cold radiator in the early part of the mission (when power is abundant), thus dispersing some of the contaminants which could have been initially trapped on the cold surface.

References

[1]Bouchez, J. P. and Gulpen, J., "The European Geostationary Communcation Satellite OTS - Two Years of Thermal Control Experience in Orbit," AIAA Paper 80-1500.

[2]Bouchez, J. P. and Howle, D. H., "The Orbital Test Satellite OTS-2 Two Years of Orbital Thermal Control Experience," ESA Bulletin 26, May 1981.

[3]Paillous, A., "Qualification des Revêtements de Côntrole Thermique aux Rayonnements Ultraviolets et Particulaires de l'Espace," Final Report ONERA/DERTS Document 79/4025/TS, July 1975.

[4]Minier, C. F., "Microscopic Examination of the Behaviour of Non-Conductive Optical Solar Reflectors after Electrical Breakdown," ESA TM 172, Sept. 1972.

[5]Gülpen, J., Konzok H. G., and Stümpel D., "OTS Thermal Subsystem Performance Evaluation based on First Orbit Data," ERNO Report AN-TD3 145/78, June 30, 1978.

[6]Bouchez, J. P., "Sensitivity Study on Winter Solstice 79 Mathematical Model," ESA Memo TST/JPB/3610/eh.

[7]Stultz, J. W., "Solar Absorptance of Second Surface Mirrors for High Angles of Incidence," AIAA Paper 74-670, 1974.

Thermal Design and Experiment Thermal Integration of the Long Duration Exposure Facility

Robert F. Green Jr.*
NASA Langley Research Center, Hampton, Va.

Abstract

The Long Duration Exposure Facility (LDEF) is a large (14 ft in diameter by 30 ft long) Shuttle transported, reusable spacecraft. The LDEF can accommodate up to 13,000 lb of experiments mounted in 86 standard trays. This paper describes the philosophy and approach used for the passive thermal design of the LDEF. Also discussed are the standardized guidelines and techniques used in the thermal design and integration of approximately 50 different thermally passive experiments. A technique for reducing multinode thermal models of experiments to two nodes is also presented. This approach allows the efficient thermal integration of large numbers of experiments with the LDEF spacecraft.

Introduction

The increasing demand for a low-cost means of utilizing the Space Transportation System (Shuttle) orbital capabilities for a wide variety of individual experimenters has led to the LDEF spacecraft concept. This concept allows the long-term space exposure of a large number of experiments at minimal, cost with retrieval and return to Earth for postflight investigations. The low-cost guideline has dictated a simple passive thermal design and the minimization of design and integration requirements on individual experimenters. The design and thermal integra-

---

Presented as Paper 82-0829 at the 3rd AIAA/ASME Joint Thermophysics, Fluids, Plasma & Heat Transfer Conference, St. Louis, Mo., June 7-11, 1982.

*Thermal Systems Engineer, Systems Engineering Division.

tion of the current 48 experiments, which have over 60 different thermal configurations, illustrate the magnitude of the problem. This concept has demonstrated that an Earth orbital space exposure capability can be provided for a large number of experiments within the temperature limits inherent in a passive thermal design.

Description of LDEF

Mission

The LDEF (see Fig. 1) is a Shuttle transported, reusable, low-cost, essentially passive spacecraft with accommodations for a wide variety of experiments requiring long-term space exposure. Many LDEF experiments are completely passive, depending on postflight laboratory investigation of the retrieved experiment.

The mission profile is illustrated in Fig. 2. The LDEF is carried into orbit aboard the Shuttle. Once in orbit the Shuttle remote manipulator system (RMS) removes the LDEF from the Shuttle bay and places it in a three-axis gravity-gradient stabilized mode with the cylindrical axis pointing toward the Earth and one row of experiments pointing toward the velocity vector of the orbit (see Fig. 3). The orbit for the first LDEF will be circular at an altitude between 250 and 296 n.mi. with an inclination of 28.5 deg. This altitude may degrade to between 200 and 275 n.mi. after 12 months in orbit. On a subsequent

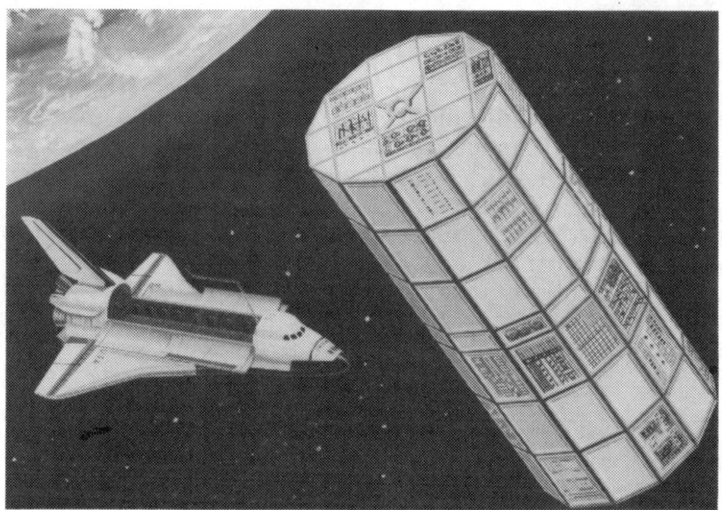

Fig. 1 LDEF in free flight.

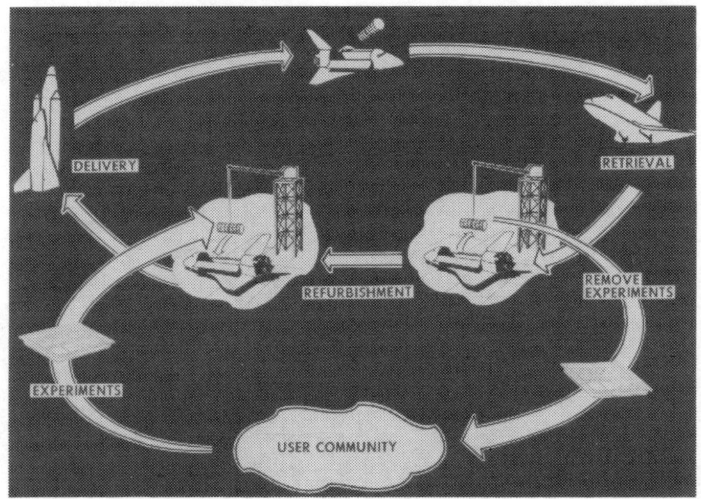

Fig. 2 LDEF mission profile.

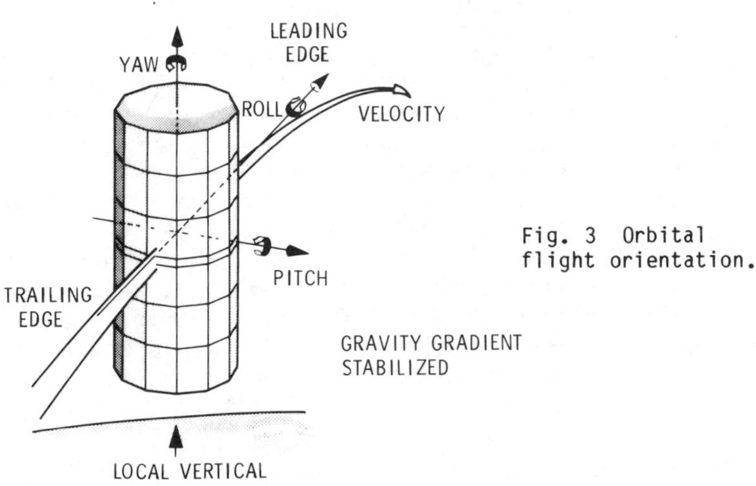

Fig. 3 Orbital flight orientation.

Shuttle flight the LDEF is retrieved and returned to Earth. The experiment trays are then removed and shipped to the experimenters for postflight inspection and analysis. The first LDEF mission is currently scheduled to fly on STS Flight 13, with subsequent missions envisioned.

Facility

The LDEF structure (see Fig. 4) is a 12-sided polygon, 14 ft in diameter and 30 ft in length. The empty facility

Fig. 4 LDEF structure.

weighs approximately 8500 lb and can accommodate trays and experiments weighing up to 13,000 lb. The structural members divide the periphery into 72 equal sized rectangular bays, each accommodating a standard size experiment tray. The end structures accommodate an additional 14 experiment trays. The structure is fabricated of standard aluminum sections using bolted and welded construction.

The LDEF provides no central power or data handling system. An initiate signal is provided to experiments when LDEF is deployed from the Shuttle, and a terminate signal is provided when the LDEF is retrieved and placed in the Shuttle bay. A viscous magnetic damper system is mounted on the facility to damp excess motions at deployment.

Experiment Trays

All experiments on LDEF are totally self-contained in standard trays furnished by NASA. The 72 interchangeable aluminum trays on the periphery are 34 in. by 50 in. long, are available in depths of 3, 6, and 12 in., and can contain experiment weights up to 180 lb. The experiment trays on the ends of LDEF are similar, but square in shape. Figure 5 shows the general construction details for the peripheral trays. The structural members forming the bottoms of the trays provide the tray/experiment attachment. The trays are clamped to the LDEF structure so as to allow for thermal expansion.

LDEF Thermal Design Requirements

The basic requirements for the LDEF thermal design are to:

1) Minimize hardware and integration costs.

2) Provide a thermal environment compatible with batteries and electronics. Temperature variations of the LDEF should be minimized.

3) Design the spacecraft for a near Earth, 28.5 deg inclination, circular orbit with gravity-gradient stabilization with one end toward the Earth. Variations in attitude of the nominal leading edge of ±30 deg in yaw as well as 180 ± 30 deg in yaw must be accommodated. A nominal 1 yr mission is planned.

4) Insure compatibility with the Shuttle environment. In addition to the orbital environment of the deployed spacecraft, the LDEF and its experiments must also be designed for the launch, orbital Shuttle bay, re-entry, landing, and postlanding environments.

5) Permit the development and thermal design of a large number of experiments in parallel with each other and with the LDEF.

6) Provide flexibility in the selection and placement of experiments on the facility. This includes maximizing the area available for experiment exposure.

7) Provide experimenters with design information, guidelines, and well-defined interfaces. These should limit constraints on experiment design and be usable by experimenters with large differences in experience and capability.

8) Simplify integration of experiment thermal models into the LDEF thermal model.

Fig. 5 LDEF trays.

A workable solution to all of the above requirements is needed for a successful thermal design.

LDEF Thermal Design Concept

The LDEF thermal design is completely passive, relying on surface coatings and internal heat paths for temperature control and equalization. The LDEF facility is a large cylindrical structure which is open on the interior with experiments mounted externally in trays on the surfaces (see Fig. 6). All interior experiment and structure surfaces are painted with a high-emissivity black paint to maximize radiation coupling across the facility and to minimize the thermal gradients around it. The gravity gradient stabilization results in a specific leading edge and allows one side of LDEF to face either the sun or space for extended periods of time.

The primary means of achieving temperature control of the average internal temperature of LDEF and of experiments is by the selection and placement of experiments and by selecting properties of thermal control coatings. Various types of experiments are placed in a checkerboard arrangement to equalize thermal properties over the surfaces of LDEF. The average LDEF interior temperature (defined as the mean average temperature of the interior surface of the

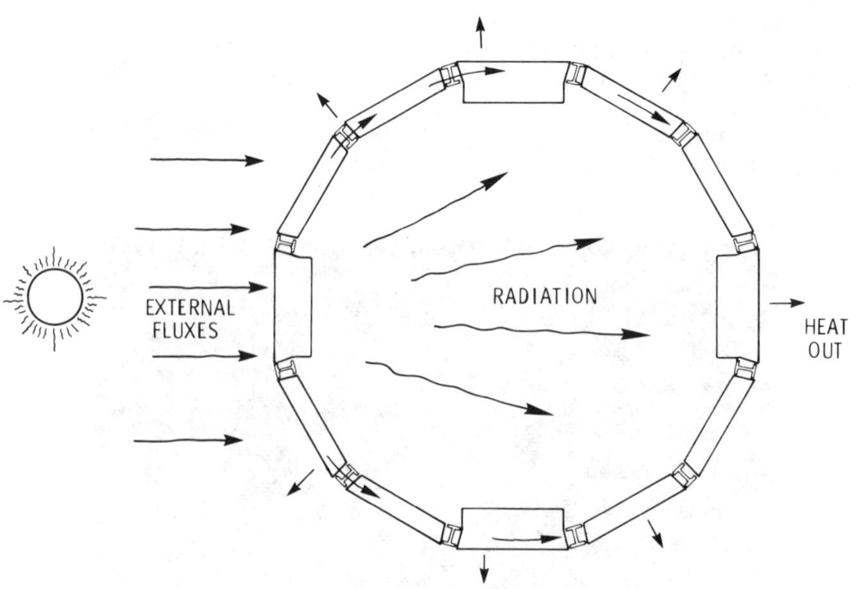

Fig. 6 Heat flow in LDEF.

experiments) is a result of the heat flow through all the experiments. The goal was to maintain this interior average temperature between 10 and 120°F to provide temperatures compatible for batteries and electronic systems.

The experimenters thermally design their experiment and tray within the LDEF supplied guidelines. The experiment thermal boundary conditions are defined in terms of the external flux, the internal average radiation temperature, and the temperature of the structure where the tray is mounted. Experiments may be subjected to different thermal environments depending upon their placement on the LDEF and by the option of coupling radiatively to the interior average temperature and/or to space. The tray is considered part of the experiment and the boundary is assumed at the LDEF structure/tray interface.

Guaranteeing the temperature average over the mission with a variety of changing or unknown experiments is a major problem. The flexibility to accommodate these experiment changes is obtained by reserving 20 to 25% of the trays for a Langley Research Center (LaRC) space debris experiment on which the thermal properties can be varied significantly after final experiment selection. These debris experiments required that coatings no thicker than 0.5 mil be used, as they reduce the quality of experiment data. A variable anodic coating for aluminum developed at Langley Research Center provides a wide range of α_S/ε (solar absorptance to total emittance ratio) with minimal solar degradation.[1] The solar absorptivity and emissivity of this chromic acid anodized coating (see Fig. 7) are selected to maintain the desired LDEF average temperature. All external structural surfaces are also anodized using this process.

Thermal verification testing is not planned for the LDEF facility; however, some components, experiments, and trays will be tested. The margin for uncertainty in analysis is therefore included in the temperature predictions.

The prelaunch, launch, and orbital environments in the Shuttle bay will be maintained at temperatures less extreme than the free-flying on-orbit environments by means of mission constraints such as time limits and attitudes. The heat soak and resultant temperature increase after landing will be minimized by preconditioning the large mass of LDEF prior to re-entry and by ground purge cooling after touchdown. Therefore the experimenter need only thermally design for a free-flying environment and be aware of the potential heat soak after landing.

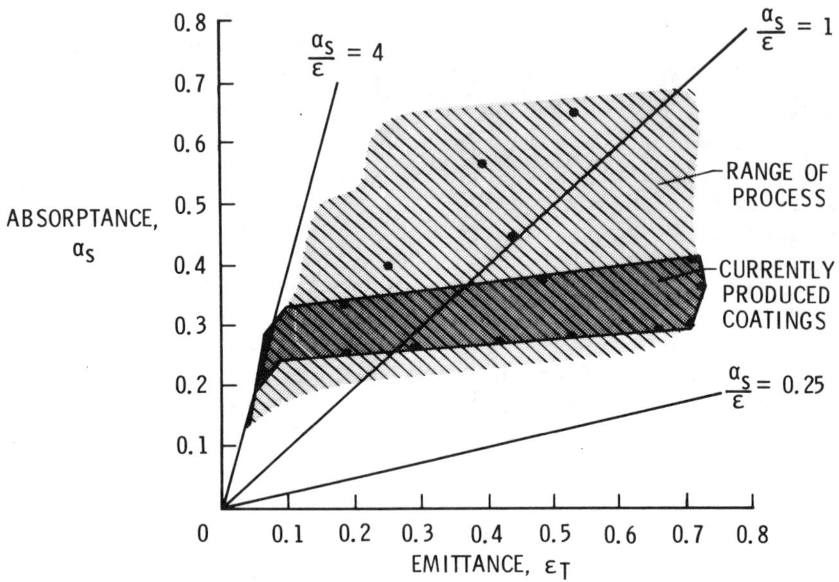

Fig. 7 Range of properties, variable anodic thermal control coating.

LDEF/Experiment Interface and Integration

The LDEF concept requires the ability to interface with, integrate, and analyze a large number of varied experiments. The following approach was developed to meet these requirements with minimal cost while allowing flexibility in manifesting experiments.

The thermal design and integration is performed in three phases. Phase I includes preliminary design, analysis, and parametric studies of the total LDEF over a range of orbital conditions, as well as with various thermally generic experiment complements. With information from these studies, an LDEF Experimenter Users Handbook is prepared to provide each experimenter with the tray thermal description and model, thermal boundary conditions, and design guidelines.[2] The thermal contents of this handbook will be covered in more detail.

Phase II involves design, review, and evaluation of all proposed experiments. Each experimenter has the total responsibility for the design and thermal analysis of his experiment and tray. Project Office personnel interface with the experimenters as well as participate in critical design reviews. Safety, experiment feasibility, and the effect on other experiments are the primary concern of the reviews.

Phase III of the integration includes receipt of the final experiment thermal models and conversion of each into a standard format. This process is simplified by the use of a standard LDEF Experiment Thermal Model Data Sheet submitted by each experimenter and containing all required thermal information. Experiment thermal models received on the first LDEF ranged from 2 to 100 nodes. Many of these experiment models were reduced to a standard two-node format prior to integration into the LDEF in order to minimize the size of the LDEF facility thermal model and permit interchangeability of experiments. A thermal model reduction program to be described later was developed for reducing the experimenter supplied models. Where two or more experimenters share the same tray, the integration and tray thermal analysis is accomplished by the LDEF Project Office.

After the equivalent two-node experiment models are calculated, these are incorporated into the LDEF facility thermal model and the facility thermal analysis is performed. Tradeoff studies on experiment location are conducted to satisfy thermal, weight, center of gravity, and experiment requirements. The process may be lengthy, as there are conflicting requirements. The flexibility and interchangeability of experiment thermal models make this tradeoff study easier. From these studies the experiment selection, locations, and the thermal properties of the anodized debris panel experiment are selected. This final configuration is then documented and forwarded to the STS office at NASA-JSC.

Shuttle thermal systems analysts incorporate this model and other payloads on the same flight into an integrated model of the Shuttle bay. A complete series of thermal analyses for all mission phases from launch through postlanding are run and the data checked to verify that all LDEF systems and experiments are within flight temperature limits during the mission.

Experimenters Handbook

To provide information and interfaces to all experimenters, an LDEF Experimenter Users Handbook has been prepared. The thermal sections of this handbook contain the thermal environments for each experiment location, guidelines and restraints for experiment thermal design, descriptions and thermal models of trays, and interface requirements. The information in the Handbook is summarized below.

General Guidelines

The experimenter has the total responsibility for his thermal design. He should perform a thermal analysis to assure the feasibility and survivability of his experiment and to satisfy the design and safety requirements. The LDEF Project Office provides a tray thermal description and model, tray boundary conditions, and guidelines which are used to design the experiment. Since the supplied boundary conditions and environments are for the tray, the experimenter must make an analysis of the combination of the experiment and the tray.

Specific Guidelines

Thermal compatibility with the overall LDEF design constitutes a major consideration in the design of an experiment. The following specific guidelines are given to experimenters to minimize integration problems with the LDEF:

1) A list of approved low outgassing thermal coatings is provided with a recommendation that an average α_S/ε of 0.30/0.15 is preferred by LDEF. However, average values of solar absorptivity (α_S) from 0.1 to 0.6 and emissivities (ε) from 0.1 to 0.80 are allowed. All surfaces facing the LDEF interior require a high-emissivity coating such as black paint.

2) The external surface of the experiment should be flush with the mounting surface to the tray, as the fluxes provided are for this flat surface.

3) No gaps are allowed that would permit sunlight to enter the LDEF interior.

Table 1 LDEF tray/experiment thermal environment

		Hot case, °F	Cold case, °F
Internal temperature-average (radiation environment)		120	10
Structure temperature		150	-10
Space end structure temperature		135	10
Earth end structure temperature		140	10
Solar radiation		408-451 Btu/ft^2-h	
Earth albedo		30-45%	
Earth radiation	458±4°R	72.9-77.4 Btu/ft^2-h	
Space sink temperature		0°R	

4) Experiment heat generation must be specified.

5) Experiments should be designed for environments at all anticipated locations on the LDEF, since integration may preclude specifying an exact tray location.

6) Experiments occupying a partial tray should be thermally isolated from adjacent experiments.

7) A conduction or radiation path through the experiment to the interior is desired by LDEF, but not required.

8) The experiment temperature should be biased as warm as allowable to help raise the facility temperature.

Environments

The orbital parameters, boundary conditions, temperatures, and fluxes that define the experiment tray environment are also furnished in the LDEF Experimenter Users Handbook. These include incident solar, albedo, and Earth emitted flux tables for each exposed surface, the average temperature of the inside of LDEF, and the structural boundary temperature where the tray is mounted. A sample of the data supplied is shown in Table 1. Data are furnished for the worst hot case and cold case orbital environments. Temperatures in the Shuttle bay while in orbit are assumed less severe than the free-flying LDEF. The LDEF will be thermally conditioned (currently between 10 and 80°F) prior to re-entry so that the heat soak to the payload after re-entry will not raise the experiment temperatures above flight limits. Fluxes and temperatures are provided for the Orbiter bay during and after re-entry for use in calculating experiment re-entry temperatures.

Tray Models

The thermal description and models of typical trays and the instructions for using these models in the experiment design and analysis are furnished to experimenters.[3] This includes a description of each tray type, a general sketch, a figure (see Fig. 8) showing how the tray was divided into nodes for thermal modeling (typically 16 nodes and four boundary nodes) and tables of the thermal model data including thermal capacitances, thermal resistances, and radiation connectors.

All tray models include boundary nodes for radiation to space, the LDEF interior, and the structure immediately surrounding the tray. These data are provided for all seven tray types. A sample experiment integrated into a tray is also modeled and included to demonstrate use of the model and to allow checkout of the model by the experimenter's thermal computer programs.

Experimenter Submittals

After completing the experiment design and thermal analysis, the experimenter must prepare and submit to the LDEF Project Office an LDEF Experiment Thermal Model Data Sheet. This describes the experiment and thermal model. It includes the following information: a sketch of the experiment, thermal coatings and surface areas, the effective solar absorptivity and emissivity of the surfaces, the weights and thermal masses, internal and external conductances, and radiation values. The experimenter also delivers a thermal model of the experiment, preferably under ten nodes.

LDEF Thermal Model Description

The LDEF thermal model consists of 290 nodes. The experiments, trays, and structural parts are modeled as shown in Fig. 9. Most experiments and trays are described by a two-node model (internal and external), with the internal node representing the tray and the external node representing the experiment. The tray node also has an external facing area around its edge.

The intercostals and longerons are grouped into 24 longeron nodes. To do this, the intercostals are divided into two halves and each half assigned to the adjacent longeron. This is done because the thermal gradients occur around the circumference. Separate nodes for the end of the longeron are used to take into account temperature dif-

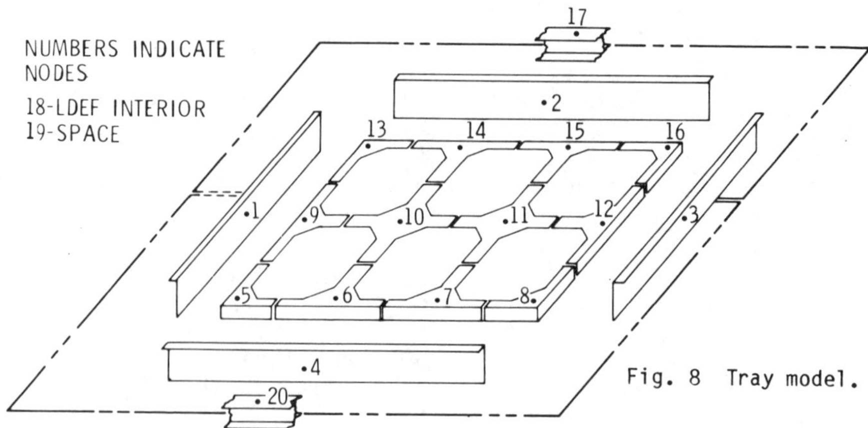

Fig. 8 Tray model.

ferences between the ends and sides of the structure. The center ring is of heavy welded construction and is modeled by 12 nodes. The end structure is modeled as shown in Fig. 10. The damper, initiate system, and supporting trunnions are also represented. The final model has 180 thermal nodes representing experiments and 110 structural nodes.

Over 20,000 internal radiation connectors are initially calculated for the open LDEF interior. The calculation of these radiation connectors includes multiple reflections and accounts for blocking by intervening surfaces. These 20,000 values are then condensed to 8000 for use in the final LDEF thermal model by dropping the very small values, connecting the deleted values to a dummy node, and then correcting the values slightly to assure that the sum of the shape factors is equal to 1 for each internal surface. The final thermal model is restricted in size to better allow the many changes and iterations required before a final experiment complement is selected.

Calculated temperature limits of the deployed LDEF and expected boundary conditions for the maximum hot and minimum cold environments are shown in Table 1. Temperatures shown include the expected variations in solar, albedo, and Earth flux, orbit parameters, spacecraft attitude, and

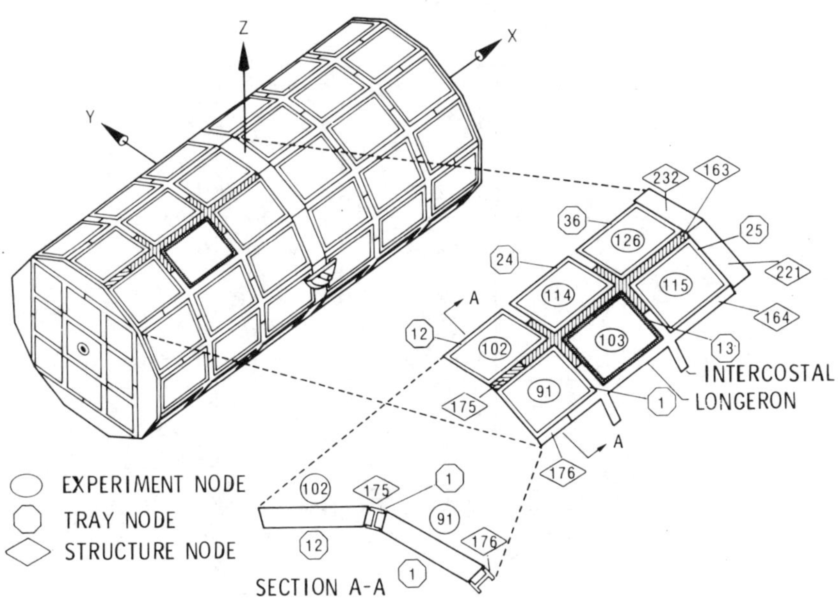

Fig. 9 LDEF thermal nodal model.

degradation of thermal coatings. An analysis uncertainty of ±20°F is included, as there is no facility thermal verification test. Figure 11 illustrates temperature distributions through the structure in a worst case hot attitude with the solar flux continuously on one side of the LDEF.

Model Reduction Program

The experiment model reduction program calculates an equivalent two-node model of a multinode experiment thermal model. The program is based on having identical steady-state energy flow through both the original and two-node model of the experiment at two given orbit conditions (see Fig. 12). The mean levels of solar, albedo, and Earth flux for the coldest and hottest orbits the experiment will experience were used as the two conditions. The two-node model matches the correct energy flow through the trays at these two orbit conditions called match points (see Fig. 13). At other intermediate conditions there is some difference in transmitted flux as calculated by the two models. Figure 13 compares the transmitted flux for two types of experiments.

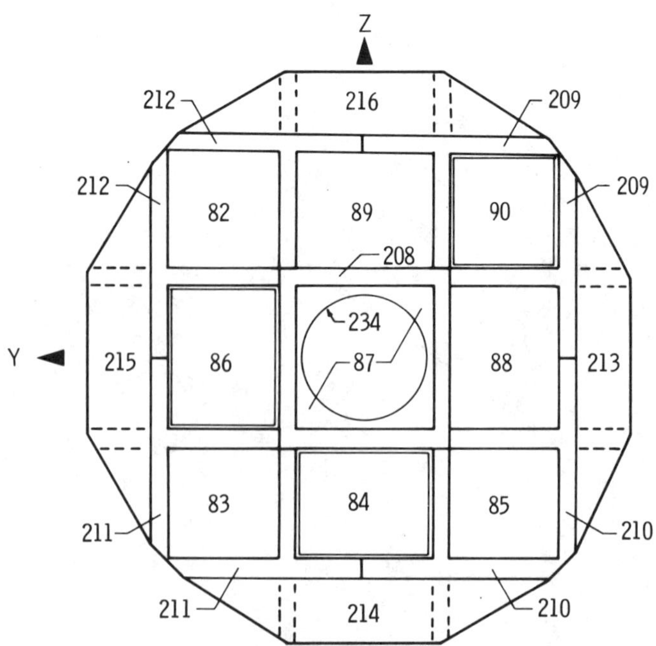

Fig. 10 LDEF thermal model - space end.

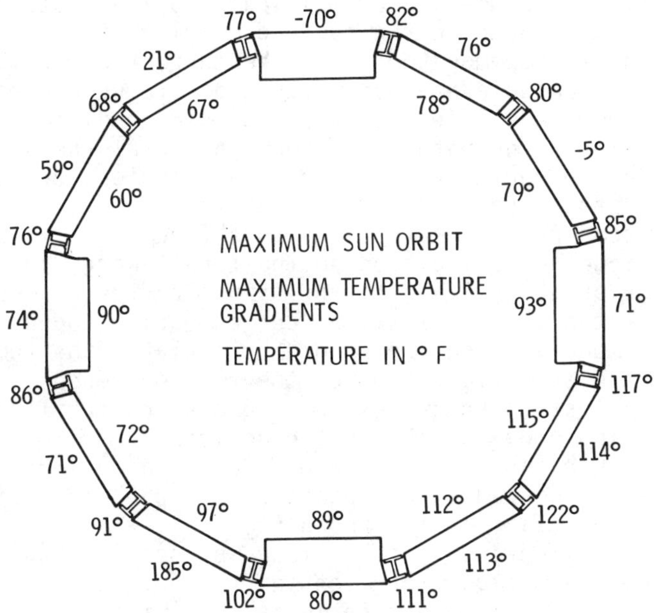

Fig. 11 LDEF temperature gradients.

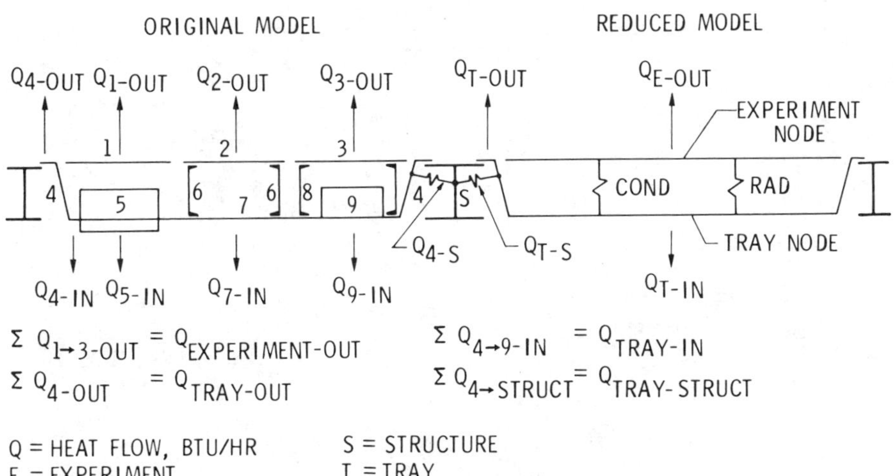

Fig. 12 Thermal model reduction program.

The following steps are taken in utilizing the program. Inputs for a hot and cold case are submitted, including LDEF internal and structural temperatures, average solar, Earth, and albedo fluxes, and the original multinode thermal model. All nodes in the original model are assigned to one of the two nodes in the reduced model. Up to 50 node models can be accommodated using the current program. The program then calculates the steady-state temperatures of the multinode model for both hot and cold conditions. It next calculates for both conditions the total energy into the tray and experiment from the LDEF interior, the adjacent LDEF structure, and to and from space.

The temperature of the tray node (see Fig. 12) of the two-node model is calculated so the total energy into the LDEF interior (Q_{t-in}) and to the structure (Q_{t-s}) equals the sum of the energy flow from corresponding nodes of the multinode model. Then the program calculates the temperature of the experiment node of the two-node model so that the energy loss to space (Q_{e-out}) equals the total loss for corresponding nodes of the multinode model.

The inner and outer nodes of the two-node model are coupled with one radiation connector and one conduction connector that satisfy the known energy flow and the nodal temperature. These values are found by solving two simultaneous equations representing the hot and cold cases. Finally, the effective absorptivity and emissivity of the

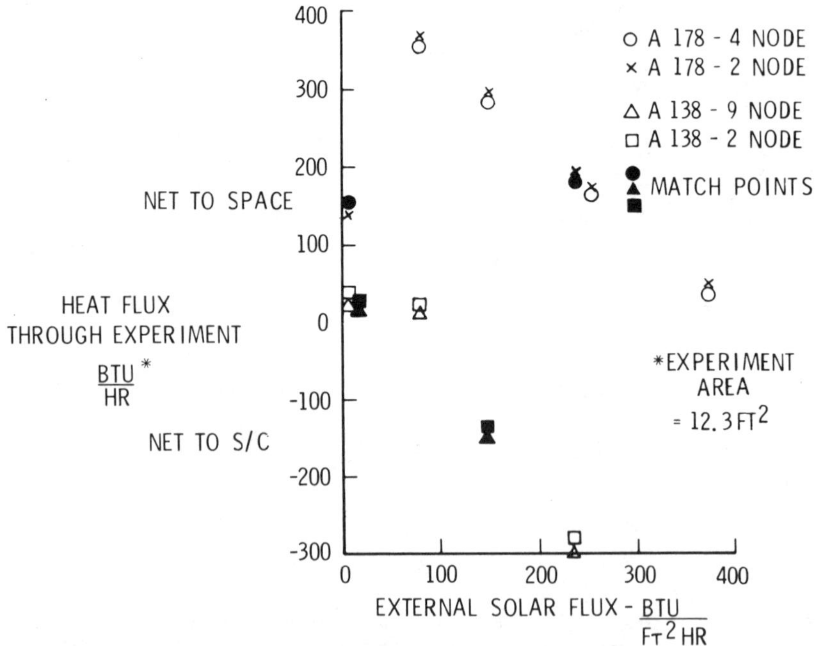

Fig. 13 Comparison of thermal models.

degradation of thermal coatings. An analysis uncertainty of ±20°F is included, as there is no facility thermal verification test. Figure 11 illustrates temperature distributions through the structure in a worst case hot attitude with the solar flux continuously on one side of the LDEF.

Model Reduction Program

The experiment model reduction program calculates an equivalent two-node model of a multinode experiment thermal model. The program is based on having identical steady-state energy flow through both the original and two-node model of the experiment at two given orbit conditions (see Fig. 12). The mean levels of solar, albedo, and Earth flux for the coldest and hottest orbits the experiment will experience were used as the two conditions. The two-node model matches the correct energy flow through the trays at these two orbit conditions called match points (see Fig. 13). At other intermediate conditions there is some difference in transmitted flux as calculated by the two models. Figure 13 compares the transmitted flux for two types of experiments.

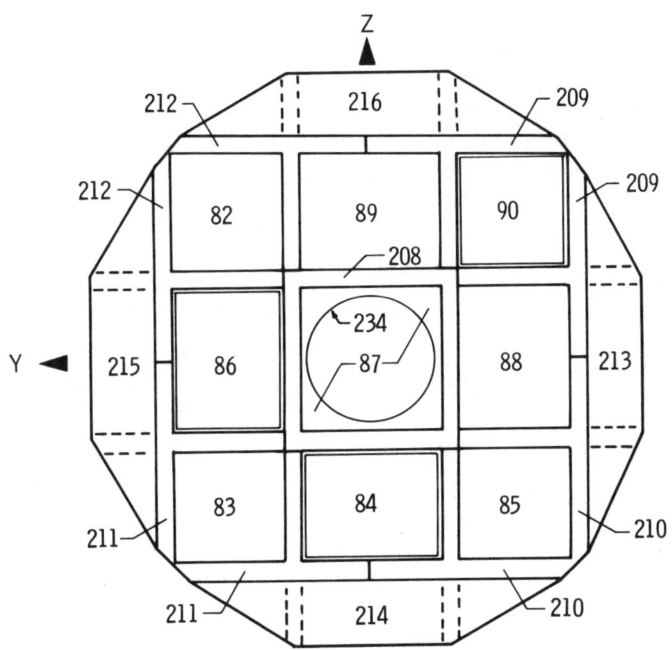

Fig. 10 LDEF thermal model - space end.

Comparison of transient temperatures for the multinode model and the corresponding two-node model for experiment A138 are shown in Fig. 14. The internal node (1) of the reduced model, which influences the overall temperature of LDEF, matches the temperature of the internal components (8 and 12) of the multinode model.

The model reduction program has worked well for the specific application of producing a two-node model with equivalent heat flow through the LDEF experiment. It has proven invaluable for incorporating the large number and variety of experiment thermal models into the LDEF thermal model.

Summary

The techniques used for the thermal integration and analysis of a large number of varied experiments from a diversified group of experimenters have proven very efficient and workable. Forty-eight different experiments from over six countries, government agencies, universities, and aerospace companies were handled in a timely manner. The author feels similar techniques and procedures will be useful in processing the large numbers of potential experiments on LDEF and other passive Shuttle payloads.

Acknowledgment

The author wishes to acknowledge the contributions and support of Harold Edighoffer, Senior Analyst with the General Electric Company, who initially developed the experiment model reduction program.

References

[1]Gilliland, C. S. and Duckett, R. J., "Variable Anodic Thermal Control Coatings," NASA Technical Brief LAR-12719, Vol. 5, No. 4, 1980.

[2]LDEF Project Office, Long Duration Exposure Facility Experimenter Users Handbook, LDEF 840-2, Change 2, April 1978.

[3]Rankin, G., "Thermal Analysis and Mathematical Modeling of Long Duration Exposure Facility Experiment Tray Thermal Model Descriptions," Final Report, NASA CR-165903, Dec. 1976.

Satellite Thermal Design and Analyses for Expendable and Shuttle Launch Environments

H. Factor* and E. A. Stipandic†
General Electric Space Division, Valley Forge, Pa.

Abstract

Designing a three-axis stabilized, geosynchronous satellite for both a conventional expendable booster and the Space Transportation System (STS) may present unusual thermal problems. The Air Force Defense Satellite Communications System (DSCS III) was originally designed and met all design requirements for launch on a Titan IIIC/Transtage. It was later required to be compatible also with the Titan 34D/IUS (Inertial Upper Stage) and the STS/IUS, with the goal of no major design changes to meet these new launch environments. Design and analyses studies will be presented only for the Titan 34D/IUS and STS/IUS launch vehicle environments. Two major concerns investigated have been heater power usage (electrical power balance) and light-weight appendage temperature responses during periods of molecular heating and during nonspinning maneuvers. Thermal worst case investigations have been made of each flight operational event by establishing scenarios consistent with launch vehicle capabilities and mission timeline constraints. Analysis has shown that excessively high heating rates of the stowed solar arrays can occur during various mission events. Preventive measures have been identified to resolve these potential temperature problems, including launch window constraints, with the goal of launching on

---

Presented as Paper 82-0828 at the 3rd AIAA/ASME Joint Thermophysics, Fluids, Plasma & Heat Transfer Conference, St. Louis, Mo., June 7-11, 1982.

*Environmental Subsystems Engineer; presently at Kulse LTD, Haifa, Israel.

†Environmental Subsystems Engineer.

Table 1 Event group naming convention with approximate nominal times

Event Group	Title	Approximate Time After Lift-Off	Approximate Duration	Comments
1	Park orbit injection	4.7 min	7.5 min	Immediately follows payload Fairing separation
2	Park orbit (roll)	12 min	55 min	Low Earth orbit; significant heating due to molecular effects
3	Transfer orbit injection	1.1 h	13.5 min	Molecular heating drops off rapidly as altitude increases
4	Transfer orbit (roll)	1.3 h	4.9-5.9 h	Duration dependent on satellite mass and other factors
5	Final orbit injection	6.7 h	20 min	Both DSCS satellites and and IUS are separated during this event group
6	Array deployment and sun acquisition	7 h	72 min	Maximum times cited for thruster enabling and solar acquisition

any day of the year. An acceptable heater power usage profile consistent with the electrical power systems capabilities of the DSCS, IUS, and STS have been developed. For the STS launch, a separate thermal control heater bus and a method to switch to different heater circuits after ejection from the Orbiter was incorporated to meet both peak and average heater power requirements. For the T34D/IUS launch, tighter launch pad environmental temperature level control was imposed to limit prelaunch heater power usage.

T34D/IUS Launch Vehicle Studies

Thermal studies for the T34D/IUS launch considered extreme sets of "cold" conditions and events which would demand maximum spacecraft (S/C) heater power and extreme sets of "hot" conditions that would yield critical upper thermal limits for the stowed solar arrays (S/A). The T34D/IUS launch thermal studies concerned themselves with the first few hours of launch, through solar array deployment and sun acquisition.

Table 1 presents approximate nominal times and the naming conventions for various event groups (EG) occurring during the T34D/IUS/DSCS II/III launch.

Following payload fairing separation (approximately 4.6 min after launch), the DSCS II/III/IUS is injected into a park orbit. Park orbit injection (EG1) is composed of three events, which include 1) payload fairing (PL/F) separation to park orbit (P/O) injection; 2) P/O injection to orientation for P/O roll; and 3) orient to P/O roll attitude. The approximate duration of EG1 is 7.5 min.

Park orbit (EG2) is a low Earth, near circular orbit, inclined 28.6 deg and approximately 55 min in duration. The IUS X (longitudinal) axis is maintained perpendicular ± 5 deg to the equatorial plane during EG2, and the IUS/DSCS II/III stack is rotated about the IUS X axis during this period for thermal considerations which will be described in a following section.

EG2 is followed by transfer orbit (T/O) injection (EG3) with an approximate duration of 13.5 min. During EG3, the IUS first-stage solid rocket motor (SRM 1) is fired and the payload is injected into a highly elliptical orbit, boosting the two DSCS satellites to a synchronous altitude. T/O injection is composed of the following events:

1) stop roll and maneuver to solid rocket motor 1, (SRM 1) attitude,

2) SRM 1 burn and coast,

3) maneuver to reaction control system 1 (RCS 1) attitude,

4) RCS 1 correction burn, and

5) orient to T/O roll attitude.

The payload is, once more, oriented with the IUS X axis perpendicular (± 5 deg) to the equatorial plane for transfer orbit (EG4). The DSCS II/III/IUS stack is rolled about the IUS X axis during T/O, which can range from approximately 4.9 to 5.9 h depending primarily on payload weight.

The IUS second-stage solid rocket motor (SRM 2) is fired near apogee of transfer orbit in order to place the DSCS spacecraft into a geosynchronous orbit. The series

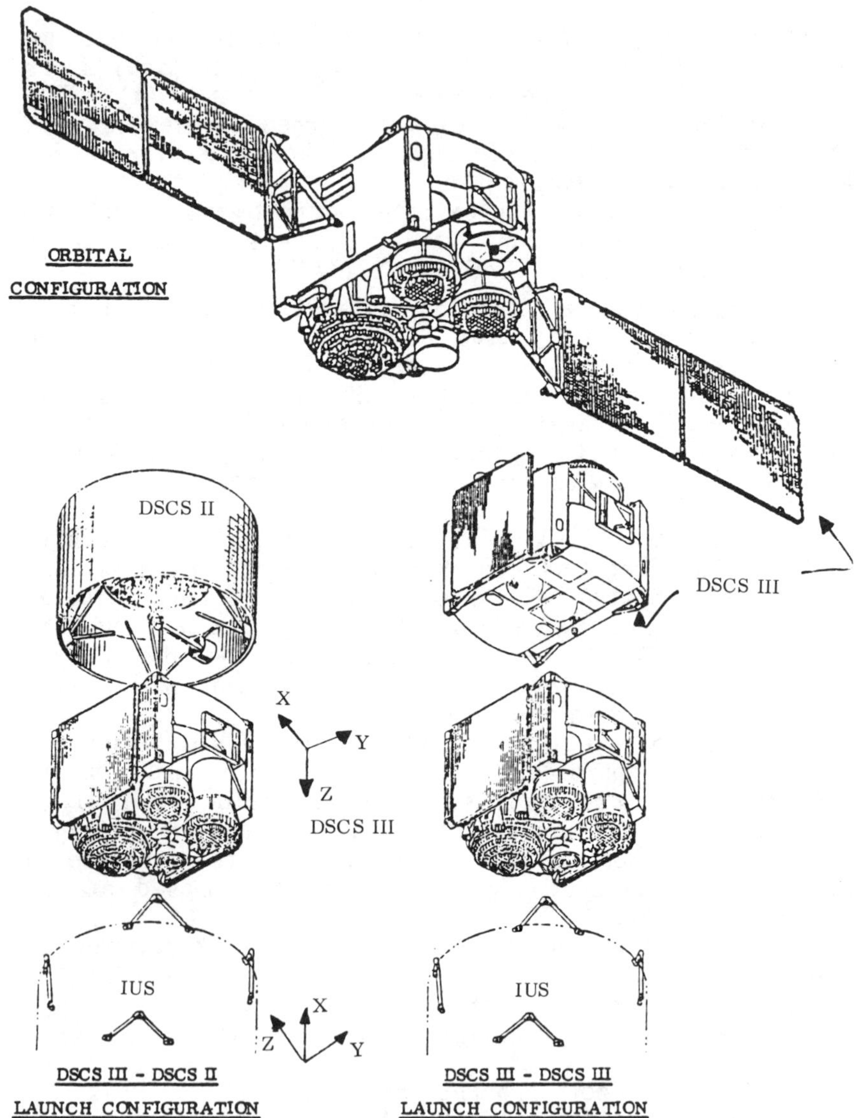

Fig. 1 DSCS III spacecraft configurations.

of events during this period is called final orbit injection (EG5). During EG5, T/O roll is stopped and the IUS first stage is jettisoned. The SRM 2 is fired after the proper attitude has been achieved. A corrective RCS 2 burn then takes place. The IUS/DSCS II/III stack is then positioned by the IUS for DSCS II release. Following separation of DSCS II, the IUS/DSCS III stack is maneuvered by the IUS to avoid collision, and subsequently positioned for release of DSCS III. All of these events, comprising EG5, have a total duration of approximately 20 min.

During EG6, the DSCS III S/C deploys its solar arrays and sun acquisition takes place. The maximum duration for EG6 is approximately 72 min. The DSCS III launch and orbital configurations are shown in Fig. 1. The major spacecraft assemblies are shown in Fig. 2.

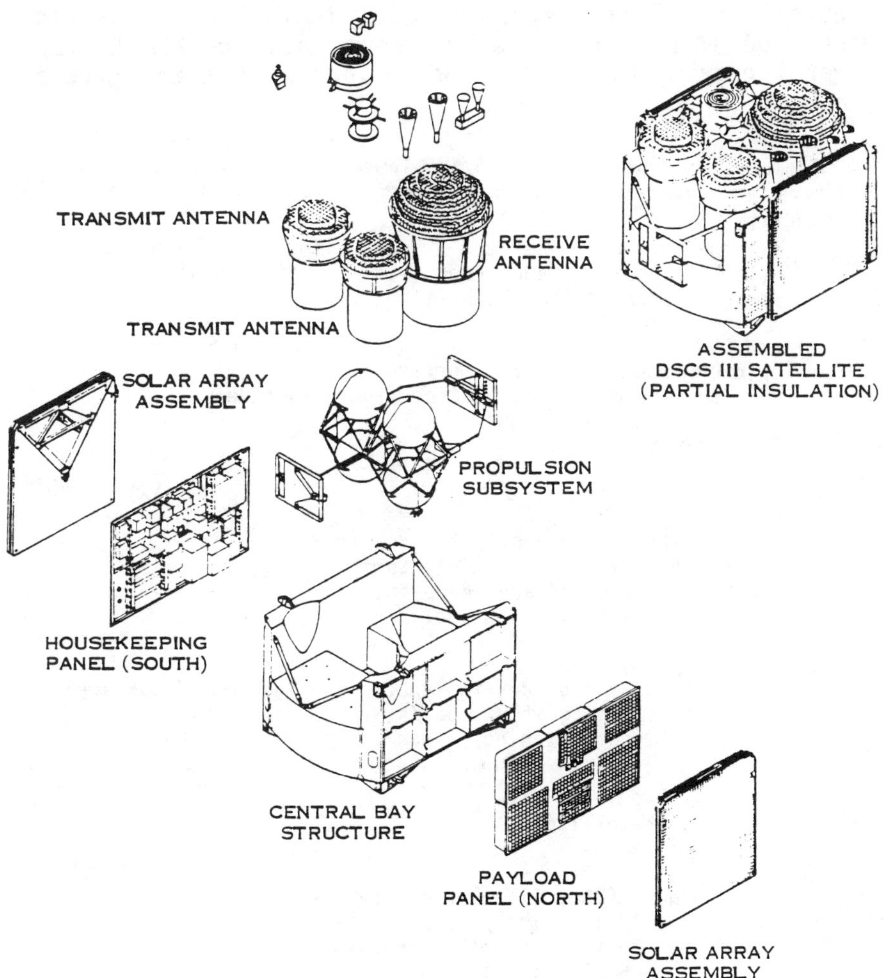

Fig. 2 DSCS III spacecraft major assemblies.

Heater Power Studies for a Cold Mission

A series of cold case analyses from prelaunch through sun acquisition were performed to define maximum heater power requirements. Conservative assumptions were applied for all satellite attitudes to yield "worst case" environments which, in turn, dictated maximum satellite heater power requirements.

The analysis for the six event groups was performed in five sections. This was done in order to more correctly model heater power requirements and to combine parts of event groups with similar environmental constraints. Three similar analytical thermal models were used in the analysis in order to calculate heater power requirements and to combine parts of event groups

Table 2 Assumption for T34D/IUS launch heater power studies

General:
- All thermal models derived from baseline 165-node model
- Satellite coatings and properties at beginning of mission
- Solar constant = 415 Btu/h-ft^2

EG1 and EG2:
- Cold environmental air temperatures
- No albedo or molecular heating fluxes included
- Satellite assumed spinning
- Total duration = 62 min

EG3:
- Duration = 13.4 min
- East satellite surfaces fully sunlit
- North satellite surfaces full-Earth-facing
- South satellite surfaces deep-space facing
- No albedo heating fluxes included

EG4:
- Satellite Z axis 60 deg to sun (no sunlight on $\pm$Z surface)
- Satellite spinning at 0.7 deg/s
- 90-min eclipse beginning 56 min after start of EG4
- Duration: nominal = 205 min; long = 354 min

EG5:
- No sun on satellite

EG6:
- "Cold" model (DSCS III with deployed solar arrays)
- Duration: 80 s (array deployment)
26 min (thrusters enabling)
45 min (sun acquisition)
- No sun on satellite

with similar environmental constraints for configurations of DSCS II/III/IUS, DSCS II/III, and DSCS III with deployed solar arrays. Three different multinode thermal models were used to represent the three different satellite and launch vehicle configurations varying between 127 and 165 nodes.

Two durations for EG4 were analyzed. A "nominal" duration was defined as 295 min; a "long" duration was 354 min. Accordingly, heater power requirements were calculated for two sets of EG5 and EG6 following the nominal and long EG4. The duration for both EG5 calculations was 19.9 min. The duration for both EG6 calculations was 72 min, which was composed of 80 s for array deployment, 26 min for enabling of thrusters, and 45 min for solar acquisition. Table 2 presents a summary of thermal assumptions, data, and analytical models used in the analysis.

Table 3 Thermal control heater power summary

Event group	Description	Duration, min	Average power, W at initial temperatures, °C (°F) of 22(72)	20(68)	15(59)	Remarks
1	Park orbit injection	6.7	10	11	20	
2	Park orbit					
	Eclipse	12.3	10	11	26	
	Sunlight	43.2	12	14	21	
3	Transfer orbit injection	13.4	14	16	22	
4	Transfer orbit					
	Sunlight	46	16	17	23	
	Eclipse	90	26	28	42	
	Sunlight	218	49	49	62	Maximum timeline
5	Synchronous injection	19.9	58	77	114	
6	Sun acquisition	72	216	216	274	
4	Transfer orbit					
	Sunlight	46	16		23	
	Eclipse	90	26		42	
	Sunlight	159	46		55	Nominal timeline
5	Synchronous injection	19.9	55		105	
6	Sun acquisition	72	188		202	

Because many spacecraft components including the telemetry transmitter are not turned on until after separation from the launch vehicle, two sets of DSCS III component thermal dissipation are used: one prior to DSCS III separation/array deployments and another following array deployment.

Table 3 presents average heater powers for the T34D/IUS following payload fairing separation. Three sets of powers are presented in Table 3 corresponding to steady-state launch pad temperatures of 22, 20, and 15°C (72, 68, and 59°F), respectively. It is readily apparent that the satellite heater power is sensitive to initial spacecraft temperature.

An important consideration in launch heater requirements is the fact that the DSCS III is switched onto internal power on the launch pad as early as approximately 7 h before liftoff. If launch environment temperatures are not at a high enough level, some heater thermostats could be activated before launch and greatly reduce the battery state-of-charge by the end of transfer orbit. Therefore it was recommended that launch pad environment temperatures be constrained to 22 ± 0.5°C. This constraint has been shown to allow an 89% state-of-charge for the DSCS III batteries at the end of a "worst case" transfer orbit scenario. This battery state-of-charge is sufficient to ensure electrical power systems margin through completion of sun acquisition.

Solar Array Studies

Another important study conducted for the DSCS T34D/IUS launch was a "hot case" solar array (S/A) response analysis. This study was of critical importance for a number of reasons:

1) In a typical flight configuration, the arrays are deployed (Fig. 1). This configuration allows heat to be radiated from both sides of the array and thereby controls its maximum temperature to approximately 55°C (131°F). During launch, however, the arrays are stowed and are effectively insulated on one side. In addition, DSCS II provides some additional localized heating and a blockage to space (Fig. 1).

2) The solar arrays are constructed of lightweight aluminum honeycomb core and lightweight aluminim face-sheets. As a result, thermal response is relatively fast.

3) The solar arrays were qualified to a temperature of 102°C based on a Titan IIIC maximum hot case environment prediction of 81°C. The 102°C limit is

very close to the structural allowable temperature of the honeycomb facesheet-to-core epoxy adhesive. As a result, special attention was given to IUS attitudes which might produce high S/A heating rates during all mission phases. (Once the solar arrays are deployed in EG6, there is no critical hot condition as previously mentioned.) Using the IUS flight operations parameters and mission timeline tolerances, worst case environmental heating rates were determined, including solar, Earth i.r., albedo, and molecular heating.

The rationale of the S/A studies initially was to identify extreme worst cases for each EG and to combine these worst cases (which individually could occur on different launch dates) in an unnatural, "inconsistent worst case" sequence. It was felt that the S/A response was satisfactory for the inconsistent worst case sequence, then any natural sequence would also be satisfactory. Unfortunately, this inconsistent worst case sequence yielded unsatisfactorily high S/A temperatures. This led to a focusing on individual EG worst cases. Table 4 presents worst cases for each event group and an abridged description of each.

For the park orbit injection sequence (EG1), various scenarios were examined involving launch dates and launch times. Efforts were concentrated on identifying launch conditions producing the worst case sun orientations in

Table 4 Worst cases for each event group

Event group	Launch date	Solar array attitude	Comments
1	May 10	Angle between sun and velocity	Low Earth orbit; molecular heating significant
2	January	S/A rolling	This launch date yields worst case initial temperature for EG3
3	January 18	S/A normal to sun	Worst possible orientation
4	Early January	S/A rolling	This launch date yields worst case initial temperature for EG5 (no Earth effects)
5	Early January	S/A normal to sun	Worst possible orientation (no Earth effects)

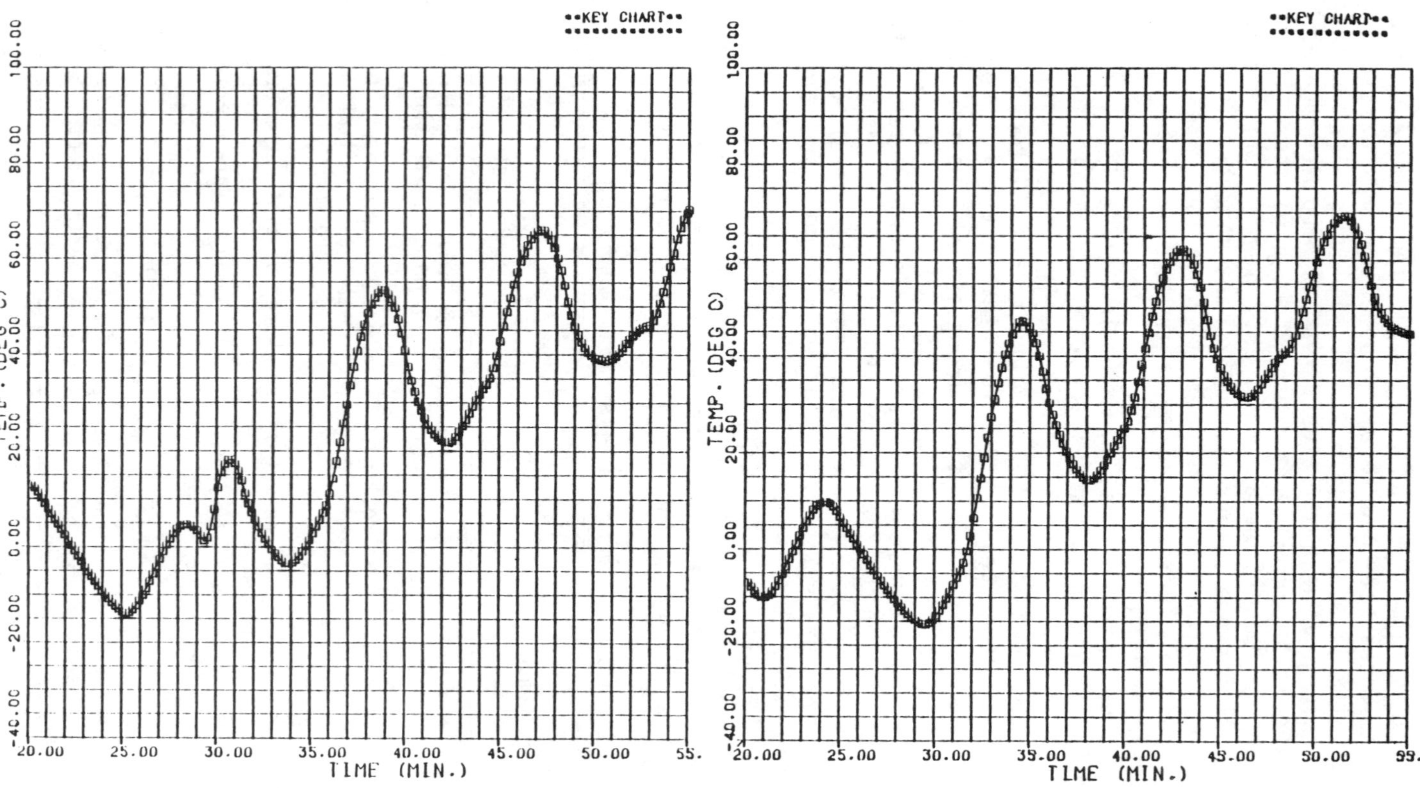

Fig. 3 Solar array thermal response in park orbit with roll reversal.

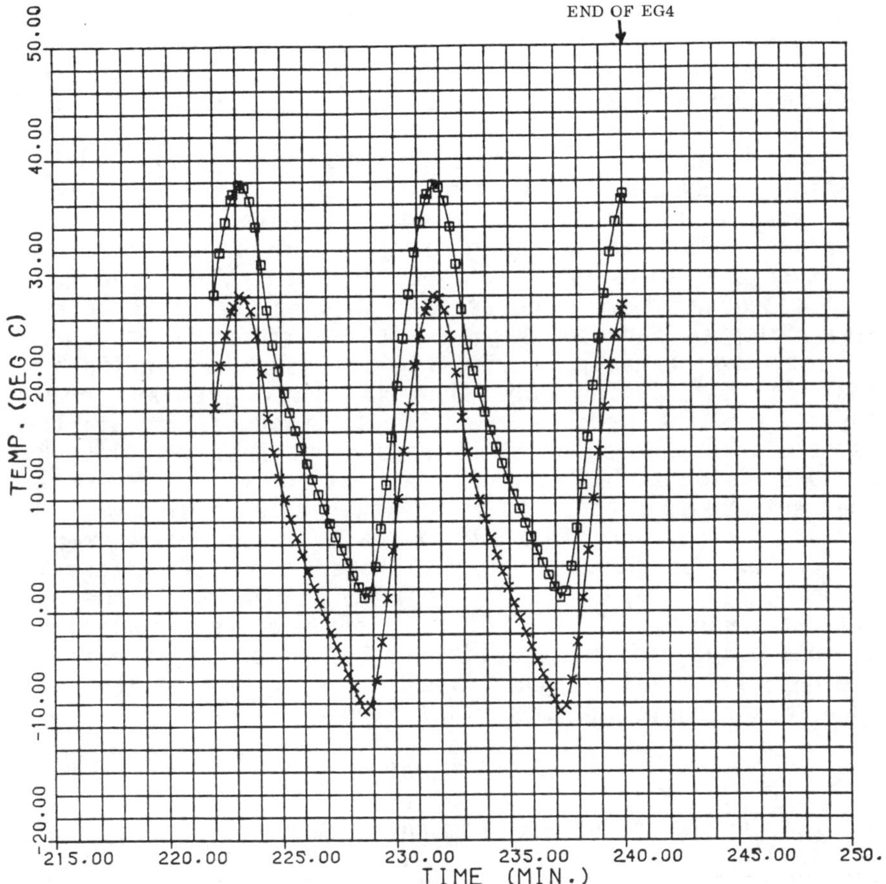

Fig. 4 Solar array thermal response in transfer orbit roll.

the early part of the EG and worst case combined molecular/sun heating during the final "stand up to roll" maneuver. Solar array temperatures in excess of 120°C were predicted for the most critical launch date identified. Analyses are in process to identify acceptable launch window times for every date of the year by adjusting launch times and limiting solar incidence angles.

Event groups 2 and 4 had no worst case that was unsatisfactory. This is because the IUS/DSCS will be rolled at a rate of 1 ± 0.3 deg/s during park and transfer orbits. These two event groups were significant, however, in that final temperatures for either event group dictated the initial temperatures for the next event group. Figure 3 shows a typical thermal response for the hottest S/A node for a minimum altitude trajectory in event group 2. Figure 4 shows the S/A maximum thermal response (repetitive) for the last 18 min of EG4.

The transfer orbit injection (EG3) was particularly critical because of length of time of nonspinning maneuvers (approximately 13.5 min) in which the sun can approach normal to a solar array. Various alternative attitude scenarios were examined, including inserting IUS roll maneuvers during the event. These alternate methods were rejected because they required either IUS hardware or software changes. Parametric analyses had indicated that for sun angles more than 45 deg off solar array normal for the entire event would produce acceptable temperatures. Based on these parametric analyses, launch windows were estimated for the entire year. Thermal analyses are in process to verify these launch windows.

Six scenarios of a worst case launch date were considered for EG5. Analysis showed that S/A thermal response could be limited by employing both IUS telemetry antennas during EG5. The solution proposed utilizing alternate IUS antennas to limit solar exposure of either the DSCS III north or south S/A by rolling out of the sun when switching to the alternate antenna.

STS/IUS Studies

The Boeing Aerospace Company (BAC), as the system integrating contractor, performed the initial STS/IUS/DSCS III/III launch analyses.[1] General Electric then used the BAC results primarily for environmental boundary conditions and initial temperature fields to further study the launch environment using more detailed DSCS III thermal models. Whereas the BAC analyses addressed the Orbiter bay/IUS and DSCS III/III, the GE analyses only addressed the DSCS satellites, with particular attention to the electrical power system's energy balance and the solar array temperatures. The BAC analyses timeline and condition flow summary are summarized in Tables 5 and 6.

Heater Power Studies

As previously discussed, DSCS III contains both control heaters (setpoints dictated by minimum operational component temperature limits) and survival heaters (setpoints dictated by nonoperational component limits).

Table 5 Launch to separation BAC thermal analysis flow

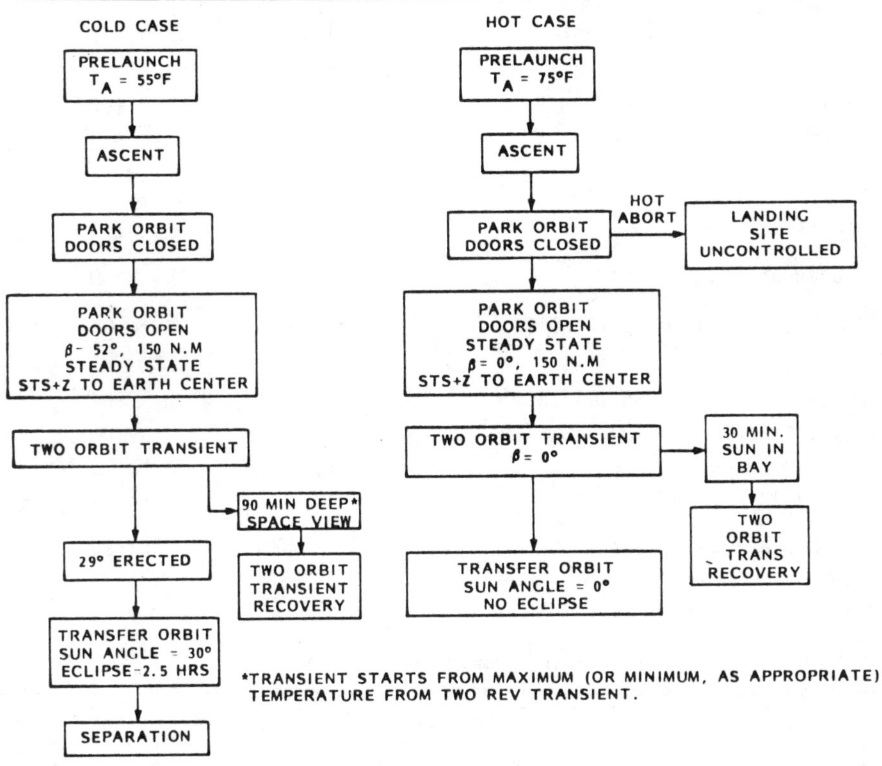

For an expendable booster, the control heaters are commanded off for launch except for the batteries and hydrazine propulsion subsystem. This ensures minimum heater power usage through sun acquisition. However, for an STS launch, this was found to be unacceptable because an extended stay in the STS park orbit could result in peak heater power demands far exceeding the STS dc/dc converter limit. Various alternative heater circuit designs and thermostat setpoint changes were investigated. A minimum design impact solution was selected in which only the primary control heaters are command engaged prior to launch. The redundant control heaters are disabled except for the batteries and propulsion subsystem. At ejection from STS, the control heaters (except for propulsion and batteries) are switched off and the satellite allowed to cool to the survival limits on the nonoperating components. This unique switching method solved the STS peak power

Table 6 STS launch analysis timeline

Mission event	Hot case, h	Cold case, h
Close Shuttle doors and DSCS III/III power up	-10:00	-10:00
IUS power up	-3:00	-3:00
Liftoff	0:00	0:00
Open Shuttle doors	3:00	1:00
Eject IUS and DSCS III/III to 29 deg	No erection	4:00
Eject IUS and DSCS III/III	6:00	7:00
SRM-1 burn	6:48	7:48
SRM-2 burn	12:08	13:08
Vehicle 2 separates	12:14	13:14
Vehicle 1 separates	12:20	13:20

Table 7 Heater power demands tabulation

Condition	Duration,h	Average power, W			Energy, W-h	Peak power, W
		Vehicle 1 aft	Vehicle 2 forward	Total		
Prelaunch for TA >16°C	Steady state	118	118	236	--	--
Park orbit, doors open, +Z Earth center	4	100	100	200	800	--
	Steady state	236	252	488	--	1199
Erected in STS	0.38	155	155	300	118	--
Transfer orbit[a]	6.0	137	189	326	1920	--
Synchronous injection through sun acquisition[b]	1.53	365	381	N/A	--	--

[a]Includes 90 min eclipse and 30 deg sun angle to roll axis normal.
[b]Assumes no sun on spacecraft for entire period.

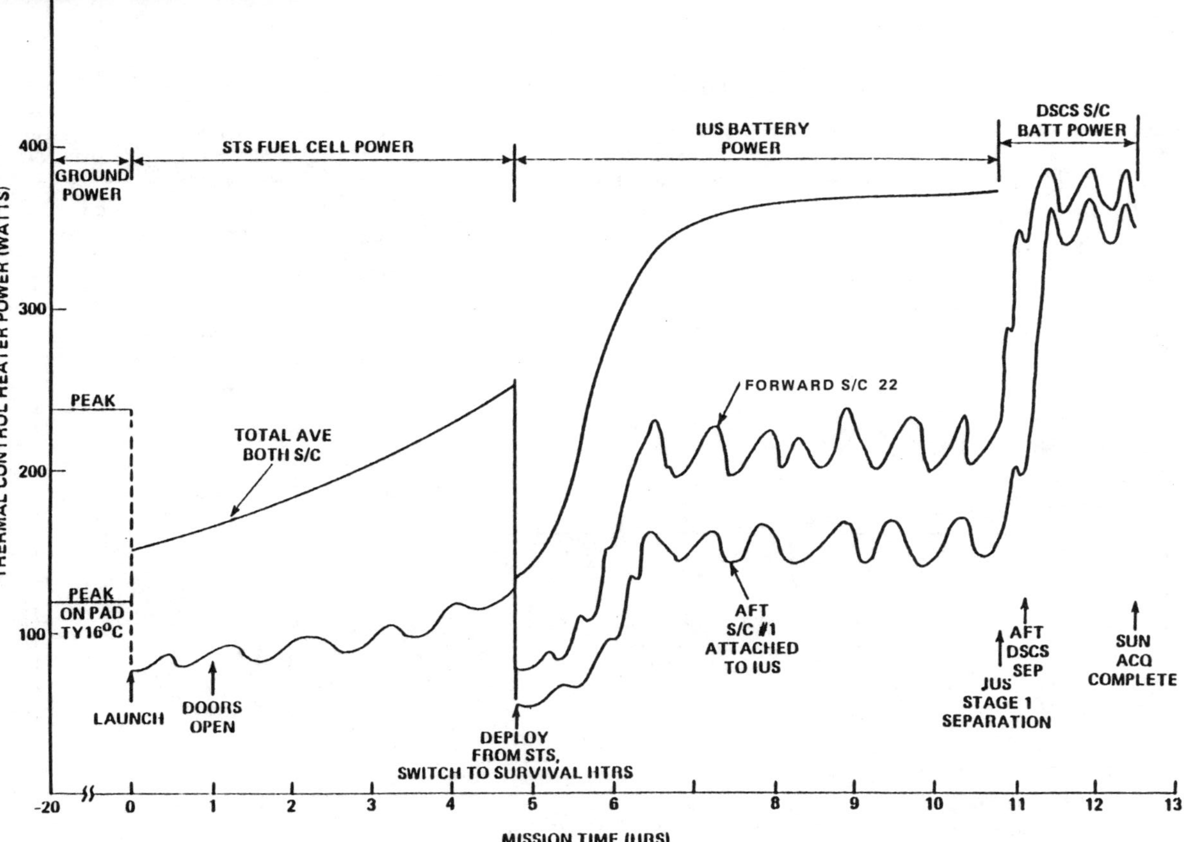

Fig. 5 DSCS III/III heater power predicted profile for STS launch.

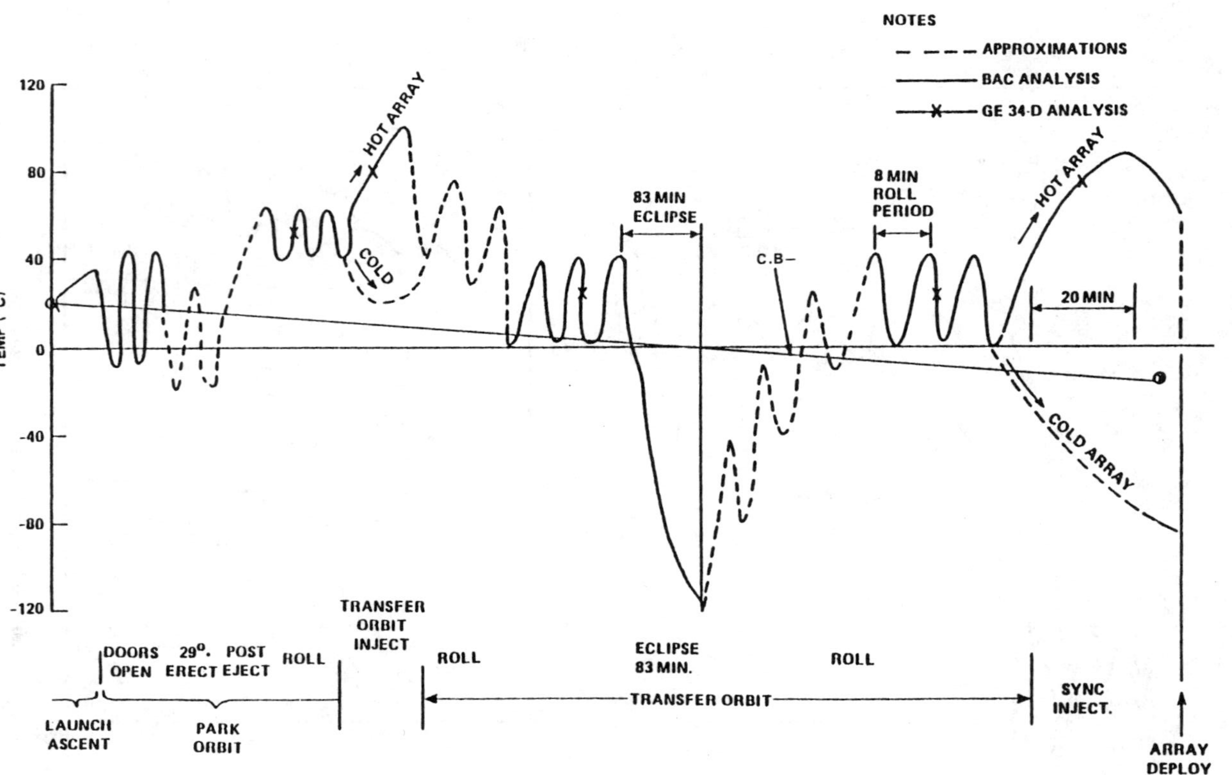

Fig. 6 Solar array temperature history local maximum/minimum.

problems and did not require additional IUS battery power during transfer orbit. A graphical representation (for illustrative purposes) of approximate heater power time history is shown in Fig. 5. Exact calculated heater power usage is shown in Table 7.

Solar Array Studies

Since the most thermally critical satellite area or component during launch is the solar array considerable attention has been paid to this item. Of particular interest were key mission phases of erection, ejection, and all IUS nonspinning maneuvers. Since the T34D/IUS studies had shown the extreme sensitivity of the arrays to direct solar impingement, an STS constraint was imposed of no sun into the Orbiter bay. This requires an STS orientation of STS + Z axis local vertical (bay toward Earth center) at all times after door opening except for star scans and IMU alignment.

During the erection and ejection phases, the DSCS III must be kept in the shadow of either the Orbiter or the Earth. Preliminary studies have been completed for the Orbiter shadow condition and indicate acceptable solar array conditions.

After STS ejection, the IUS mission profile for nonspinning maneuvers is similar to T34D/IUS. The approximate solar array temperature response is shown in Fig. 6 for an STS mission in which T34D analyses are used to approximate the actual response.

Conclusions

The conclusions derived from the launch vehicle studies are as follows:

1) Solar array high-temperature problems were identified by examining in detail the worst case attitudes/timelines for each mission phase.

2) Parametric study results produced limiting vehicle attitudes and timelines. These data were then used to establish launch window constraints.

3) The spacecraft heater power circuitry design was not adequate for an STS launch because both peak and average heater power requirements were exceeded. A separate heater electrical bus and a method of switching to different circuits were incorporated.

4) For the Titan launch, a tighter launch pad environmental control was necessary to limit prelaunch

heater power usage and satisfy the spacecraft battery state-of-charge requirements.

Reference

[1]Hartung, R. and Goo, S., "SSV/IUS/DSCS III/III Integrated Thermal Analysis Report," Rept. D290-10557-1, Boeing Aerospace Co., Kent, Wash., May 1980.

Chapter II. Subsystem and Components

Thermo-Mechanical Design and Analysis System for the Hughes 76-in. Parabolic Antenna Reflector

S.C. Clark*
Therma-Tech, Inc., Pasadena, Calif.
and
G.E. Allen†
Hughes Aircraft Company, El Segundo, Calif.

Abstract

This paper describes the methods and materials used to control the temperatures on a dichroic, multibeam antenna reflector for a communications satellite. It illustrates the combination of analysis and test programs in both thermal and mechanical disciplines used to predict thermal distortions and verify the final design.

Nomenclature

α = solar absorptance
ε = thermal emittance
τ = transmittance

Introduction

The communication performance of the Hughes Series 376 satellite is strongly dependent on the thermo-mechanical design of its antenna reflector. This component shapes, polarizes, and directs microwave beams for transmission between the spacecraft electronics and specific stations on the ground. Reference 1 discusses the microwave performance requirements for such large, multibeam antenna systems. The effects of various error tolerances are also described, and a list of needed technologies, including thermal control, is presented.

Presented as Paper 82-0864 at the 3rd AIAA/ASME Joint Thermophysics, Fluids, Plasma & Heat Transfer Conference, St. Louis, Mo., June 7-11, 1982.

*Consultant.
†Senior Project Engineer, Thermophysics Department.

Reference 2 focuses on the mechanical and thermal considerations involved in the design of the reflectors used with these systems. Diurnal and seasonal solar variations can cause temperature extremes and gradients which mechanically distort the reflector and degrade these communications signals. This paper describes the evolution of the Hughes HS 376 reflector thermal design, a process which required a combination of analysis, solar-vacuum testing, and thermal-atmospheric testing.

The configuration of the satellite and its antenna systems is shown in Fig. 1. The reflector receives microwave signals from Earth stations and focuses them on the feed array. Transmitted signals trace the same path in the reverse direction. The reflector, the construction of which is detailed in Fig. 2, consists of two circular paraboloids oriented one behind the other and joined by two cylindrical rings. Each of these four structural elements is a formed sandwich of honeycomb core and enclosing facesheets. Woven Kevlar fabric, impregnated with epoxy resin, is the primary structural material. An epoxy foam is used at selected locations in the honeycomb to provide additional rigidity, and the major parts are epoxy-bonded together. As these materials are radio-frequency transparent, copper polarizing bands are bonded to the dish surfaces to create the rf reflective properties. Horizontally polarized signals are reflected by the front dish. Vertical signals, unaffected by the front dish, are reflected at the rear.

The limits of solar variation are summarized in Fig. 3. They are typical for all synchronous orbit satellites. Diurnally, the sun line sweeps out a right-hand circle about the spacecraft -Z axis, so that each 24-h day the reflector is cycled from full front sun to rear sun and back again. The solar azimuth angle β, which defines this variation, is measured from the +Y axis. At the equinox seasons this sun line rotation takes place in the X-Y plane, so that the solar input is perpendicular to the spacecraft spin axis. At other seasons the diurnal sun line path traces out a cone at an elevation angle which depends on the time of year. At summer solstice the cone is elevated to its maximum of 23.5 deg above the plane. During this season the forward surfaces of the vehicle are illuminated and add reflected and reradiated energy to the direct solar load on the reflector. This produces the highest reflector temperatures, nearly 200°F at some locations. During the winter season the solar elevation becomes negative, as much as 23.5 deg below the X-Y plane at winter solstice. At this

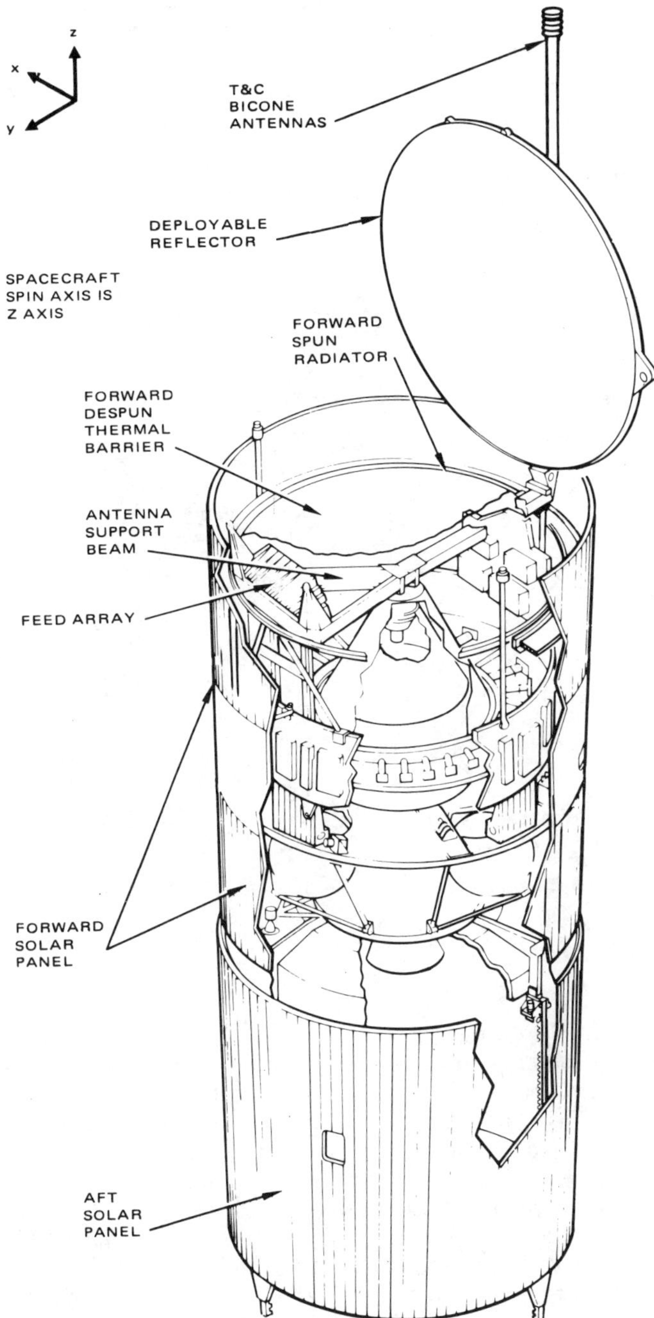

Fig. 1 HS 376 spacecraft.

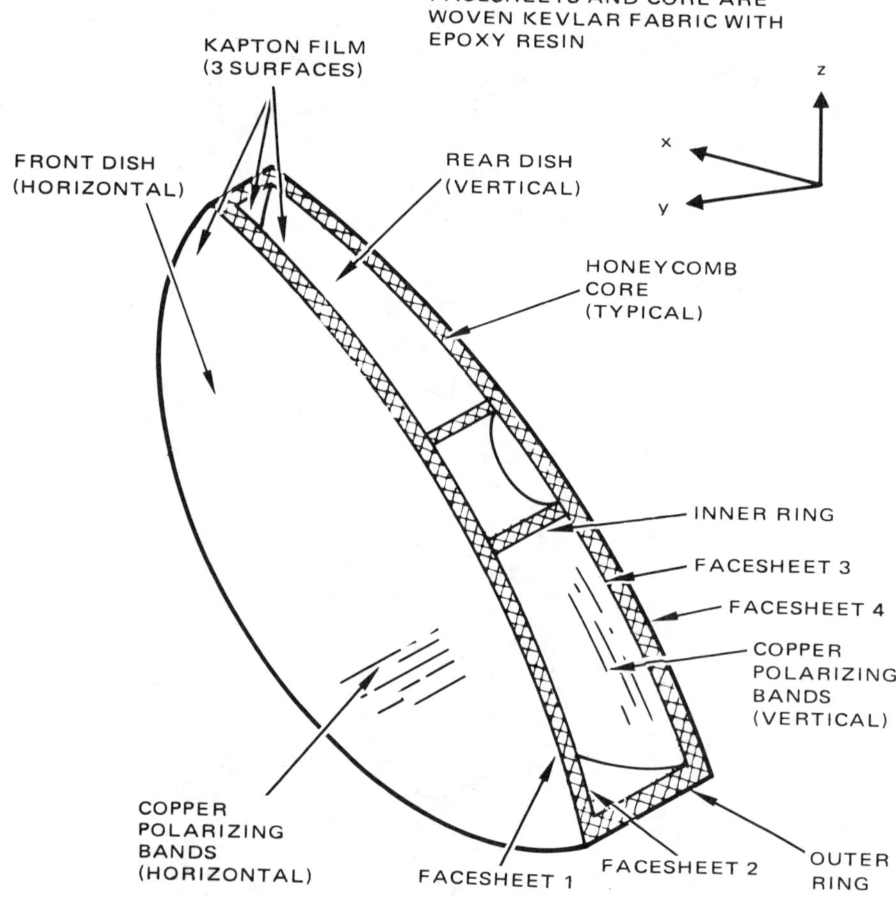

Fig. 2 Reflector construction.

time the geometry of the spacecraft and the low sun angle create a lower-half shadowed condition at the front sun position. This position can create temperature differences of hundreds of degrees on an unprotected reflector. Finally, at equinox, the satellite experiences a total solar eclipse once daily for periods of up to 72 min. These eclipses interrupt periods of full front sun illumination and produce very rapid temperature changes. Furthermore, substantial temperature differences are produced by the differential response between the inner and outermost sections of the reflector.

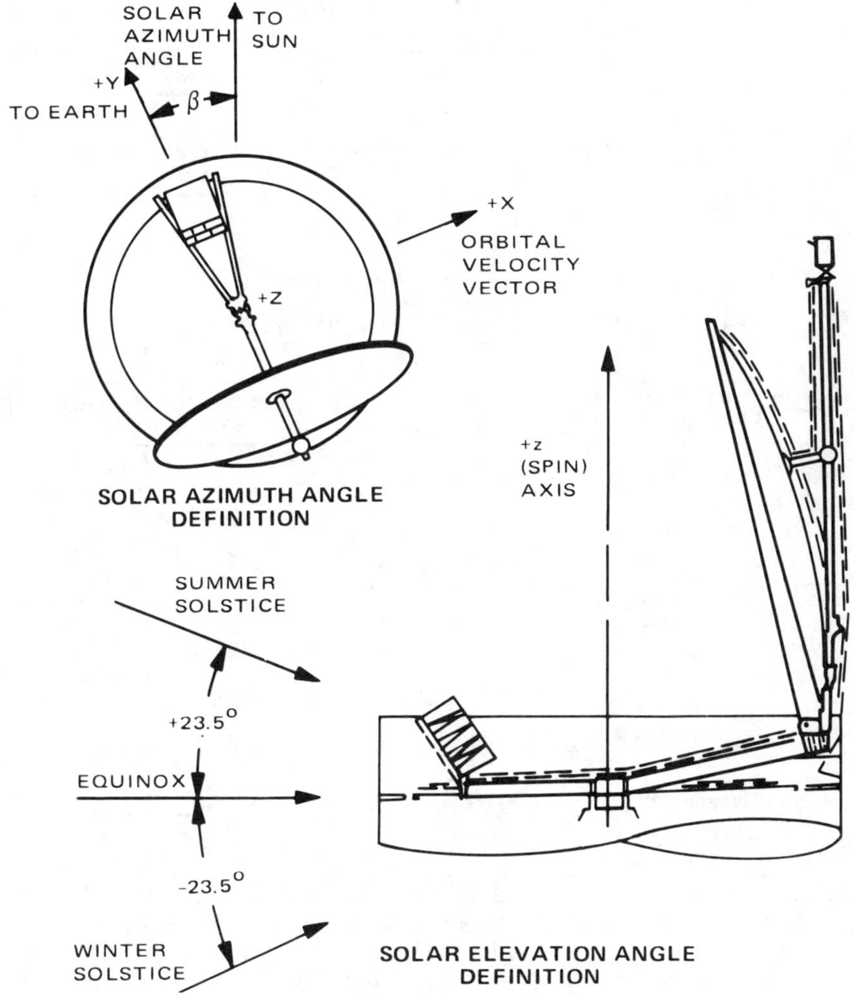

Fig. 3 Environmental characteristics.

Analytical System

The evaluation of thermal design alternatives for the antenna reflector required a systems approach to the analysis. Three analytical models are required to predict 1) antenna structural temperatures (thermal model), 2) structural distortions (mechanical model), and 3) microwave optical performance (communications model). Figure 4 shows the evaluation loop using these models to develop "production phase" thermal designs. A distinct

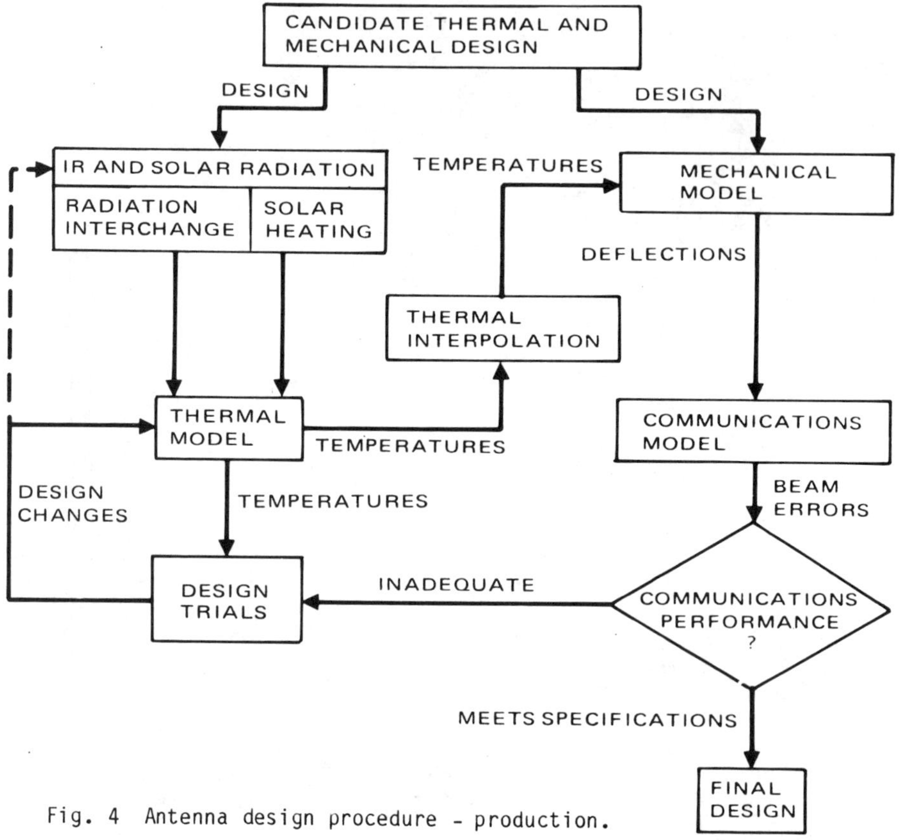

Fig. 4 Antenna design procedure - production.

technical discipline is required for each step, and large quantities of data must be transmitted at the interfaces. A thermal design is acceptable only if absolute temperatures remain within the allowed extremes for the various materials and if the temperature-induced deflections do not cause unacceptable communications performance degradation.

The prototype analysis system developed much greater complexity, primarily owing to the need to verify the mathematical models. Figure 5 shows the path which evolved during the development of the first HS 376 antenna design. Two test programs, described in detail below, were performed and resulted in model changes and a recycling of the evaluation loop. The solar-thermal-vacuum test was performed 1) to evaluate the performance of the final reflector thermal design under simulated orbital conditions and 2) to determine the ability of the mathematical model to predict temperature performance. The distortion test was

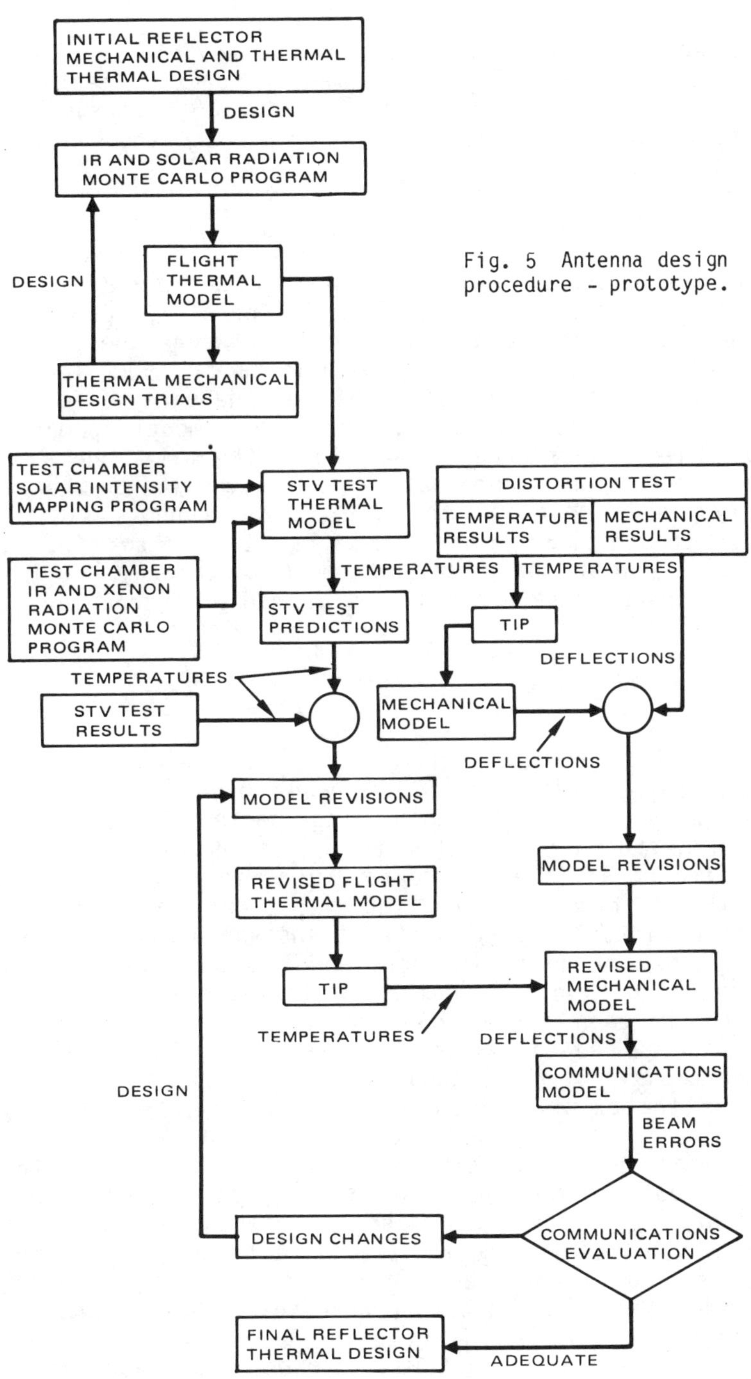

Fig. 5 Antenna design procedure - prototype.

done to obtain improved correlations between temperature and dimensional changes for adjustments to the mechanical model. The remaining discussion will emphasize the analytical and experimental features of the thermal-mechanical design.

Mathematical Models

In the systems design of complex structures, it is common to encounter differences in the nodal representation and the level of detail required between thermal and structural models. In this instance 500 elemental temperatures were needed to define adequate distortion maps. A representation of this mechanical model appears in Fig. 6. Yet, experience shows that thermal models can become unwieldy and inordinately expensive to run when the number of nodes exceeds the 100 to 200 range. This is because the number of connectors tends to increase exponentially with the number of nodes. For example, in an enclosure in which each nodal area is radiately coupled with all others,

$$\text{connectors} = \frac{(\text{nodes})^2 - (\text{nodes})}{2}$$

This equation follows from a basic relationship of combinatorial analysis. If 100 nodes are to be used, it theoretically will be required to define 4950 radiation connectors. Furthermore, the smaller the nodal volumes, the smaller the nodal time constants. The significance is that the smallest time constant in the model determines the computation interval used by the transient analysis. Thus, as the number of nodes is increased the computational cost is increased both by the number of connectors that must be processed at each inteval and by the increased number of intervals required to traverse a given orbital time period.

Additionally, the mechanical model required minimum detail in the axial dimension; this model assigns a single layer of elements to each section of two facesheets and honeycomb core. Since certain solar exposures create large gradients across the core, the thermal model required two layers of nodes for each dish or presumably double the number of nodes needed for mechanical analysis. These problems were reconciled by development of separate, specialized models and a temperature interpolation processor to perform the interface function.

The thermal model that was developed is illustrated in Fig. 7. It reflects the tradeoff between the informational

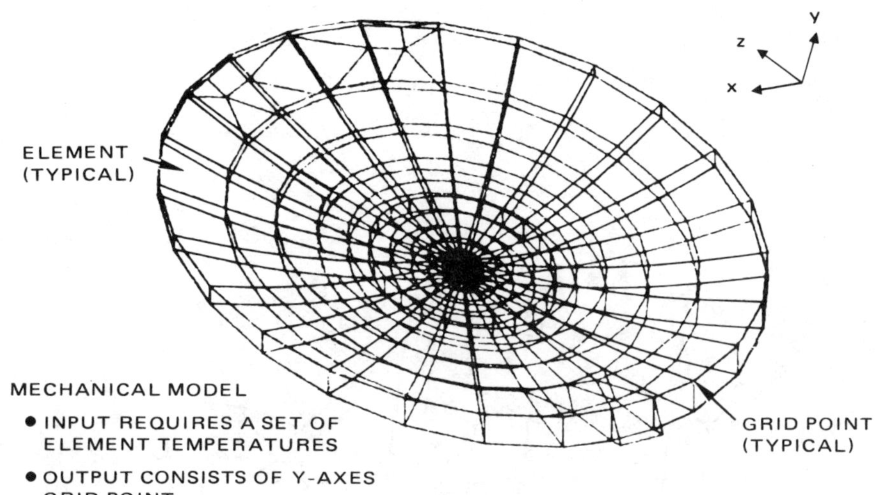

Fig. 6 Mechanical math model.

value of copious detail and the costs of complexity. Each facesheet is represented by 14 concentrically arranged nodal areas. The two central nodes cover 1.5 ft^2 each, and the middle and outer groups are about 2.5 ft^2 per node. Each "plane" simulates one facesheet and one-half of the attached core. When a thermal shield and a multilayer blanket were added to the front and rear surfaces of the reflector, this 14 node pattern was repeated in their representation.

The structural outer ring is represented by six pairs of nodes with inner and outer facesheet surfaces. Simulation of the inner ring, joining the two reflectors at about a 12-in. radius from the center, was deleted from the thermal model when it was found to have negligible effect on temperatures. In addition to the reflector itself, the current thermal model also contains representation for the support structure, feed components, and forward spacecraft surfaces, although these do not appear in the nodal diagram of Fig. 7. This model can reveal all of the major temperature patterns which affect the performance of the antenna.

A large proportion of thermal model input data pertains to the radiation interchange network. This information is descriptive of the geometric configuration and the optical properties of the surfaces involved. Because most materials treat the solar spectrum differently from longer wavelength thermal radiation, two separate radiation analyses are needed to obtain 1) the direct and reflected solar radiation

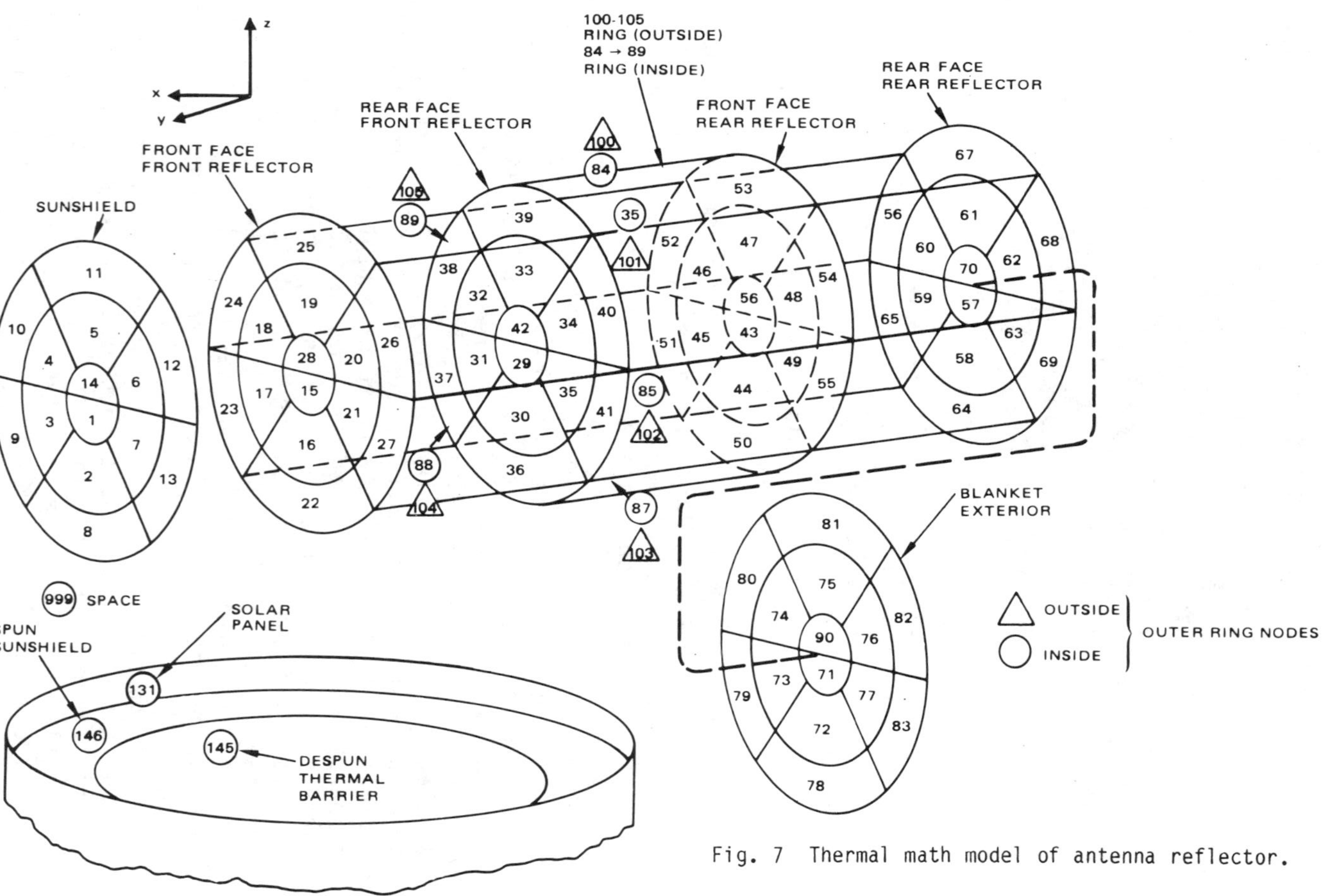

Fig. 7 Thermal math model of antenna reflector.

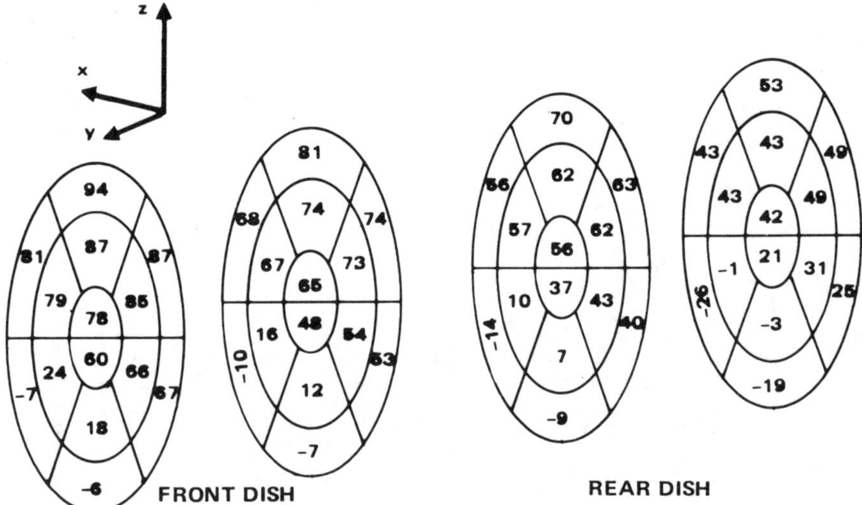

Fig. 8 HS 376 reflector temperature distribution (^{o}F).

absorbed by each node and 2) the thermal coupling between pairs of nodal surfaces at moderate, i.e., nonsolar, temperatures. A ray tracing program using Monte Carlo techniques was chosen for this task. The program had a proven ability to reliably handle a variety of shapes, specular reflections, and shadowing effects, all of which were present in the analysis (though the parabolic dishes were modeled as spheres).

Interface Programs

To expedite the transfer of data between the thermal and mechanical models, the thermal interpolation program (TIP) was created. It is a specific routine for this problem which constructs a temperature distribution for the mechanical model by interpolating geometrically among the sparse thermal nodes and associates these temperatures with the element locations in the format of the mechanical model. This allows the thermal evaluation of a variety of thermal designs over all environmental conditions at a relatively low cost. When a distortion analysis is required, the thermal program can be directed to output the data needed to run the TIP for one or more orbital positions. Figures 8 and 9 show part of a temperature distribution for a winter, oblique-front sun situation as

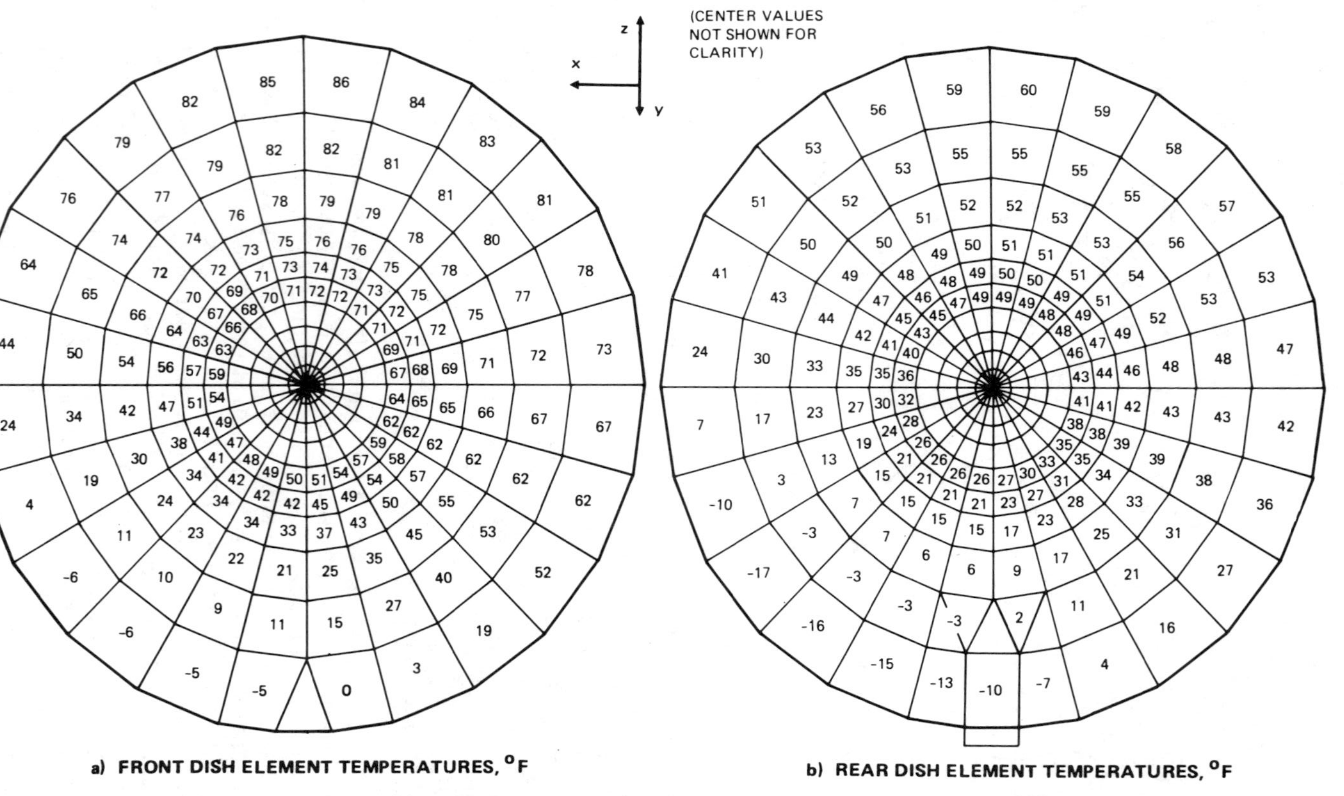

Fig. 9 TIP program output winter β = 320 deg.

input to and output from the TIP. Non-facesheet temperatures and the center temperatures of the output are omitted for the sake of clarity, and it should be noted that, in addition to the average element temperatures shown, the mechanical model receives axial (Y-direction) temperature gradient information at each elemental location. Thus the bending moments due to these gradients are accounted for in the deflection analysis.

The output of the distortion analysis consists of a set of deflections, measured perpendicularly to the aperture plane, for each of the two dish surfaces. The displacement of each grid point of the mechanical model is placed in a data set for submittal to the beam error program for the evaluation of communications performance.

TIP is not the only special processor used in this task. Computation of radiation connectors with the radiation program generates a great deal of data which must be placed in a format compatible with the thermal model. Solar heat inputs, also generated by the Monte Carlo program, require similar treatment. Small programs were coded in FORTRAN to accomplish these tasks.

Design Selected

A variety of thermal control approaches was evaluated by this process. Bare and painted reflectors were rejected for temperature extremes and large temperature differences. Figure 10a is an example of such a design exposed to the half-shadowed condition of a winter front sun. Distortion analysis showed that acceptable performance could not be obtained under this condition. The use of a sunshield over the reflector aperture moderates the gradients and low-temperature extremes, but most rf transparent materials also transmit substantially in the solar waveband. Early sunshield designs produced very high temperatures during front sun conditions because of the greenhouse effect.

The final design chosen uses a multicomponent sunshield over the reflector aperture and a multilayer insulation blanket over the convex rear surface, as illustrated in Fig. 11. The sunshield is Kapton film with a germanium coating on the outside and a layer of vacuum deposited aluminum (VDA) with an etched grid pattern on the inside surface. The germanium increases solar reflectance and provides an electrical grounding function. The gridded VDA blocks 95% of the solar spectrum while transmitting microwave communication signals. It is highly reflective to infrared radiation from the inside. This property is the

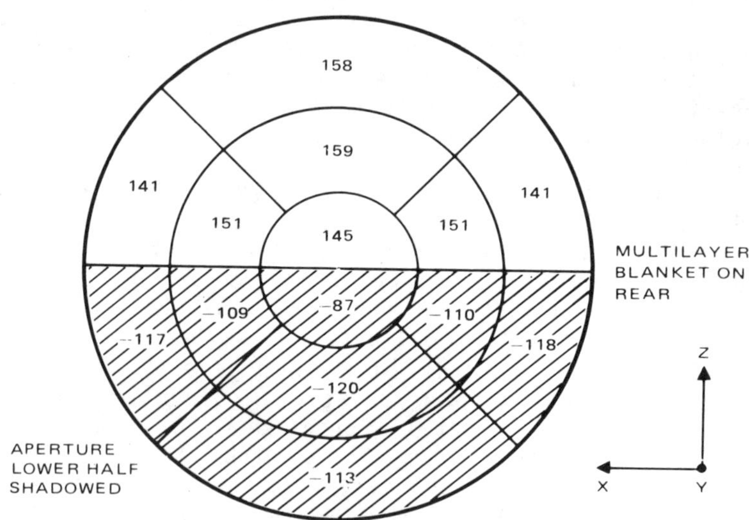

a) PAINTED FRONT SURFACE (ABSORPTANCE = 0.45)

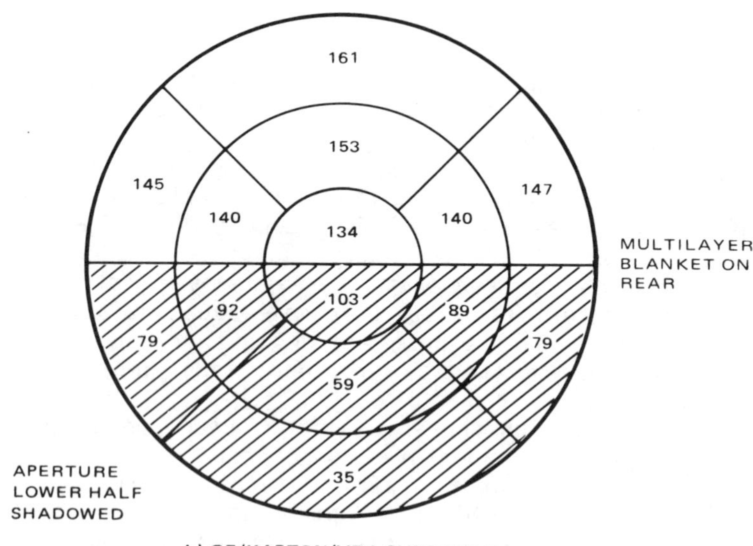

b) GE/KAPTON/VDA SUNSHIELD

Fig. 10 Front facesheet temperature distributions for winter solstice -- front sun, ^{o}F.

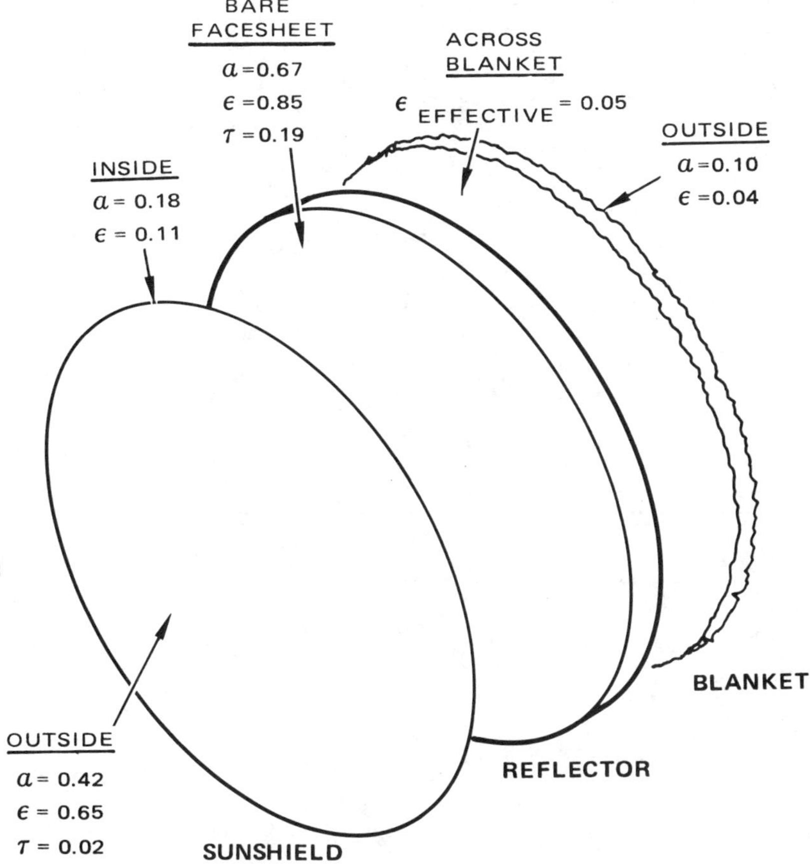

Fig. 11 Final design.

primary control of low-temperature extremes when the aperture is shaded, as in the case of side sun and eclipse conditions. Performance is also improved by this property when the reflector is partially shadowed. The high internal reflectance spreads energy from high- to low-temperature areas, reducing the differences and the distortions. Figure 10b shows the improvement obtained by the use of this sunshield under the same half-shadowed condition applying to Fig. 10a. The maximum temperature difference was reduced from 270 to 126°F.

The rear of the reflector is covered with a two-layer Kapton-VDA blanket. This blanket serves a similar function by controlling the rate of heat loss and gain by the rear dish. The aluminum side is outermost because the high α/ε ratio of this material is needed to maintain proper bulk temperature levels when the sun is on the back of the dish.

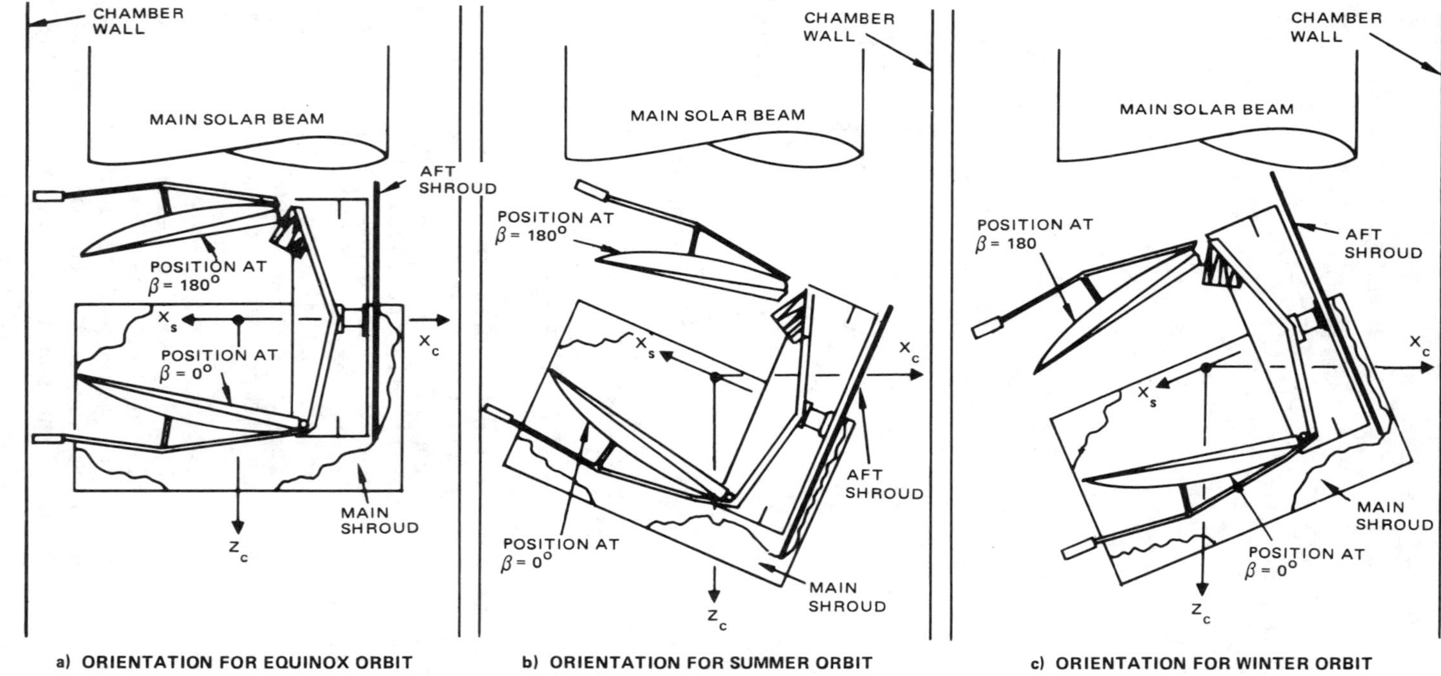

Fig. 12 Test article in solar-thermal-vacuum chamber for synchronous orbit: a) orientation for equinox orbit; b) orientation for summer orbit; c) orientation for winter orbit.

Model Verification

A solar vacuum test of the antenna was conducted at Hughes to check the validity of the thermal model. The test simulated a summer transfer orbit case with the antenna in stowed position followed by the three basic seasonal orbits (i.e., equinox, summer solstice, and winter solstice). Conditions in the test chamber are substantially different from those in space, and the test article differed in certain details from the flight design; a specific STV version of the thermal model was developed to provide a basis for applying test data to the flight model.

Both the radiation and the thermal analyzer programs figured prominently in this test effort. The solar heating and infrared interchange values which were calculated included all important chamber surfaces and accounted for the diurnal rotation of the test article in the chamber. The only effects which could not be handled directly were a slight decollimation and uneven intensity distribution of the solar beam. The decollimation effect was tolerated because of its generally minor effect on temperatures. A special mapping routine was created to account for intensity variations, however. This was possible because a detailed set of thermal mapping data for the chamber was available. The mapping data were placed in the thermal program as a three-dimensional matrix. Also input were arrays which described the location, in chamber coordinates, of each of the reflector nodes at the equinox, front sun position of the test article. A special relative solar intensity routine was developed to compute the nodal locations and, by interpolation of the mapping data, the corresponding relative solar intensity for each node around the orbit. This instantaneous relative intensity was used as a multiplicative component of the solar heating value impressed on the node.

The test article, shown in Fig. 12, consisted of the development model of the reflector combined with a mocked-up structure simulating the forward end of the spacecraft. Instrumentation included approximately 250 temperature sensors. Those installed on and in the reflector were placed to coincide with the location of the thermal model nodal centers.

When the test was completed and the results were evaluated, the reflector predictions were found to be in acceptable agreement, with one minor exception. The calculated transverse conductance value across the honeycomb core of the dishes appeared to be too large. That is, the

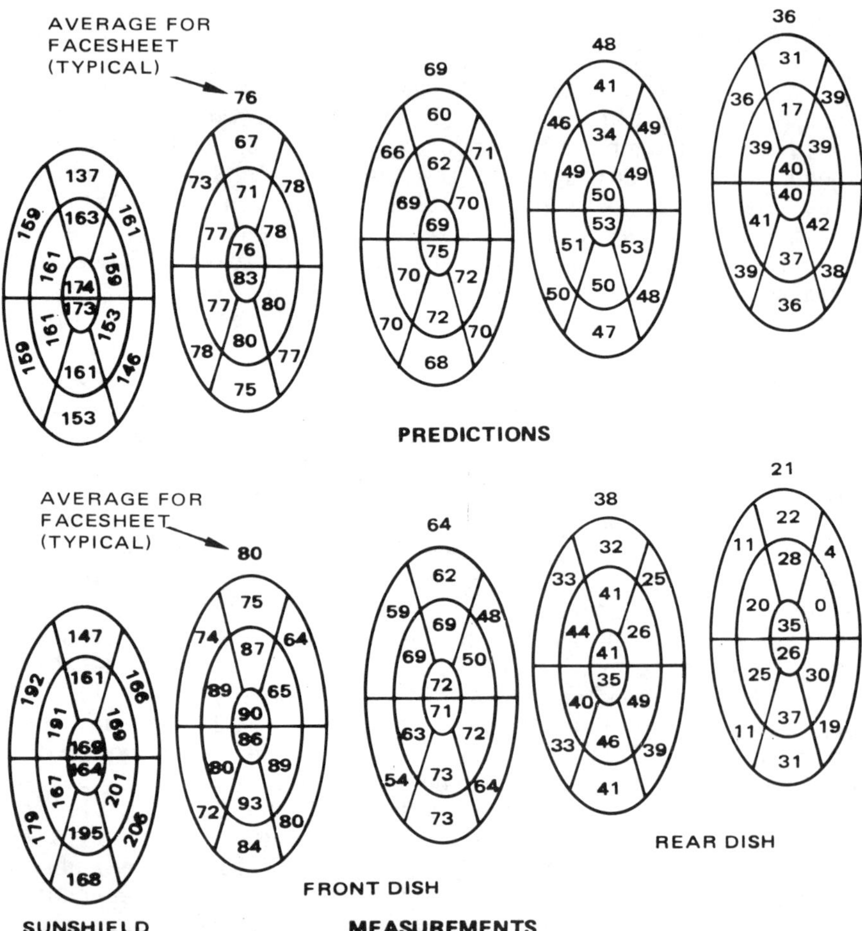

Fig. 13 Antenna STV test.

measured differences between the front and rear facesheet of each dish were significantly larger than calculated. The effect is most pronounced just after eclipse. Figure 13 shows the predicted and measured temperature distribution for the time, about 15 min from eclipse exit, when the temperature difference from front to rear facesheets is maximum. The comparison shows that the model predicts a substantially better coupling through the dish than is evidenced in the test results. Figure 14 shows the summer

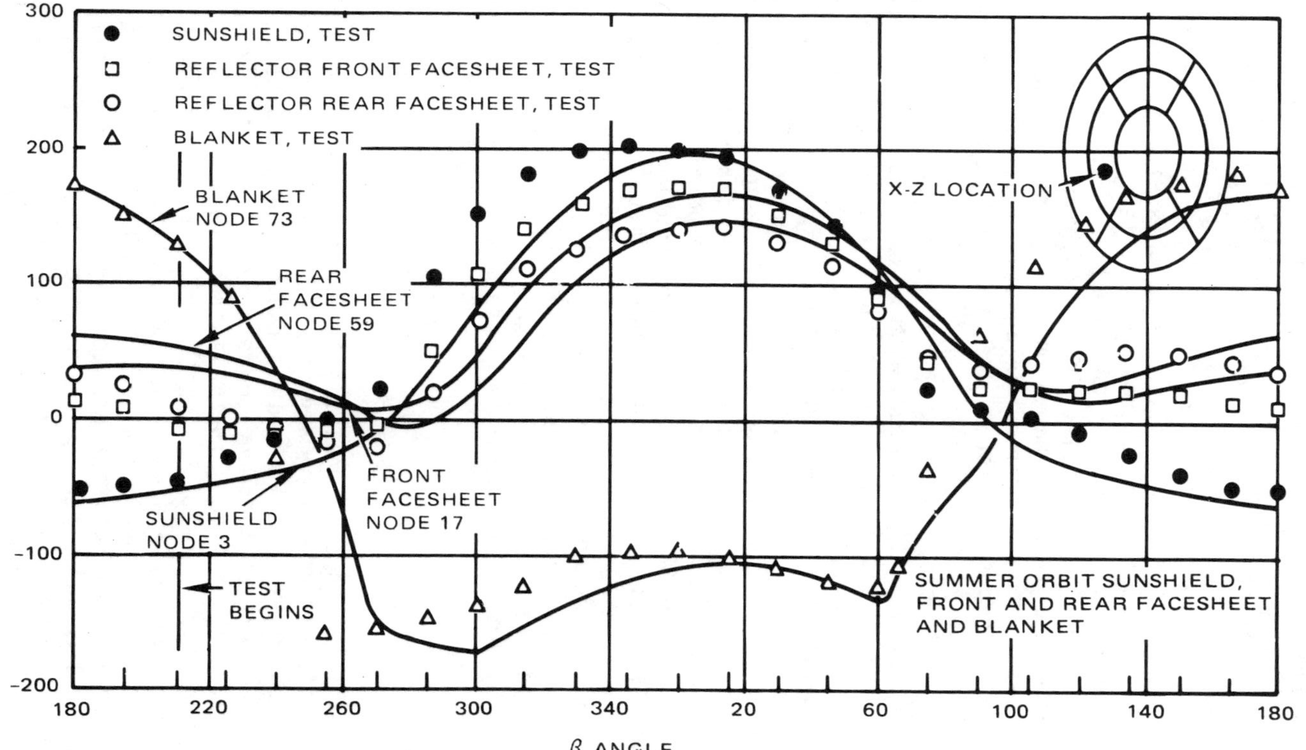

Fig. 14 STV predictions and test results.

orbit diurnal temperature history for a group of axially separated thermal model nodes. Here again the measured reflector gradient was somewhat greater than predicted. This figure also typifies the overall accuracy of the anlytical techniques employed, however. The honeycomb conduction factor was reduced in the flight version, and the model was considered to be thermally verified.

Subsequently, structural considerations implicit in the mechanical model led to another test program. The need for this much simpler task was based on an uncertainty about the correct coefficient of thermal expansion to use for certain areas of the composite dish. The test was conducted in a normal laboratory environment.

The test article, the same reflector used earlier in the STV test, was illuminated with lamps and cooled with cold GN_2 to create specific temperature distributions. These distributions were selected to simulate the most likely distortion cases as indicated by previous analysis and the STV test results. The reflector itself was instrumented with strain gages for deflection readings and the temperature probes emplaced for the earlier STV test.

At the completion of the test the temperature readings were entered into the TIP interpolation routine just as had been the analytical data from the thermal model. The output of TIP was then input to the mechanical model. The deformations produced at this step were then compared with the strain gage readings recorded during the test. The results led to certain refinements of the mechanical model. This refined model is in active use on current applications.

Conclusion

Primary reliance on analysis to predict the performance of spacecraft structures remains the current practice, since verification of mechanical and communication performance in a thermal-vacuum environment is still an area of extreme difficulty. Therefore, thermal-structural-rf design of high-performance parabolic reflectors for communication satellites requires accurate and cost-efficient analytical tools which can evaluate numerous design tradeoffs, predict results for thermal verification testing, and perform flight analysis. The analytical process which led to the HS 376 satellite reflector design represents one approach which filled these requirements.

Future effort should be directed toward combining the numerous steps involved in the analysis process as well as expanding the breadth of accurate data bank material information (over a wide temperature range) on the lightweight, low thermal expansion composites now being used.

Acknowledgments

The authors wish to thank R. K. Miyakawa and G. D. Wolodkin, without whose support and encouragement this paper could not have been prepared. Special thanks are also due R. P. Bobco for invaluable editorial advice.

References

[1]Foldes, P. and Dienemann, M. W., "Large Multibeam Antennas for Space," Journal of Spacecraft and Rockets, Vol. 17, July-Aug. 1980, pp. 363-371.

[2]Archer, J. S., "High Performance Parabolic Antenna Reflectors," Journal of Spacecraft and Rockets, Vol. 17, Jan.-Feb. 1980, pp. 20-26.

Thermodynamic Optimization of a Cryogenic Storage System for Minimum Boiloff

G.R. Cunnington*
Martin Marietta Denver Aerospace, Denver, Colo.

Abstract

This paper establishes a thermodynamic basis for reducing boiloff losses in cryogenic systems. Implementation of the second law of thermodynamics shows the proportionality between boiloff and entropy production (thermodynamic irreversibility). The method minimizes entropy production and boiloff by optimizing the location, temperature, and number of cooling stations in the system. Applying this method to the insulation and mechanical supports of a particular liquid hydrogen storage system results in a significant reduction in boiloff losses. Comparison with the best possible reduction shows that the actual reduction can closely approach the lowest possible boiloff.

Nomenclature

A = area
E = system energy
h = enthalpy
k = thermal conductivity
L = support length
m = mass
$\dot{m}$ = flow rate
M = system mass
P = pressure
$\dot{Q}$ = heat input rate
s = entropy

Presented as Paper 82-0075 at the AIAA 20th Aerospace Sciences Meeting, Orlando, Fla., Jan. 11-14, 1982.

* Senior Engineer, Propulsion.

S = system entropy
t = insulation thickness, time
T = temperature
u = internal energy
x = coordinate position

Subscripts

C = cold
e = exit
f = saturated liquid
g = saturated vapor
gen = generation
H = hot
o = no cooling
opt = optimum
S = shield
x = support location

Introduction

Recent NASA and DOD mission models have indicated future needs for orbital cryogenic storage and supply systems. Cryogens required will include hydrogen, oxygen, methane, argon, helium, deuterium, and nitrogen triflouride. Tank sizes will vary from 0.6 m^3 (22 ft^3) to 37.4 m^3 (1320 ft^3), with orbit durations varying from a few hours for such missions as Orbit Transfer Vehicle Low-Earth Orbit to Geosynchronous Equatorial Orbit transfer and resupply, to several years for missions such as station-keeping and space-based laser systems.[1]

Reduction of boiloff losses can be accomplished by using vent fluid to intercept insulation heat leak on a cooled shield and to intercept structural heat leak.[1,2] However, careful thermal design is necessary to achieve the minimum boiloff within the system constraints. Traditional thermal design and analysis uses only conservation of mass and energy, and requires iterative minimization. Iterative minimization is costly, time consuming, and is subject to arbitrariness and inaccuracies. An analytical technique to directly reduce the heat leak from each storage system component to the minimum possible would be very useful, both in terms of saving cost and in achieving the minimum possible boiloff. Establishing a thermodynamic basis for a method to reduce boiloff loss to the minimum possible by minimizing heat input to the cryogen from each heat leak path separately, and applying it to a particular cryogenic storage system is the motivation for this paper. The method used is based on minimizing the local rate of entropy production (thermodynamic irreversibility[3]) in each storage system component by optimizing the location, temperature, and number of cooling stations at which heat is intercepted along each heat leak path.

The traditional approach to thermal design and analysis overlooks the second law of thermodynamics and its usefulness. The second law of thermodynamics provides a statement about the loss of useful energy in any process. The loss of available energy is quantified by the production of entropy or irreversibility. The thermodynamic state of any real system always changes irreversibly, reducing available energy in a system, thus resulting in an increase in entropy. The usefulness of the second law is that the direction of a process is constrained, and the loss of availability can be determined. As shown in this paper, by using the second law of thermodynamics it is possible to determine the optimum thermal design of a system directly, as opposed to the traditional iterative approach to thermal optimization.

The remainder of the paper is devoted to establishing the minimum boiloff thermodynamic basis, and its application to a generic cryogenic storage system. The paper includes a description of the system to be optimized, the derivation of the proportionality between entropy production and boiloff, and minimization of the insulation system entropy production and structural entropy production, along with the associated boiloff.

Cryogenic Storage System to be Optimized

The cryogenic storage system described here is for orbital long-term storage of subcritical liquid cryogens. The system consists of a pressure vessel containing the saturated liquid cryogen, a structural support system, multilayer insulation (MLI), a vapor-cooled shield (VCS) with a heat exchanger, and an outflow line along with other penetrations such as fill lines, drain lines, and electrical cables. Figure 1 is a schematic representation of such a storage system. Other system components, such as the liquid acquisition device and instrumentation have been ignored for the purposes of this paper. A detailed cryogenic storage system description can be found in Refs. 1 and 2.

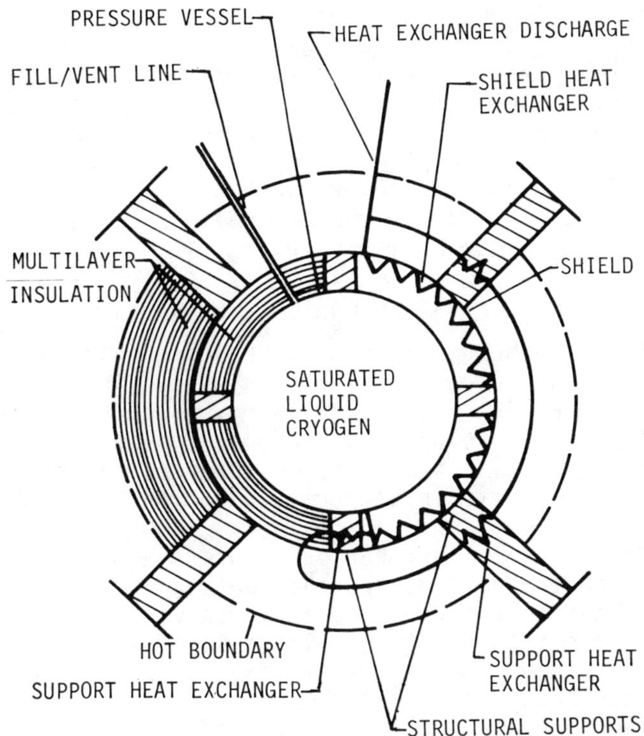

Fig. 1 Schematic diagram of a typical long-term orbital cryogenic storage system.

The pressure vessel typically consists of a thin-walled (0.080 cm) aluminum tank. The support system consists of low-thermal conductance high-strength tubes or straps designed to minimize heat leak while providing for the launch loads. Thermal control is facilitated, in addition to structure, using MLI in conjunction with a shield embedded in the MLI. The shield is cooled with vented vapor. In addition to intercepting heat with vapor on a shield, the vapor vent line can also be used to intercept support heat leak. Finally, outflow lines, fill and drain lines, and electrical cabling are all necessary for various ground and orbital operations.

The MLI consists of layers of 0.00635 cm aluminized mylar sheets, separated by dacron net spacers, 0.0178 cm thick.[1] Typical layer densities are 24 layers/cm. The MLI vacuum thermal performance can be modeled by an effective thermal conductivity,[+] proportional to $T^{0.6}$.[4] The VCS is typically a thin aluminum shell, with the same shape as the pressure vessel. A small diameter vent line is wrapped, in intimate thermal contact, around the shield. In addition, the vent line may be wrapped around the support(s), to intercept heat leak. The support thermal conductivity is assumed to be proportional to temperature.[4] With these thermal models the cryogenic storage system boiloff rate can now be minimized.

Boiloff is Proportional to Entropy Production

Boiloff can be diminished by reducing entropy production within the cryogenic storage system. This can be shown by performing a mass and energy balance on the overall system, and combining the result with the second law of thermodynamics. Thermodynamic irreversibility is produced within the system at a rate proportional to the boiloff rate.

To show that boiloff can be reduced by minimizing entropy production, consider the control volume shown in Fig. 2. Heat is entering and cryogen vapor is venting, both at the hot temperature. The mass inside the control volume is assumed to consist of fixed spacecraft mass plus the cryogen mass, which can change with time. The cryogen is assumed to be saturated liquid, and the spacecraft is assumed isothermal. All processes are assumed isobaric. Finally, all potential and kinetic energy changes are negligible.

Conservation of mass and energy requires, respectively,

$$\frac{dM}{dt} = -\dot{m} \tag{1}$$

and

$$\frac{dE}{dt} = \dot{Q} - \dot{m}h_e \tag{2}$$

Combining the two conservation relations yields

$$\dot{Q} = \dot{m}(h_e - u_f) \tag{3}$$

Note that this equation neglects the accumulation of vapor in the tank, which can be shown to be small, and will not change the validity of the statements to follow.

+The effective thermal conductivity is obtained by applying Fourier's Heat Conduction Law to experimentally measured MLI performance.

It is interesting that for pool boiling of a fluid the energy change is not the enthalpy of vaporization, because there is no flow work entering. Table 1 shows the comparison between the enthalpy of vaporization and the actual energy change during pool boiling for several fluids. It appears that the error in assuming the enthalpy of vaporization increases with pressure, i.e., as the boiling process is closer to the critical point. The error will tend to overestimate the boiloff rate, since $h_{fg} < h_g - u_f$.

The second law of thermodynamics for the control volume is

$$\dot{S}_{gen} = \frac{-\dot{Q}}{T_H} + \dot{m}s_e + \frac{dS}{dt} \geq 0 \qquad (4)$$

Combining this definition for the production of entropy, $\dot{S}_{gen}$, with the energy and mass balance results yields the following expression between the overall system entropy production rate and the boiloff rate

$$\dot{S}_{gen} = \dot{m}\left[(s_e - s_f) - \frac{1}{T_H}(h_e - u_f)\right] \qquad (5)$$

For specified boundary conditions the quantity in brackets, which consists of thermodynamic state parameters only, is a constant, and the boiloff rate becomes proportional to the rate of aggregate system entropy production. Therefore, minimizing the rate of entropy production brings about the minimization of the cryogenic boiloff rate.

The aggregrate system entropy production rate is the sum of the entropy production rates in each cryogenic storage system component.[5] Therefore, when each component contribution of $\dot{S}_{gen}$ to the aggregrate entropy pro-

Table 1 Comparison of enthalpy of vaporization with pool boiling enthalpy change for several cryogens

	h_{fg}(kJ/kg)		$h_g - u_f$(kJ/kg)	
Cryogen	P = 1 atm	P = 0.9P_C	P = 1 atm	P = 0.9P_C
Hydrogen	445.84	223.47	447.28	244.73
Helium	20.42	10.28	21.23	12.37
Nitrogen	198.74	62.22	198.88	66.43
Oxygen	213.02	67.80	213.11	74.80
Methane	386.00	175.50	390.10	191.10
Argon	161.92	40.32	161.99	46.52

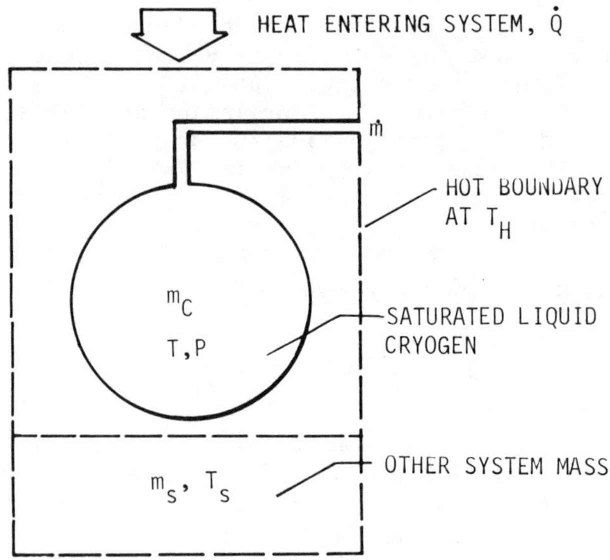

Fig. 2 Control volume for minimum boiloff analysis.

duction rate is minimized, the truly minimum boiloff rate will be achieved.

For the cryogenic storage system described previously, the heat leak to the pressure vessel is the dominant source of thermodynamic irreversibility. The heat leak consists chiefly of heat input through the insulation and structure. Consequently, when the insulation entropy production rate and the structure entropy production rate, are each minimized, the minimum rate of cryogen loss will be achieved.

Minimization of Entropy Production in Multilayer Insulation

Multilayer insulation, consisting of layers of metallized thin plastic sheets separated by net spacers, can be modeled by an effective thermal conductivity given by

$$k = k_1 T^{0.6} \tag{6}$$

where $k_1 = 4.6 \times 10^{-8}$ W/cm $K^{1.6}$ over the range of 20 K to room temperature.[4]

Reduction of MLI thermodynamic irreversibility can be achieved by controlling the temperature distribution (and the heat leak) through the insulation blanket, in an optimum manner, which yields the minimum entropy production rate. Radiation shields within the MLI, cooled with the boiloff vapor, will control the temperature throughout the MLI. Bejan[3] showed that minimum thermodynamic irreversibility is achieved based on continuous cooling throughout the insulation. For the MLI model chosen, the optimum temperature distribution through the MLI is

$$T_{opt} = \left[T_C^{0.33} + \frac{x}{t}(T_H^{0.33} - T_C^{0.33})\right]^{3.33} \tag{7}$$

The optimum heat removal rate is given as

$$\dot{Q}_{opt} = 3.31\frac{k_1 A}{t}(T_H^{0.33} - T_C^{0.33})T_{opt} \tag{8}$$

Thus, with these relations the optimum heat removal rate through the MLI can be determined.

Continuous MLI cooling is generally not practical. However, it is thermodynamically important because it leads to the greatest possible reduction in entropy production and should be compared to other more practical cooling schemes. For continuous cooling, the reduction in thermodynamic irreversibility reduces the boiloff rate to 0.429, relative to no cooling.

A more practical boiloff reduction scheme is to use a single, or at most, a few cooled shields. Consider a single shield whose position and temperature need to be optimized for minimum entropy production. The control volume to be analyzed is shown in Fig. 3. Conservation of energy leads to

$$\dot{Q}_H - \dot{Q}_C - \dot{Q}_S = 0 \tag{9}$$

The second law of thermodynamics for the control volume is

$$\dot{S}_{gen} = -\frac{\dot{Q}_H}{T_H} + \frac{\dot{Q}_C}{T_C} + \frac{\dot{Q}_S}{T_S} \geq 0 \tag{10}$$

Combining these two expressions with the effective thermal conductivity model leads to this expression for the rate of entropy production in the MLI,

$$\dot{S}_{gen} = \frac{k_1 A}{1.6(t-x)}\left[\frac{T_H^{1.6} - T_S^{1.6}}{T_H - T_S}\right]\left[\frac{T_H}{T_S} + \frac{T_S}{T_H} - 2\right] + \frac{k_1 A}{1.6t}\left[\frac{T_S^{1.6} - T_C^{1.6}}{T_S - T_C}\right]\left[\frac{T_S}{T_C} + \frac{T_C}{T_S} - 2\right] \quad (11)$$

For the case of no cooling the entropy production rate is

$$\dot{S}_{gen,o} = \frac{k_1 A}{1.6t}\left[\frac{T_H^{1.6} - T_C^{1.6}}{T_H - T_C}\right]\left[\frac{T_H}{T_C} + \frac{T_C}{T_H} - 2\right] \quad (12)$$

The ratio of $\dot{S}_{gen}$ to $\dot{S}_{gen,o}$ provides an objective function for determining the minimum boiloff for a single shield embedded within the MLI. For a given location the shield temperature can be optimized to give the minimum boiloff. Figure 4 shows the ratio of boiloff with cooling to boiloff without cooling, as a function of the dimensionless distance from the MLI cold boundary. Figure 5 is a plot of optimum shield temperature against dimensionless distance from the tank. From these two figures it can be seen that the minimum boiloff occurs when the shield is operated at 87 K, and located at a distance of x/t = 0.44.

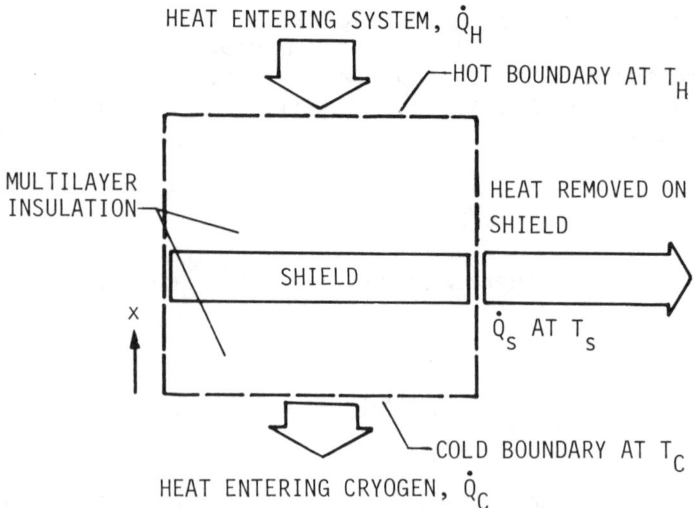

Fig. 3 Control volume of insulation analysis.

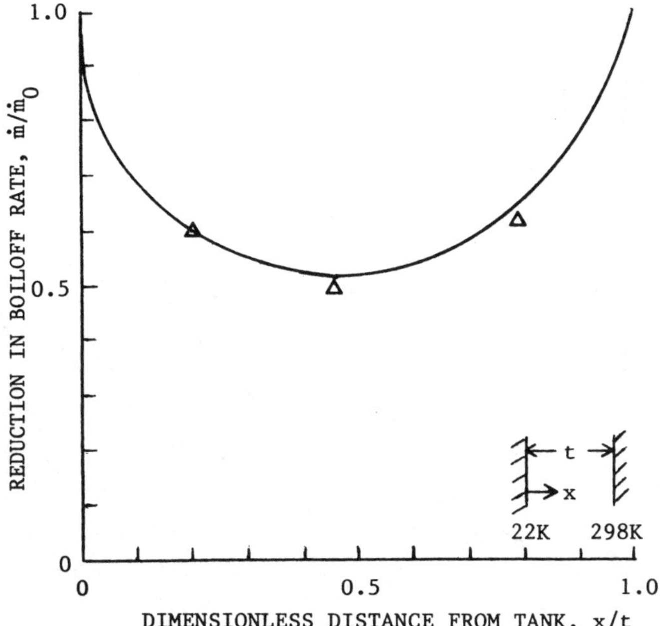

Fig. 4 Variation in boiloff reduction with distance from tank.

An optimized single-shield design reduces the boiloff by 45% as compared to the best possible reduction of 57% (continuous cooling).

A computer model of a cryogenic storage system was set up to numerically determine the minimum boiloff for a single shield. The model consisted of liquid hydrogen surrounded by one inch of MLI, with a variable position VCS embedded within the MLI. All structural heat leak was zeroed out, so that only MLI heat leak contributed to the boiloff. The vented vapor was then used to intercept heat on the VCS. In order to determine the optimum VCS location, it was positioned at several different locations within the MLI blanket, with the minimum boiloff rate iteratively calculated at each position. The boiloff rate with the VCS inoperative was also calculated and used to normalize the other boiloff rates, to see the reduction in boiloff rate as the VCS was moved around. The results of the iterative numerical calculation are shown in Fig. 4 as the triangular points. The agreement between the iterative numerical calculation and the minimum entropy production technique is very good.

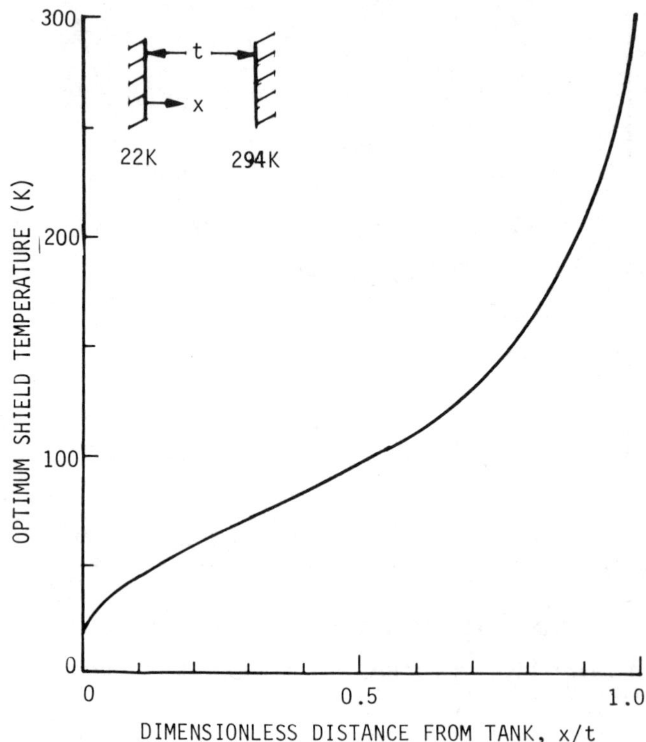

Fig. 5 Optimum shield temperature variation with distance from tank.

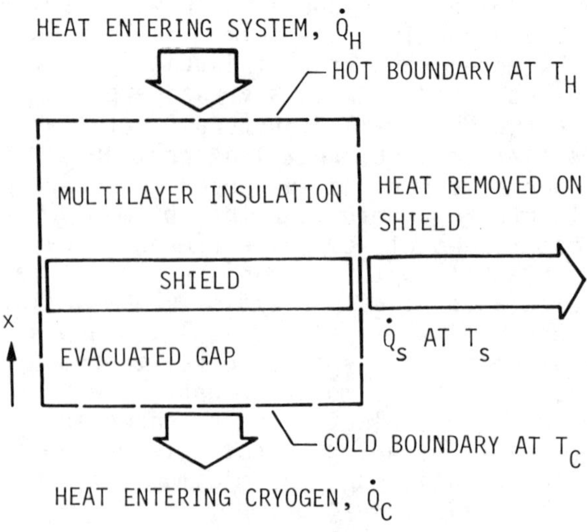

Fig. 6 Single shield system with a gap.

A special case of the single shield concept occurs when the shield is located between the tank and MLI, as shown in Fig. 6. In this case radiation is the heat-transfer mode from the insulation to the tank. Applying the second law of thermodynamics to the energy balance leads to an expression for the ratio of entropy production with cooling divided by the entropy production without cooling. For this case the location is fixed, but the hemispherical total emissivity is temperature dependent. A shield temperature of 50 K reduces the boiloff by 30%.

Adding additional shields will further reduce the boiloff. Using the control volume of Fig. 3, except that n shields are located within the MLI, the expression for the entropy production rate becomes

$$\dot{S}_{gen} = \frac{k_1 A}{1.6 x_1}\left[\frac{T_1^{1.6} - T_C^{1.6}}{T_1 - T_C}\right]\left[\frac{T_1}{T_C} + \frac{T_C}{T_H} - \frac{T_1}{T_H} - 1\right]$$

$$+ \frac{k_1 A}{1.6}\left(\sum_{i=2}^{n-1} \frac{T_{i+1}^{1.6} - T_i^{1.6}}{(T_{i+1} - T_i)(x_{i+1} - x_i)}\right.$$

$$\left[\frac{T_{i+1}}{T_i} + \frac{T_i}{T_H} - \frac{T_{i+1}}{T_H} - 1\right] - \frac{T_i^{1.6} - T_{i-1}^{1.6}}{(T_i - T_{i-1})(x_i - x_{i-1})}$$

$$\left.\left[1 - \frac{T_{i-1}}{T_i} - \frac{T_i}{T_H} + \frac{T_{i-1}}{T_H}\right]\right) \tag{13}$$

with shield enumeration outward from the cold boundary. Optimizing this expression, normalized with $\dot{S}_{gen,o}$, is accomplished numerically. The results of adding a second shield are reported in this paper.

Figure 7 shows the reduction in boiloff, relative to no cooling, vs the boundary temperature ratio T_H/T_C, for a single shield, two shields, and continuous cooling. The top three curves are for a single shield that is, successively from the top, (i) temperature optimized at $x/t = 0.5$, (ii) position optimized with the shield temperature at the average of the hot-cold boundary temperatures, and (iii) optimized with respect to both temperature and position. The next curve (iv) is for two shields, both position and temperature optimized. The bottom curve (v) is for continuous cooling.

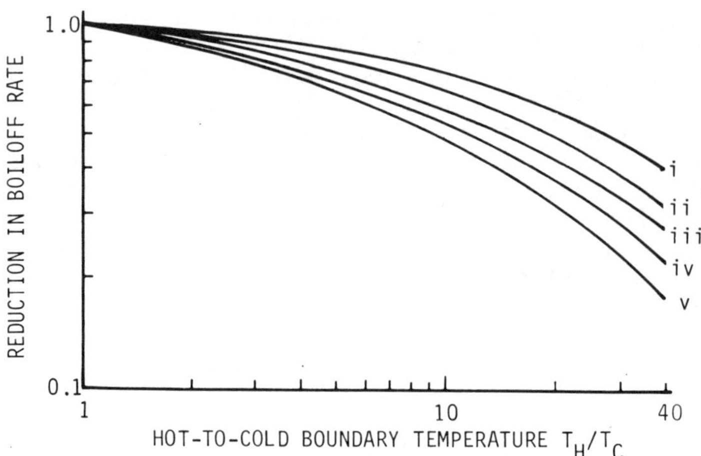

Fig. 7 Reduction in boiloff with hot-to-cold boundary temperature ratio.

A typical value of T_H/T_C for liquid hydrogen storage is about 13. For continuous cooling, the boiloff is reduced to 43% of the uncooled case. Two shields reduce the boiloff to 49% of the uncooled case, only 6% less than the continuous case. Thus, two shields are quite good, relative to the best achievable boiloff reduction. A single, completely optimized shield reduces the boiloff to 55% of the uncooled case, about 6% above the two shield case, and 12% above the best possible case. Note that for the two shields, the first should be located at $x/t = 0.3$ and at a temperature of 60 K, and the second shield should be located at $x/t = 0.6$ and at 131 K.

Figure 7 shows that substantial reductions in boiloff are realized for complete optimization of a single shield. For a temperature ratio of 13, temperature optimization with the shield midway through the MLI reduces the boiloff by 31%, while position optimization with the shield at the average of the hot and cold temperatures reduces the boiloff by 40%. Thus, an additional 14% reduction in boiloff could be achieved if the shield is optimized with respect to both temperature and position, rather than temperature only. A 5% decrease is possible if the shield is completely optimized rather than only position optimized.

Minimizing Entropy Production in the Structure

Structural supports can be modeled thermally as leaky insulation. A typical tank support is made of glass/

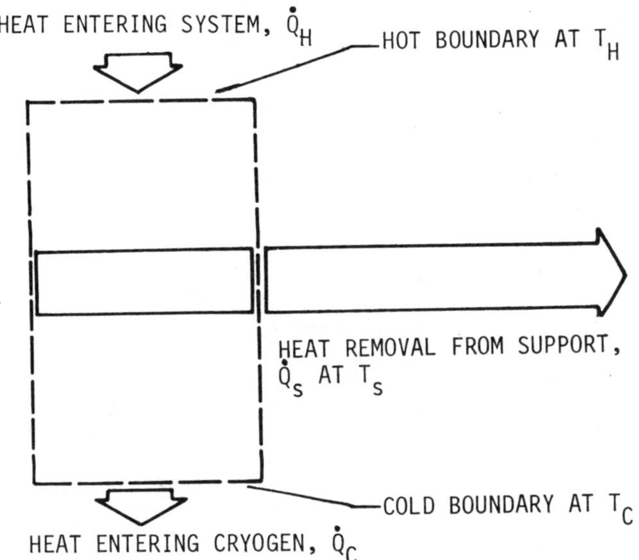

Fig. 8 Control volume of support analysis.

epoxy, with a thermal conductivity typically given as

$$k = k_2 T \tag{14}$$

For S-glass/epoxy, $k_2 = 3.1 \times 10^{-4}$ W/cm-K^2.[4]

A support can be cooled by placing a heat exchanger tube in intimate thermal contact with it. Figure 8 is a schematic diagram of a support and the control volume for analysis. Following a procedure identical to that used for the insulation optimization yields an expression for the support entropy production rate of

$$\dot{S}_{gen} = \frac{k_2 A}{2(L-x)} (T_H^2 - T_x^2)(\frac{1}{T_x} - \frac{1}{T_H}) + \frac{k_2 A}{2x} (T_x^2 - T_C^2)(\frac{1}{T_C} - \frac{1}{T_x}) \tag{15}$$

The uncooled support entropy production rate is

$$\dot{S}_{gen,o} = \frac{k_2 A}{2L} (T_H^2 - T_C^2)(\frac{1}{T_C} - \frac{1}{T_H}) \tag{16}$$

Taking the ratio of $\dot{S}_{gen}$ divided by $\dot{S}_{gen,o}$, and minimizing the result with respect to temperature and location

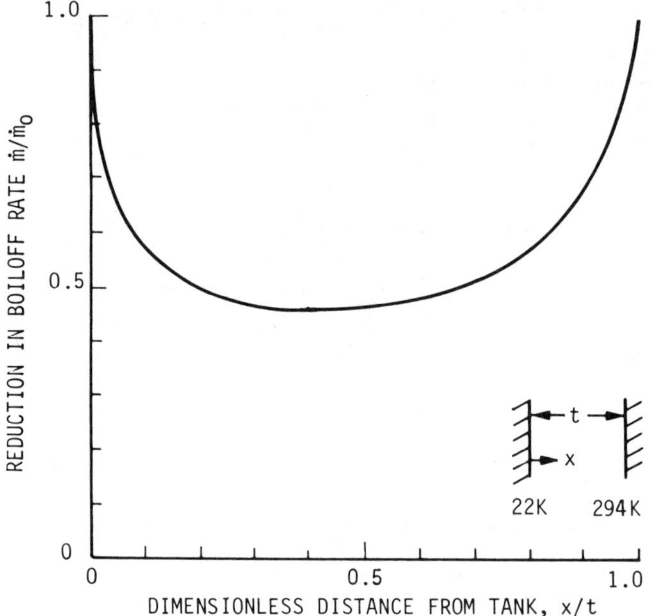

Fig. 9 Reduction in boiloff with distance from tank.

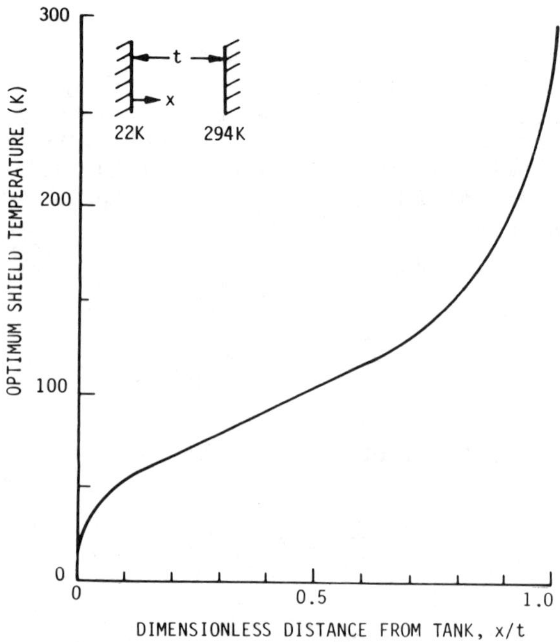

Fig. 10 Optimum support temperature variation with distance from tank.

leads to the minimum boiloff design. Figure 9 shows the reduction in boiloff, plotted against dimensionless distance from the cold end. The minimum boiloff occurs when the support heat exchanger is located at x/t = 0.40. Figure 10 shows the support optimum temperature. At x/t = 0.40, the heat exchanger temperature is about 78 K, the same as the support temperature at that point.

Conclusions

Consideration of the second law of thermodynamics in the operation of a cryogenic storage system showed that the cryogenic fluid boiloff rate was proportional to the rate of overall entropy production within the system. Consequently, by minimizing the contribution of each component to the aggregate production rate of thermodynamic irreversibility of the system, it was possible to achieve a design with the minimum boiloff rate.

This approach was used to minimize the boiloff rate for a particular multilayer insulation system. The minimum boiloff possible was found to occur for continuously cooled insulation in accordance with the general principle found in Ref. 3. A single cooled shield was not as effective as continuous cooling in boiloff reduction. Two shields were more effective than one shield, but not as effective as continuous cooling.

The boiloff caused by irreversibility produced through structural heat leak was also reduced using the same approach. The optimum location and temperature for intercepting structural heat leak were determined.

The results presented here demonstrate the usefulness of the second law of thermodynamics in thermal design. Although the same results can be achieved using the traditional system mass and energy balances, determining the system's thermal and hydraulic optimma requires an iterative procedure which can be time consuming and expensive. This approach zeroes in on the system optimums quickly and easily.

Acknowledgment

The author would like to express his appreciation to Professor Adrian Bejan of the University of Colorado, Boulder, for his guidance in preparing this paper.

References

[1] Eberhardt, R. N., Cunnington, G. R., and Johns, W. A., "Conceptual Design and Analysis of Orbital Cryogenic Liquid

Storage and Supply Systems," Martin Marietta Denver Aerospace, Denver, Colo., Final Rept. MCR-81-546, May 1981.

[2] Eberhardt, R. N., Fester, D. A., and Bailey, W. A., "Cryogenic Fluid Management Experiment," Martin Marietta Denver Aerospace, Denver, Colo., Final Rept. MCR-81-597, July 1981.

[3] Bejan, A., "A General Variational Principle for Thermal Insulation System Design," International Journal of Heat Mass Transfer, Vol 22, Feb. 1979, pp. 219-228.

[4] Gille, J. P., "Development of Advanced Materials for Integrated Tank Insulation System for the Long-Term Storage of Cryogens in Space," Martin Marietta Denver Aerospace, Denver, Colo., Final Rept. MCR-69-405, Sept. 1969.

[5] Reynolds, W. C. and Perkins, H. C., Engineering Thermodynamics, McGraw-Hill, New York, 1977, p. 233.

Molecular Absorption Cryogenic Cooler for Liquid Hydrogen Propulsion Systems

G.A. Klein* and J. A. Jones†
Jet Propulsion Laboratory,
California Institute of Technology, Pasadena, Calif.

Abstract

A light-weight, long-life molecular absorption cryogenic cooler system is described which can use low-temperature waste heat to provide cooling for liquid hydrogen propellant tanks for interplanetary spacecraft. Detailed tradeoff studies were made to evaluate the refrigeration system component interactions in order to minimize the mass of the spacecraft cooler system. Based on this analysis, a refrigerator system mass of 31 kg is required to provide the 0.48 W of cooling required by a 2.3-m-diam liquid hydrogen tank.

Introduction

Recent investigations have indicated that a substantial increase in available spacecraft payload mass can be realized by using liquid hydrogen rather than Earth-storable, i.e., room-temperature, fuels as a propellant for interplanetary missions. One of the primary obstacles to the use of hydrogen as a fuel for long-duration missions has been the fact that it must be actively refrigerated in order to maintain a temperature near its normal boiling temperature of 20 K. With the introduction of new cryogenic

---

Presented as Paper 82-0830 at the 3rd AIAA/ASME Joint Thermophysics, Fluids, Plasma & Heat Transfer Conference, St. Louis, Mo., June 7-11, 1982.

*Member of Technical Staff, Control and Energy Conversion Division.

†Member of Technical Staff, Applied Mechanics Divison.

refrigeration technology, however, it appears likely that a reliable, light-weight, long-life molecular absorption cryogenic cooler (MACC) can be built to provide the necessary cooling.

The cooler has essentially no moving parts and can use low-grade radioisotope thermoelectric generator (RTG) waste heat or direct solar heat as an energy source for the compressor. This refrigerator is currently being developed at JPL for spacecraft sensor cooling in addition to the present application.

The objective of the effort summarized in this report was to design a complete, spaceborne, sorption compressor, Joule-Thomson (JT) cooler system to provide long-term refrigeration and storage of a liquid hydrogen fuel system.

Cooler Operation

The cooler operates by using a hydride power ($LaNi_5$) to absorb large quantities of hydrogen at temperatures around 290 K and 1 atm pressure. When heated to about 390 K, however, the hydrogen pressure increases to about 4 MPa (40 atm). The pressure from the hydrogen gas then activates a diode check valve and the hydrogen flows through a series of space radiators and heat exchangers, as shown in Fig. 1. When the gas reaches the JT valve, it is at about 30 K. When expanded to 1 atm through the JT valve, it is further cooled and partially condensed. Heat from the hydrogen fuel tank then vaporizes the condensed hydrogen, and the cycle is closed back through the counterflow heat exchangers, heating the returning hydrogen while prechilling the high-pressure hydrogen.

An attractive method for getting heat into and out of the hydride compressors is by direct RTG radiation heating and cooling of the compressors (Fig. 2). This type of heating cycle can be accomplished by alternating the orientation of radiant louvers. During heating, the compressors view the RTGs and are shielded from the cold of deep space; during cooling, the RTG heat is shielded, and the compressors radiate directly to space.

Refrigeration Cycle Analysis

The refrigeration load, Q_L, is directly related to the mass flow rate through the JT valve as well as the enthalpy difference between the gas leaving the low-pressure side of the JT counterflow heat exchanger and the gas exiting the precooler which enters the high-pressure side of the JT counterflow heat exchanger. Studies indicate that the

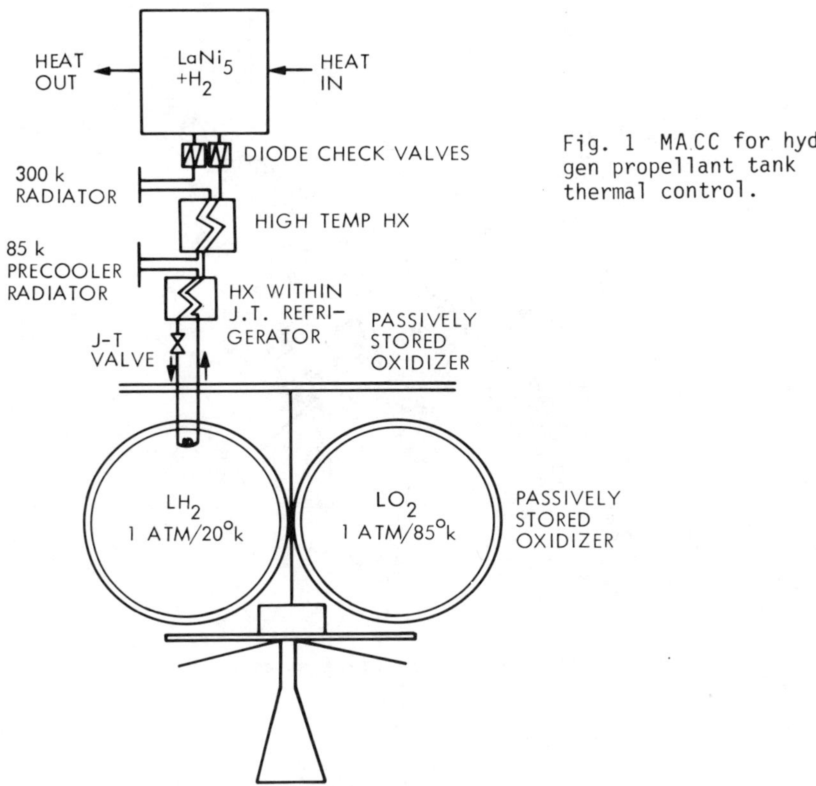

Fig. 1 MACC for hydrogen propellant tank thermal control.

refrigeration mass flow rate for any given cooling rate requirement is strongly influenced by the pressure ratio and the precooler temperature. As the pressure ratio increases, the mass flow decreases. The mass flow rate corresponds directly to the $LaNi_5$ mass in the compressor. A high-pressure requirement would add structural mass to the system. A lower precooling temperature would also add mass (corresponding to additional radiative surface area requirements) to the system. An optimization analysis is required to determine the point where the hydride mass required due to a reduced hydrogen mass flow, is offset by the additional mass of structure in the precooler and the heat exchanger. The results of such an analysis will be presented later in this report.

Further analysis of the refrigeration cycle indicates that, as the precooling temperature is decreased, the lower limit for the heat exchanger effectiveness, ε, can be decreased. A lower value for the effectiveness, which is a function of the temperature drop across the JT refrigera-

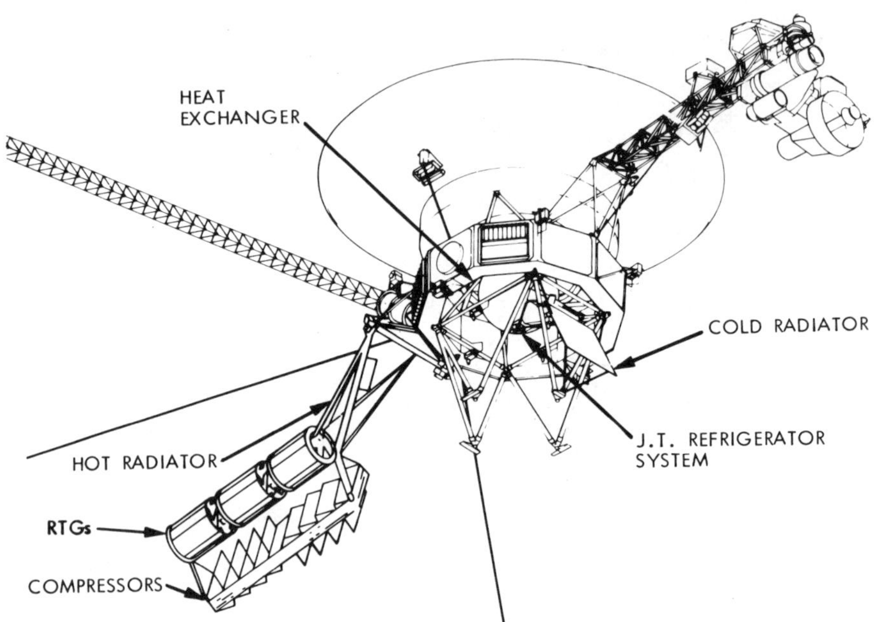

Fig. 2 Conceptual installation of absorption system on interplanetary space vehicle.

tor, ΔT_{JT}, provides more tolerance for the inefficiency of the JT heat exchanger. A practical limit to the temperature difference is 5 K. A low precooling temperature is necessary in order to keep the mass flow at a reasonable level (see Fig. 3) without placing a strain upon the operating limits of the advanced precooling radiator.

Gas Absorption Compressor Design

For the purpose of this study, a three-stage sorption compressor was examined.[1] This compressor consists of three $LaNi_5$ beds cycled in a properly phased sequence so that at least one of the stages is always absorbing low-pressure hydrogen from the external load while at least one is always desorbing high-pressure hydrogen to the load. So far as the refrigerator is concerned, there is a constant supply of high-pressure hydrogen and a constant return of low-pressure hydrogen.

For simplicity of analysis, two generalized processors were used to represent the basic compressor cycle, as shown in Fig. 4, i.e., 1) absorption/desorption of hydrogen along a plateau with flow into/out of the bed (while removing/adding the heat of absorption), and 2) temperature change due to heating/cooling of the bed with accompanying reduction/increase in concentration and no gas flow.

Analytical compressor performance studies indicate that for a given cooling load, as the inlet temperature to the JT refrigerator increases and the JT heat exchanger effectiveness decreases, the system mass flow rate increases and the compressor heat input rate requirements increase. In addition, it is possible for the system components to show similar performance characteristics for two entirely different sets of design requirements.

Figure 5 shows the sensitivity of the compressor heat rate requirements to changes in the hydrogen tank cooling requirements as a function of system pressure ratio. The system becomes increasingly sensitive to changes in the pressure ratio as the system pressure ratio decreases.

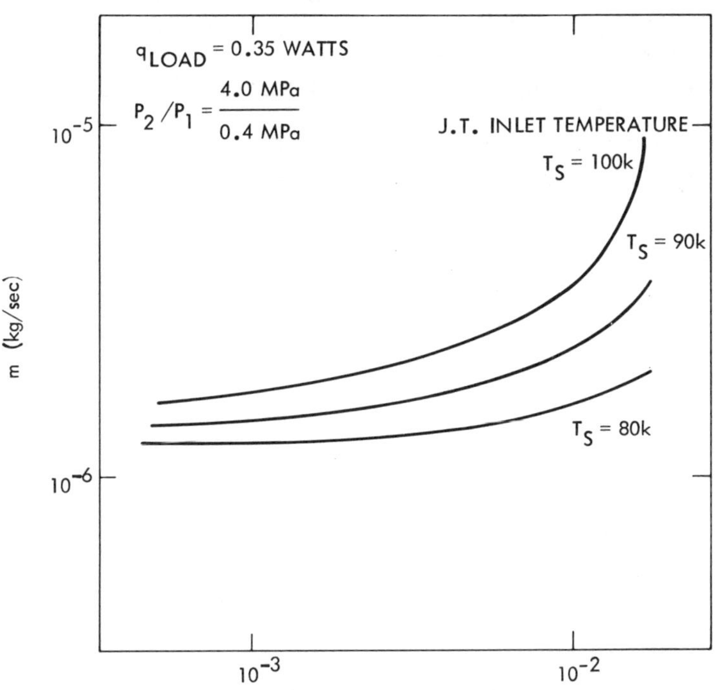

Fig. 3 Mass flow vs effectiveness of heat exchanger at the JT refrigerator as a function of inlet temperature to the refrigerator.

Thus the system pressure ratio should be maintained as large as is practical, being limited by the maximum temperature under which the compressor can be safely operated without encountering a burnout condition. For the compressor examined in this study, this value has been selected to be P_2/P_1 = 4.0/0.4 MPa; however, higher values may be selected for operation in the future.

Heat Rejection Radiators for the Compressor

The function of the waste heat rejection radiator is to dissipate the compressor system's thermal losses to space. The total heat load was assumed relatively constant and resulted in constant inlet and outlet coolant temperatures for a radiator of fixed orientation. For this analysis, the three compressor modules were reduced into component modules. Each module can be obtained by dividing the total $LaNi_5$ bed into beds of smaller volume, while maintaining constant the total $LaNi_5$ powder mass for that particular module. The modules consist of multiple $LaNi_5$ beds separated by fin-type radiators whose thickness and length were taken equal to the compressor diameter and length, respectively.

The minimum number of "component modules" required to effectively remove the compressor heat was determined

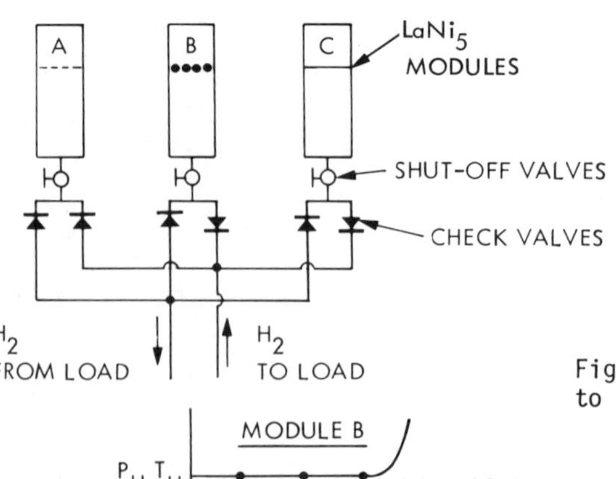

Fig. 4 Phasing three modules to form a sorption compressor.

through a heat balance performed upon the combined compressor and fin-system geometry for a given sorption time. It was assumed that heat flowed by conduction from the compressor into its adjacent fin, while heat flowed out of the fin by radiation from the surface. The compressor system mass, including radiators, is shown in Fig. 6 vs the compressor heat input requirements as a function of sorption time. For fixed compressor heat rate requirements, as the sorption time increases, the system mass requirements also increase as shown.

High-Temperature Radiator

The high-temperature radiator operates between the compressor outlet temperature (341 K) and the inlet tem-

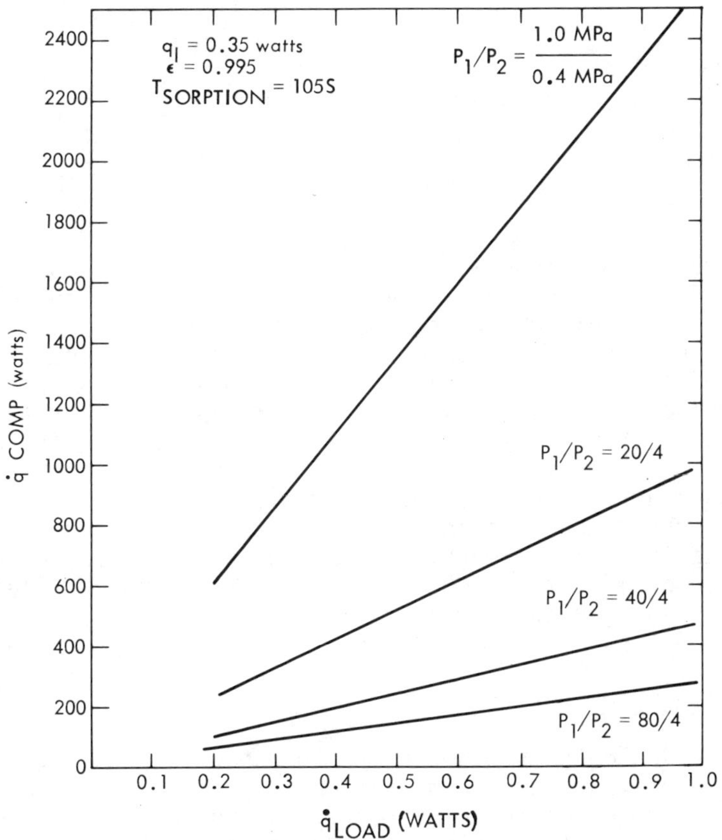

Fig. 5 Compressor heat rate requirements vs hydrogen tank cooling requirements as a function of average system pressure ratio.

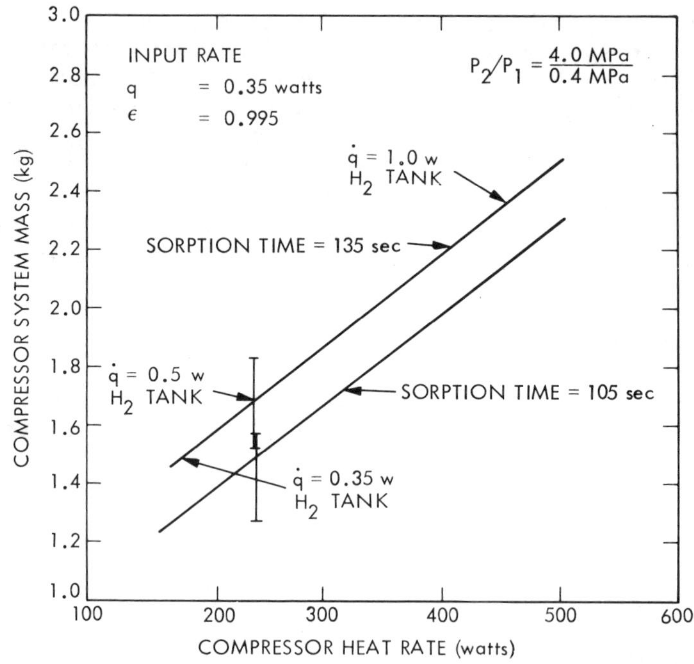

Fig. 6 Compressor system mass requirements vs compressor heat.

perature to the high-temperature heat exchanger. Initial design for the radiator is based on the design of the high-temperature radiator on the Voyager spacecraft, which was used to dispose of the spacecraft's excess electrical power. The radiator consists of an aluminum honeycomb design, and its surface is covered by electrical resistors. For the purpose of this study, the resistors have been eliminated and the hydrogen fluid loop is designed to make a series of passes across the back of the radiator.

In the present design, the radiator was scaled in size. The size factor was roughly the ratio of radiative heat transfer desired to the radiative heat transfer obtained from the Voyager spacecraft radiator, which weighed approximately 6.8 kg.

Precooling Radiator

The Joule-Thomson cooler relies upon the cooling of gases when expanded at constant enthalpy, assuming the gas has been precooled below its inversion temperature. The hydrogen flow-rate requirements depend strongly on the precooling temperature achieved before the throttling operation.

The data used in the analysis of the precooling radiator were based on an AIAA paper[2] which parametrically presents the radiator system weight as a function of operating temperature for variations in the heat removal requirements. For very low temperatures, the radiator must be shielded from the sun, the Earth, and warm spacecraft components. Considerable reductions in system weight may be realized in the future by employing an advanced cryogenic radiator design which is currently under development at JPL.[3] The temperature at which the radiator should operate is a tradeoff between its efficiency and the conductive loss from the high-temperature heat exchanger on one hand, and the JT effect and possible ΔT_{JT} on the other. Specifically, as the radiator temperature, T_R, increases, the radiator operates more efficiently and the high-temperature heat exchanger loss decreases; however, with a high T_R, both the JT effect and the ΔT_{JT} would decrease, requiring a larger mass flow ($\dot{m}$) to produce the desired refrigeration. It is apparent that the radiator

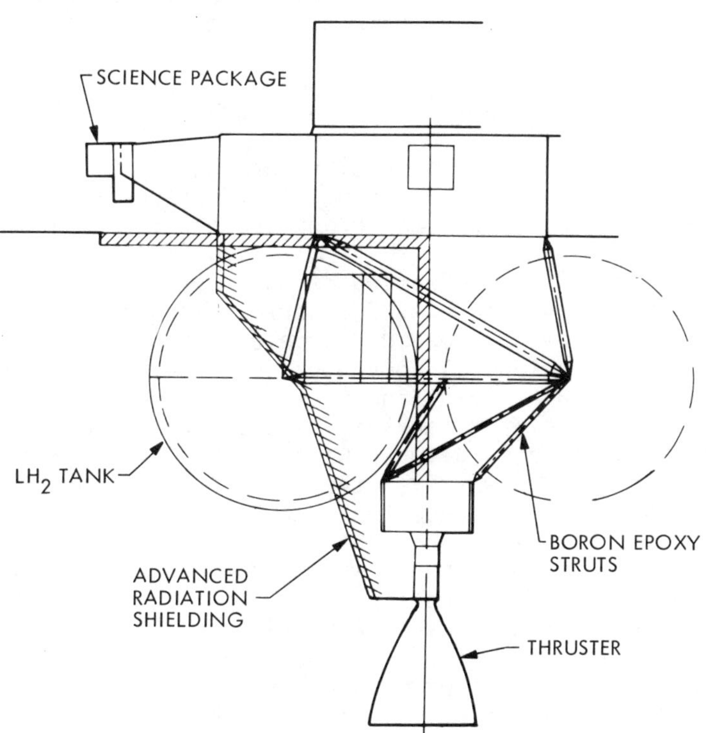

Fig. 7 LH_2 tank configuration.

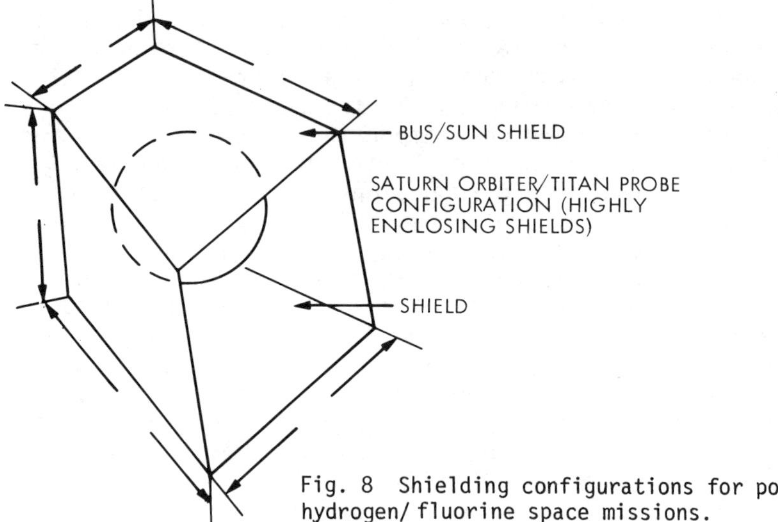

Fig. 8 Shielding configurations for potential hydrogen/ fluorine space missions.

performance has a substantial impact upon the overall system design, and a compromise between these conflicting requirements is therefore necessary.

H_2 Tank Cooling Requirements

The spacecraft configuration for the LH_2/O_2 tanks was based upon the thermal design of the tank support structure for the Saturn Orbiter/Titan Probe spacecraft configuration.[4]

The general configuration of the LH_2/O_2 propulsion spacecraft is shown in Fig. 7, where the propellant tanks are surrounded by a five-sided shield (Fig. 8).[5] The spacecraft/science package is above the bus shield, the LH_2 tank is behind the LH_2 shield, and the rocket thrusters are beneath the propellant tanks. The spacecraft is flown with the longitudinal axis aligned with the sun vector and with the payload facing away from the sun. A large sunshield must be positioned between the rocket thrusters and propellant tanks. Finally, a bus shield is necessary to shield the LH_2 tank from the upper, warmer spacecraft structure and scientific package region.

An advanced multilayer insulation configuration, shown in Fig. 9, was recommended for the shielding construction. The gap in this design permits a large portion of the incident energy to be reflected out the sides of the shields, thereby reducing the amount of heat that is transmitted through the MLI.

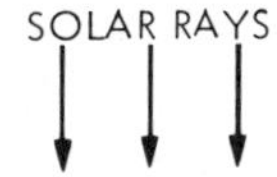

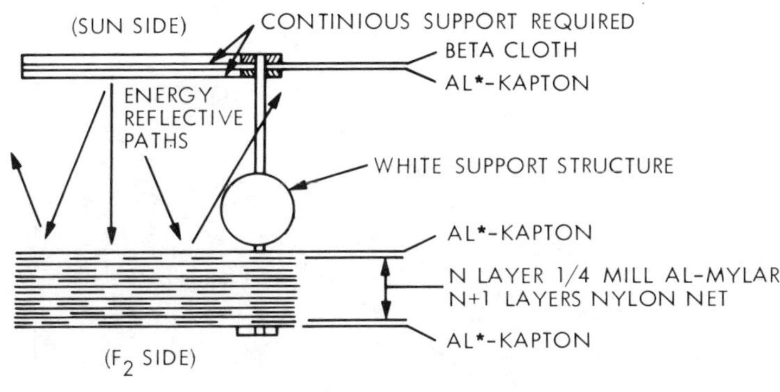

Fig. 9 Advanced configuration MLI radiation shield.

Detailed thermal computer models of the shield and tank configurations were developed to determine the anticipated LH_2 tank temperature for steady-state operation of the spacecraft in a high Earth orbit. The thermal environment was selected to give an upper bond on the maximum heat load required for removal by the JT refrigerator.

By surrounding the tank with low-emittance multilayer insulation, the tank's absorption of the thermal radiation from the inner shield is minimized. For tank sizes ranging from 1.5 to 2.3 m, the heat load into the tank from both conduction and radiation varied from 0.32 to 0.48 W.

All analyses in this report have been based on diffuse, rather than a specular radiation, in order to save both engineering and computer time. An increase in the predicted interchange factor to space and thus an additional safety margin can be anticipated for the shield configuration using a specular shield model. This study did not take into account thermal transients which occur during the boost phase of the trajectory.

Cooler System Numerical Model and System Analysis

A numerical model incorporating the analysis of each cooler component was developed to examine the performance of the various refrigerator components. These components

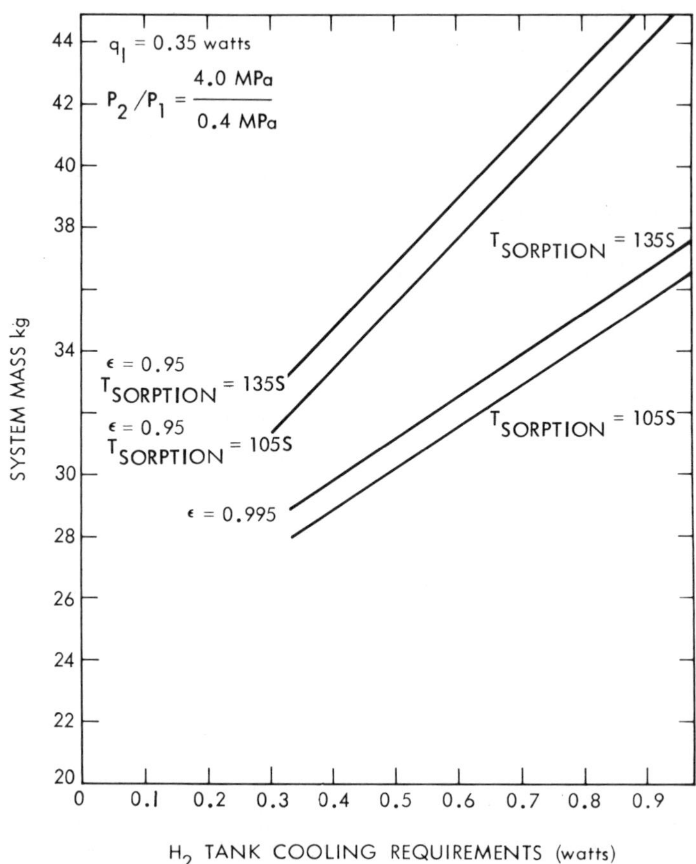

Fig. 10 Total system mass vs hydrogen tank cooling requirements.

are the JT refrigerator, consisting of JT valve, H_2 tank and heat exchanger; the compressor and its associated radiators; the high-temperature radiator and heat exchanger situated on the high-pressure return leg; and the precooling radiator located just above the JT refrigerator. The system mass is highly sensitive to the performance of the system components. As an example, the system mass is plotted in Fig. 10 vs the hydrogen tank cooling requirements as a function of compressor sorption time and effectiveness of the heat exchanger at the JT refrigerator. For fixed H_2 tank cooling requirements (i.e., fixed system mass flow rate), and for fixed compressor total heat input requirements, as the sorption time decreases, the compressor system mass decreases, thereby reducing the total system mass (the mass of all other system components remaining constant). The lower limit of sorption time is determined by how fast heat can be pumped in and out of the $LaNi_5$ powder. As the JT heat exchanger (HX) effectiveness (ε) decreases, the system mass flow rate increases, and the mass of each cooler system component will increase proportionately. The specific cooling requirements of the hydrogen tank were determined for two different tank sizes for an Earth-orbital mission. The cryogenic cooling system was sized to satisfy these heat load requirements. A detailed weight breakdown is provided in Table 1 for a case that typifies requirements of an acceptable spacecraft cooler design.

In addition, system studies indicate the influence of pressure ratio upon the performance of the JT refrigerator and the compressors. Since the compressor system comprises a substantial fraction of the total system weight, the sys-

Table 1 Design requirements and weight breakdown for typical spacecraft propellant tank cryogenic cooler

Parameter	
q	0.325 W
P_2P_1	4.0/0.4 MPa
m	0.45×10^{-5} kg/s
Sorption time	135 s
Q_{AVE}	167.8W
T_{in} JI HX	80° K

Component	Mass, kg	Mass fraction,%
Compressor and radiator system mass	4.5	15.4
High-temperature radiator	0.3	1.0
Precooling radiator	11.99	40.6
High-temperature HX (2)	0	
JT refrigerator HX (2)	0	0.0
JT valve	0	
Filters	0.4	1.4
Plumbing (piping joints and fittings)	2.0	6.8
Louvers	1.0	3.4
Insulation (for piping & hot and cold HX)	3.2	10.9
Supports and miscellaneous	6.0	20.5
Total	29.39	100.0

tem mass will increase in proportion to the compressor system's sensitivity to changes in the pressure ratio, as reflected in Fig. 5.

JPL Absorption Compressor Development and Model Verification

A $LaNi_5$ molecular absorption compressor has been tested at JPL and has provided favorable compressor characteristics. The compressor tube is heated by an electrical resistance surface heater and is cooled by water flowing in a water jacket surrounding the tube. A specially designed copper heat-transfer matrix enhances heat transfer to the $LaNi_5$ powder, and provides a heat-up/desorption time of 2.5 min. With 150 W of power, 7.5 standard liters of hydrogen gas are thus liberated at 4.0 MPa of pressure. During a 2.5-min cool-down, all of the hydrogen gas is then reabsorbed at 0.4 MPa of pressure. With LN_2 prechilling of the gas before the lower-stage heat exchanger (Fig. 1), the resultant cooling capacity has been estimated to be 170 mW at 27 K for 75 W of power as averaged over the entire 5- min cycle. This is consistent with the theoretical prediction curve for P_1/P_2 = 4.0/0.4.

A continuously operating, three-unit compressor is presently under construction at JPL and provides 650 mW of cooling at 27 K.[6] The imput average power to each compressor is 83.3 W for a total of 250 W. This predicted value is also consistent with the prediction curve for P_1/P_2 = 4.0/0.4 in Fig. 3. Contamination due to hydrogen gas impurities has prevented long-term compressor operation in excess of 30 h per test period. After contamination, however, the $LaNi_5$ power is quickly reactivated by heating the powder while under high vacuum. The predicted lifetime for very clean hydride systems below 100 C (373°K) is hundreds of thousands of cycles,[7] although some degradation may eventually occur.[8] Use of $LaNi_{4.7}Al_{0.3}$ in place of $LaNi_5$ could potentially increase hydride life even further, since $LaNi_{4.7}Al_{0.3}$ has exhibited no measurable degradation even after thousands of cycles.[8]

Due to the lack of moving parts, other than self-operating, long-life valves, the JPL refrigerator's predicted life (with clean hydrogen) is extremely long. In fact, no mechanical failures have occurred during the entire test periods.

Results and Conclusions

An analytical model has been developed to estimate the sizes of the components of the molecular absorption cryogenic cooler (MACC) for a liquid hydrogen propellant thermal control application. Detailed tradeoff studies were made using this model to evaluate the system component performance and interactions. Weight estimates of the spacecraft cooler requirements have been obtained for various design requirements. In addition, hydrogen tank cooling requirements were determined for two different tank sizes for a Saturn Orbiter spacecraft mission. For a system pressure ratio $P_2/P_1 = 4.0/0.4$, the overall cryogenic cooling system weight needed to satisfy the tank cooling requirements varied from a minimum value of 29 kg to a maximum value of 31 kg. This weight estimate was determined for tank sizes ranging from 1.5 to 2.3 m, corresponding to heat loads ranging from 0.325 to 0.48 W. The resulting system mass assumes a fixed mass for all structural and plumbing components. Thus final masses are not directly proportional to the required cooling loads.

The results of this study thus indicate that liquid hydrogen can be maintained in deep space at 20 K by means of a long-life molecular absorption cryogenic cooler (MACC) that has essentially no moving parts.

References

[1]Lehrfeld, D. and Boser, O., "Absorption-Desorption Compressor for Spaceborne/Airborne Cryogenic Refrigerators," Air Force Flight Dynamics Laboratory, Wright Patterson AFB, Dayton, Ohio, AFFDL-TR-74-21, March 1974.

[2]Haskin, W. L. and Dexter, P. F., "Ranges of Application for Cryogenic Radiators and Refrigerators on Space Satellities," 17th Aerospace Science Meeting, New Orleans, La., Jan., 1979.

[3]Bard, S., Stein, J., and Petrick, S. W., "Advanced Radiative Cooler With Angled Shields," AIAA Paper 81-1100, 16th Thermophysics Conference, Palo Alto, Calif., June 1981.

[4]Stultz, J. W. and Grippi, R. A., "A Summary and Update of the Configuration and temperature Control Studies for a Fluorine-Hydrazine Propulsion System," 1979 JANNAF Propulsion Meeting, Anaheim, Calif., March 1979.

[5]Jones, J. A., "Spacecraft Mission and Shielding Thermal Study for a Space-Storable Fluorine-Hydrazine Propulsion System," 16th Thermophysics Conference, Palo Alto, Calif., June, 1981.

[6]Jones, J. A., "$LaNi_5$ Hydride Cryogenic Refrigerator Test Results," Dec., 1982.

[7]Cohen, R. L. and Wernick, J. H., "Hydrogen Storage Materials Properties and Possibilities," Science, Vol. 214, No. 4525, Dec. 4, 1981.

[8]Goodell, D., International Nickel Co., White Plains, N. Y., Personal Communications, Nov., 1982. Publication pending in Journal of Less Common Metals.

Chapter III. Heat Pipes

Design, Fabrication, and Test of Liquid-Metal Heat-Pipe Sandwich Panels

A. Basiulis*
Hughes Aircraft Company, Torrance, Calif.
and
C. J. Camarda†
NASA Langley Research Center, Hampton, Va.

Abstract

Integral heat-pipe sandwich panels, which synergistically combine the thermal efficiency of heat pipes and the structural efficiency of honeycomb sandwich panel construction, were fabricated and tested. The designs utilize two different wickable honeycomb cores, facesheets with screen mesh sintered to the internal surfaces, and potassium or sodium as the working fluid. Panels were tested by radiant heating, and the results indicate successful heat-pipe operation at temperatures of approximately 922 K (1200°F). These panels, in addition to solving potential thermal stress problems in an Airframe-Integrated Scramjet Engine, have potential applications as cold plates for electronic component cooling, as radiators for space platforms, and as low-distortion, large-area structures.

Introduction

Design studies of the NASA Langley Airframe-Integrated Scramjet Engine[1] have indicated potential thermal stress problems. Excessive thermal stresses result from large transient temperature gradients across the honeycomb sandwich walls of the engine structure during engine startup and shutdown. Conventional heat-pipe panel designs can reduce the thermal gradients. However, inherent in these designs are problems associated with

---

Presented as Paper 82-0903 at the 3rd AIAA/ASME Joint Thermophysics, Fluids, Plasma & Heat Transfer Conference, St. Louis, Mo., June 7-11, 1982.

*Manager, Thermal Devices, Electron Dynamics Division.

†Aerospace Research Engineer.

bonding the heat pipes to the honeycomb panels, the resultant thermal gradients due to contact resistances, and the probability of a substantial increase in panel mass. An alternate solution to these problems is the development of an integral heat-pipe sandwich panel[2] that synergistically combines the thermal efficiency of heat pipes with the structural efficiency of sandwich construction, with only a negligible increase in mass. A preliminary evaluation of such a concept was reported by Peeples.[3]

In addition to the above application, heat-pipe sandwich panels have potential as cold plates for electronic and circuit card cooling, as radiators for space platforms, and as low distortion, large-area structures (e.g., space antennas). To verify the feasibility of a heat-pipe sandwich panel, a program was initiated

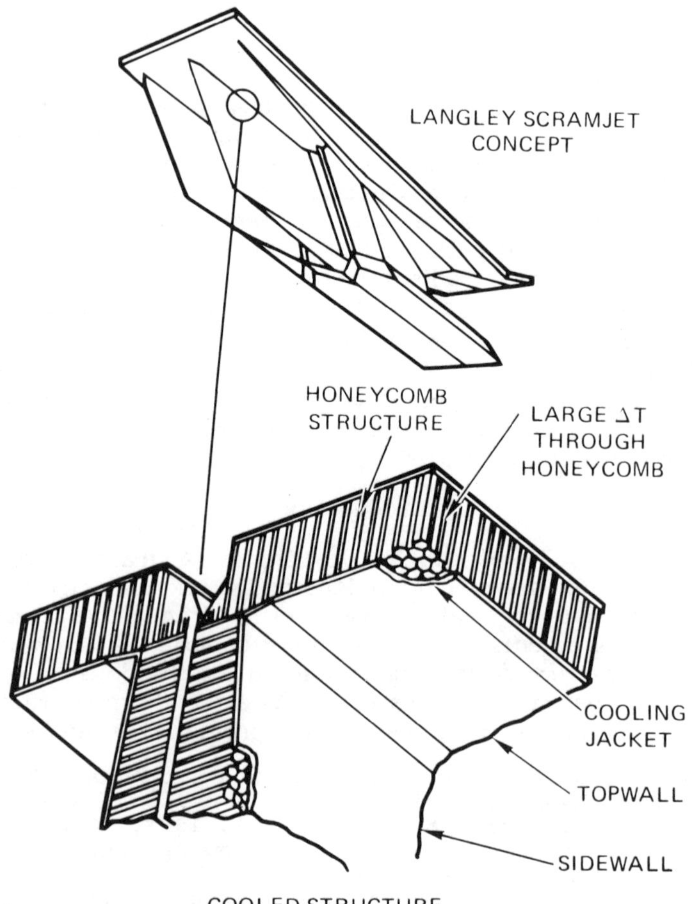

Fig. 1 Features of cooled scramjet structure.

(NASA Contract NAS1-16556) to fabricate several low-mass liquid-metal heat-pipe honeycomb panels.

This paper describes the thermal environment that led to the investigation of a heat-pipe sandwich panel, illustrates the preliminary design considerations and testing, describes manufacturing and fabrication details, discusses preliminary performance testing, and comments on potential future applications.

Design of Heat-Pipe Sandwich Panels

NASA Langley Research Center has been involved in a research program for the development of Airframe-Integrated Scramjet Engine concepts.[1] Results of that study indicate that an all-honeycomb primary structure, illustrated in Fig. 1, has less deflection and complexity than beam and honeycomb combinations of equal mass. Hence an all-honeycomb configuration was chosen as the best structural concept. All internal and external engine surfaces exposed to aerodynamic flow are cooled regeneratively by the circulation of hydrogen fuel (prior to injection) through a cooling jacket. Inconel 718 was chosen for the honeycomb primary structure, with Hastelloy-X or Nickel-200 chosen for the cooling jacket. The honeycomb front facesheet is 0.15 cm (0.06 in.) thick, the back facesheet is 0.13 cm (0.05 in.) thick, and the honeycomb cell is a 0.64-cm (0.25-in.) hexagonal arrangement constructed of 0.008-cm (0.003-in.) foil-gage ribbon.

Environment

Temperature gradients through the honeycomb walls during transient operation (i.e., engine startup or shutdown) may very

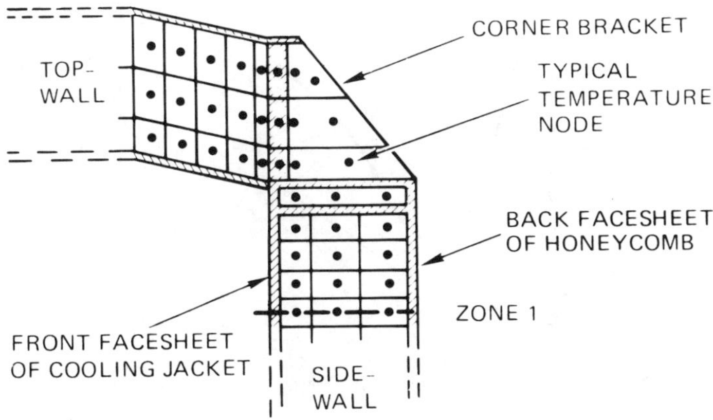

Fig. 2 Mathematical model for transient thermal analysis of honeycomb topwall-sidewall corner section.

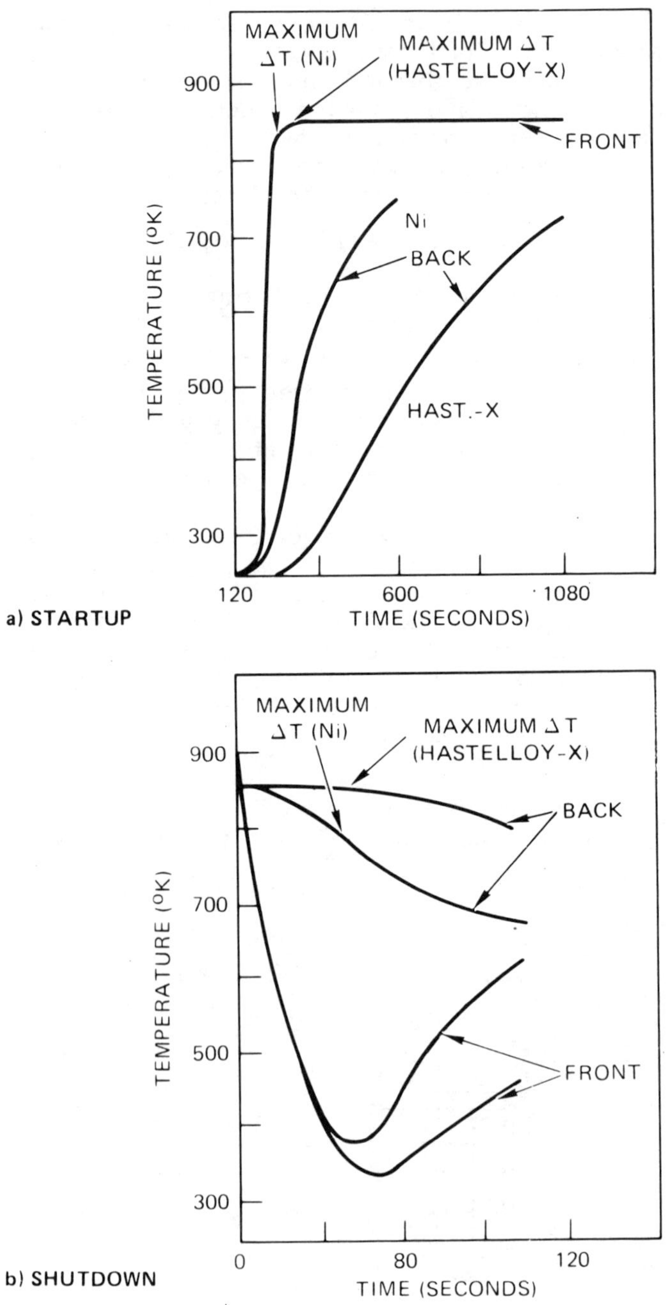

Fig. 3 Honeycomb temperature histories at zone 1.[1]

well control the structural design. A mission profile of a research-type vehicle was used by Buchmann[1] to predict the thermal/structural response quantities. A finite-difference analysis model of a section of the sidewall-topwall (Fig. 2) was used to calculate the transient temperatures shown in Fig. 3. Note from Fig. 3a that, at startup, the front facesheet quickly rises to 890 K (1140°F), resulting in a front-to-back ΔT of 667 K (1200°F) for a Hastelloy-X core. At engine shutdown, whether caused by normal occurrences or an abnormality such as a flameout, the temperature relationships of the front of the cooling jacket and the back of the honeycomb are reversed, as shown in Fig. 3b. The front-to-back ΔT developed is somewhat less than at startup--on the order of 556 K (1000°F) for the Hastelloy-X core. These thermal gradients result in excessive thermal stresses and premature fatigue failure.[1] Solutions to this problem noted in Ref. 1 result in concepts that are either more complex or heavier, or both.

Concept

The basic idea for the heat-pipe sandwich panel emerged as a solution to the above problem. The heat-pipe sandwich panels fabricated in the past met unique requirements of uniform tem-

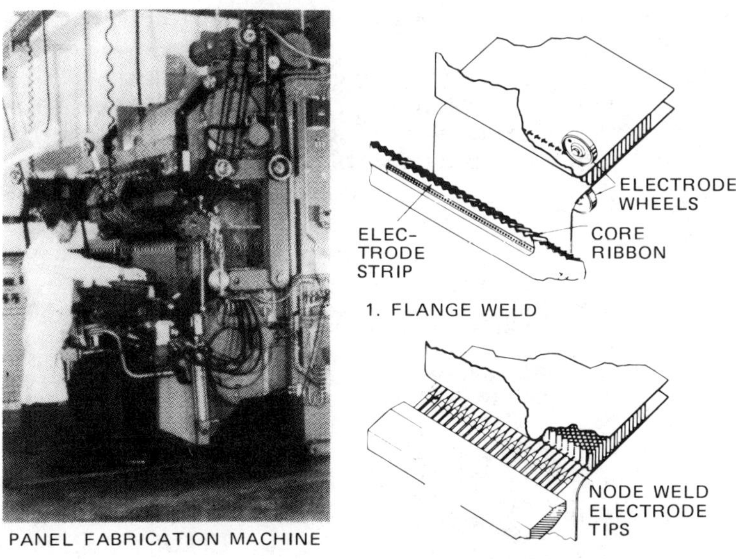

Fig. 4 Honeycomb panel welding machine and manufacturing technique (courtesy of Astech).

perature over a large surface area. These panels have demonstrated operation in a 0-g field,[4] uniform temperature over a large area[5] [0.5x6 m (1.64x19.7 ft)], and an isothermal surface[6] [within 0.01 K (0.02°F)]. However, all these panels were built by welding or furnace brazing by highly skilled technicians and, although they met all the technical requirements, they were very costly to manufacture.

The objective during this program was to design and fabricate a cost- and mass-effective sandwich panel using existing manufacturing techniques and equipment. The upper and lower ends of the core have flanges that enable spot welding to the faces. The entire sandwich panel can be constructed by simultaneously spot welding the core ribbons to each other and to the faces using the manufacturing technique illustrated in Fig. 4. The spot welds are so close together that they form an almost continuous bond. Since the entire panel is spot welded, this eliminates the need for bonding and possible materials compatibility problems.

The primary objective was to fabricate a heat-pipe honeycomb sandwich using a wickable honeycomb core, appropriate working fluid, and wickable internal faces that would enhance the transverse heat-transfer capability of the honeycomb. During operation, heat would be absorbed at the heated face by the evaporation of working fluid. The heated vapor flows, due to a pressure differential, to the cooler face, where it condenses and gives up its stored heat. The cycle is completed with the return flow of liquid condensate back to the heated face by the capillary pumping action of the wickable core. A schematic of the heat-pipe sandwich panel concept is shown in Fig. 5. A screen is sintered to the internal faces of the sandwich to allow intracellular liquid flow by capillary action. This design also allows the

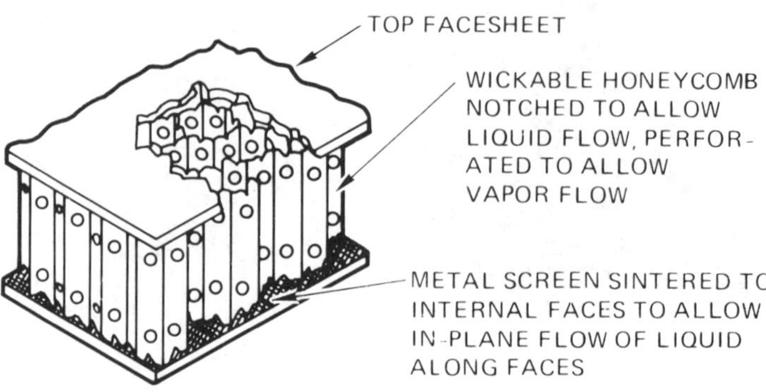

Fig. 5 Heat-pipe sandwich panel concept.

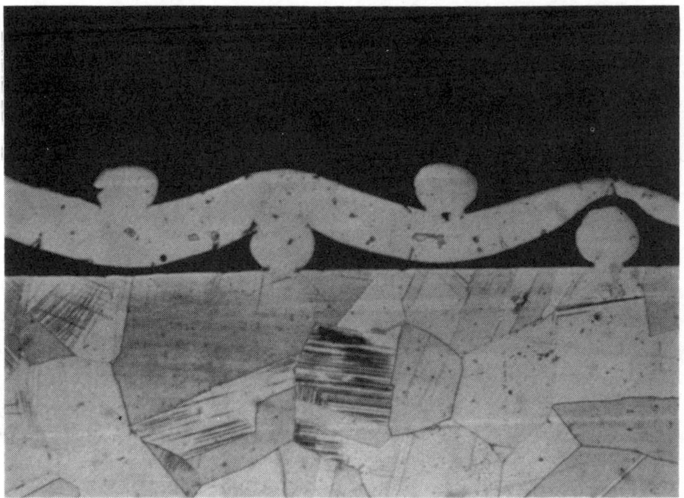

Fig. 6 Photomicrograph showing diffusion bonding of screen sintered to facesheet.

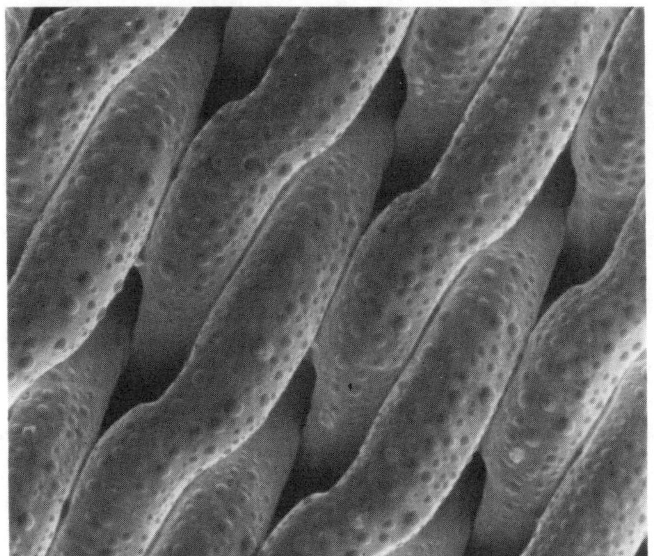

Fig. 7 Photomicrograph of Regimesh K sintered screen.

entire surface of the facing to be wetted by liquid and thus aid in evaporation and also help to reduce thermal gradients in the faces. The wickable honeycomb core could be a foil-gage woven mesh screen or a screen sintered to foil ribbons; this allows face-to-face liquid flow. The honeycomb is notched at each end to allow intracellular liquid flow and perforated to enable intracellular vapor flow. Although the primary mode of heat transfer is in the transverse direction (face-to-face) for the present application, the choice of other design alternatives can enable varying degrees of in-plane heat transfer.

Critical Element Evaluation

To accommodate the heat-pipe sandwich panel requirements, the structure must consist of two facings with internally wickable faces bonded to a perforated, wickable honeycomb core material (as shown in Fig. 5). Several techniques were considered for internal facesheet wicking: sintering a screen to the facing, spot welding a screen to the facesheet, and grooving or roughening the facesheet by grid blasting. Grooving and roughening were rejected because of facesheet warping and the poor surface left for subsequent welding. Sintering the metal screen to the facing

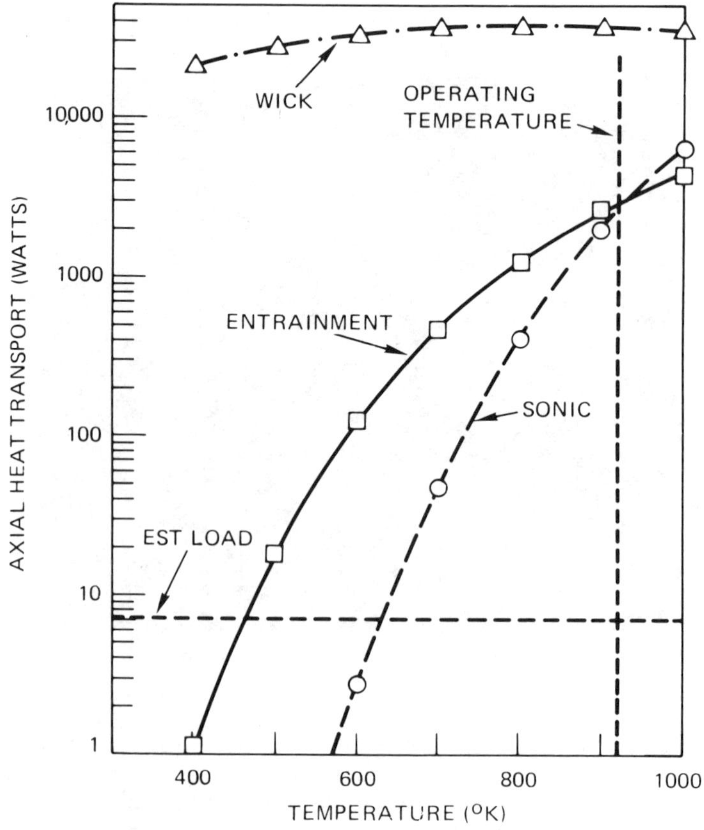

Fig. 8 Performance limits vs temperature for Regimesh K and potassium fluid.

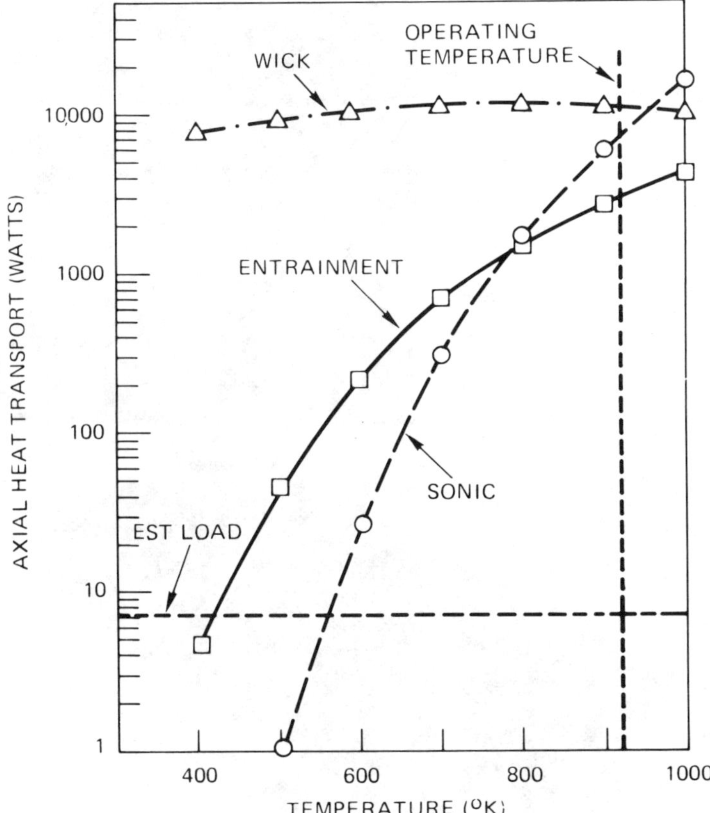

Fig. 9 Performance limits vs temperature for Regimesh K and potassium fluid.

was chosen as having more structural integrity than spot welding. Figure 6 shows a photomicrograph of the sintered screen facesheet.

Two designs for the honeycomb core were considered: a foil-gage sintered screen material (shown in Fig. 7) and a metal screen sintered to foil-gage stainless steel material. Both designs met structural and wicking requirements, with the former offering better wicking and the latter providing a stronger structural design. Figures 8 and 9 show performance limits for a heat-pipe honeycomb sandwich panel constructed with Regimesh K material for sodium and potassium working fluids.

Sample honeycomb ribbons were formed by Astech (a division of TRE Corporation, Santa Ana, Calif.) using standard equipment, and test samples were fabricated for evaluation. Both samples, sintered screen and screen on foil, met strength and wicking requirements.

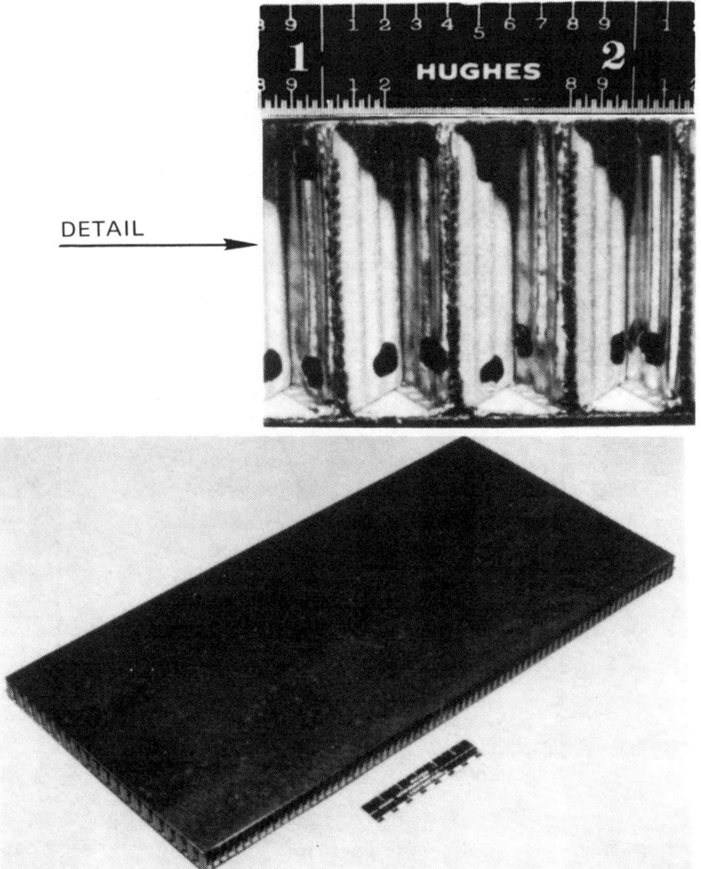

Fig. 10 Completed honeycomb panel prior to processing and final assembly.

Three different designs of honeycomb sandwich panels were fabricated: a resistance-welded core assembly for proof-pressure testing; a handmade, spot-welded core assembly for process testing and preliminary performance testing; and a machine-assembled, resistance-welded prototype for delivery and final testing.

The proof-pressure test specimen was assembled, vacuum leak checked, pressure tested up to 3.45 MPa (500 psi), and vacuum leak checked again. During and after testing, the honeycomb panel assembly retained structural and vacuum integrity. A hand-built, spot-welded core assembly was fabricated, processed with potassium working fluid and, after a preliminary test at 1075 K (1475°F), was delivered to NASA Langley Research Center for further testing.

Fig. 11 Completed heat-pipe assembly prior to processing.

Fabrication of Test Models

Sandwich panels were fabricated by Astech using an automated procedure for simultaneously resistance welding honeycomb ribbons to the facesheets. Completed sandwich panels were delivered to Hughes for further processing. Figure 10 shows the completed honeycomb panel. To eliminate potential contamination, panels were degreased, then fired in dry hydrogen at 1173 K (1652°F). At this point the sidewalls and processing tube were welded in place, completing the heat-pipe assembly. Figure 11 shows the complete heat-pipe assembly. The panel was then fired in dry hydrogen at 1173 K (1652°F) to remove oxides which were formed during final assembly. After leak check, the panel was placed in a vacuum chamber and heated to 1273 K (1832°F) for final cleaning and outgassing. After final leak check, the panel was charged with working fluid and processed. During preliminary tests, the heat-pipe panel was isothermal over the active surface, but did show some excess fluid in the processing tube. Figure 12 shows the heat-pipe panel during preliminary test.

Preliminary Radiant Heat Tests

Two prototype panels, one empty and the other containing potassium as the working fluid, were heated simultaneously by radiant heat lamps, as shown in Fig. 13. The heaters are quartz

Fig. 12 Heat-pipe panel during preliminary testing.

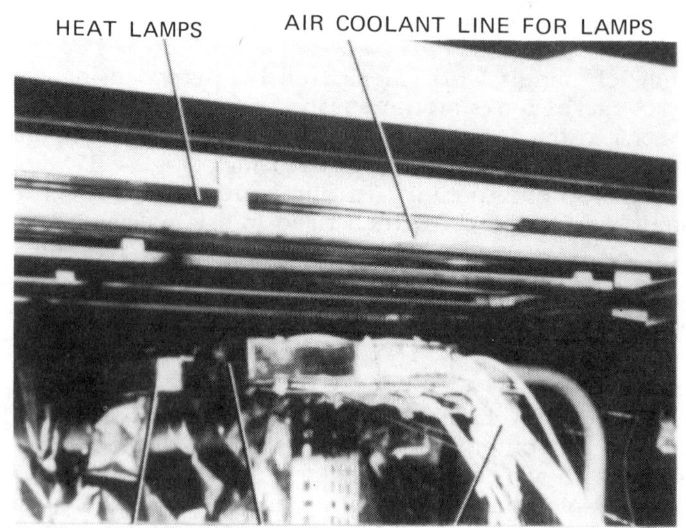

Fig. 13 Sandwich panels in position under radiant heat lamps.

lamps with a heated length of 6.35 cm (25 in.) and having a rated power of 2500 W (2.37 Btu/s) at 500 V. Each lamp bank contains eight lamps. Six lamp banks were energized for each test. One of the panels was located directly under one lamp bank and the other

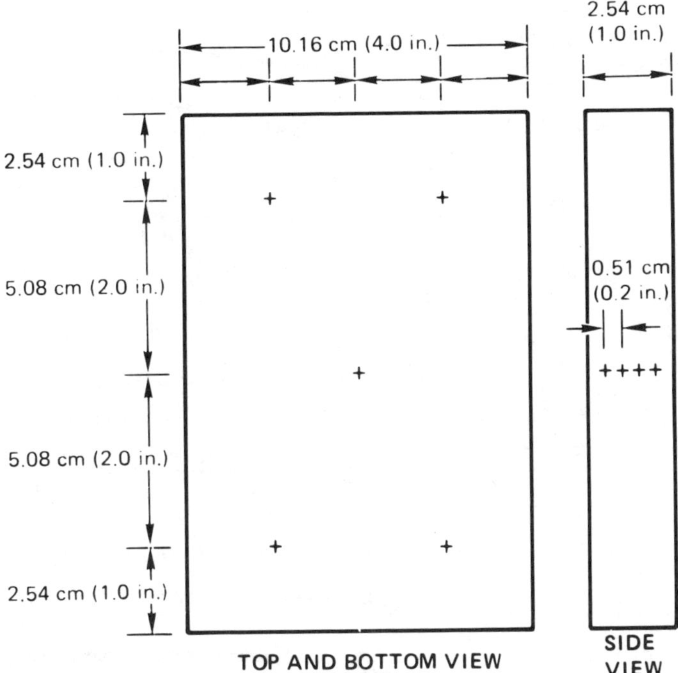

Fig. 14 Thermocouple locations.

panel was located the same vertical distance from the heaters but under another lamp bank. The distance of the panels from the lamp banks and the voltage to the lamps were varied. Power was applied as a step voltage input to the lamps. Power was applied for approximately 5 to 10 min and then abruptly shut off. Five thermocouples were located on the top and five on the bottom surfaces of the panels to measure temperature gradients, and four thermocouples were located along one side to study heat-pipe startup performance. Thermocouple locations are shown in Fig. 14. The panels were tested with and without insulation covering the bottom and sides of the panels. The insulation prevents heat loss by free convection and simulates the adiabatic boundary conditions described in Ref. 1.

The panels were tested nine times, and results of those tests are summarized in Table 1. Comparisons of temperature histories of a heat-pipe and non-heat-pipe sandwich panel with insulated and uninsulated surfaces are shown in Fig. 15. For the insulated panel tests shown in Fig. 15a, the temperatures of the back face of the non-heat-pipe panel and the heat-pipe panel temperatures continue to rise and slowly approach the temperature of the front facesheet of the non-heat-pipe panel as expected. Results of the

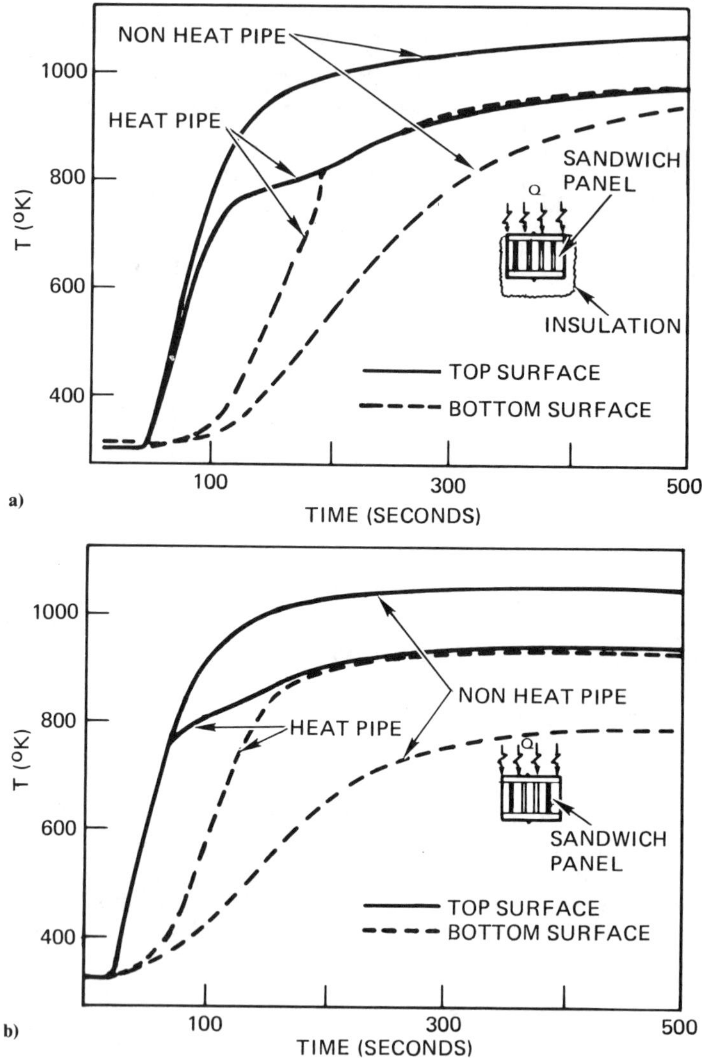

Fig. 15 Comparison of temperature histories of a heat-pipe and non-heat-pipe sandwich panel with and without insulation: a) with insulation; b) without insulation.

uninsulated panel tests (Fig. 15b) indicate that all temperatures level off and appear to reach a steady-state condition. As mentioned in Refs. 8 and 9, during heat-pipe startup from the frozen state, a nearly constant temperature continuum region propagates from the evaporator to the condensor section of the heat pipe. As shown in Fig. 15b, once this continuum front

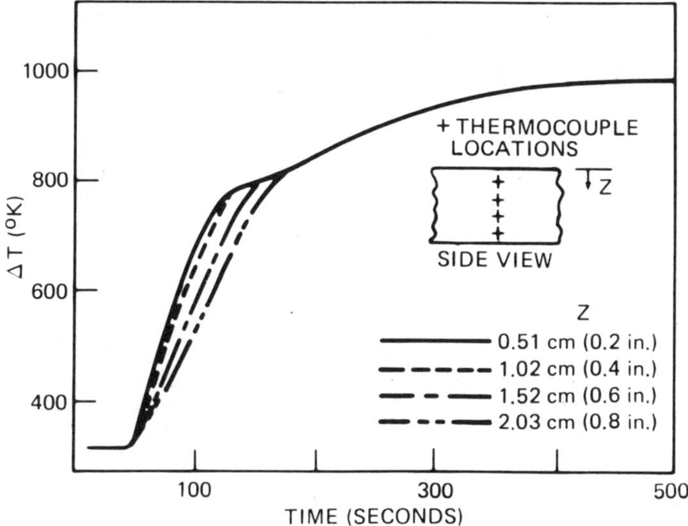

Fig. 16 Temperature histories along the side of the heat-pipe panel, illustrating startup performance.

Table 1 Summary of radiant heat tests of heat-pipe sandwich panel

Run No.	Insulated	Distance From Heater cm (in.)	Voltage V	Max ΔT K (°F) Heat pipe	Max ΔT K (°F) Non-Heat pipe	Max T K (°F) Heat pipe	Max T K (°F) Non-Heat pipe
1	No	10.5 (4.125)	380	338 (609)	403 (725)	891 (1145)	926 (1208)
2	No		460	352 (634)	474 (854)	916 (1189)	970 (1287)
3	No	5.1 (2.0)	250	313 (564)	407 (733)	803 (986)	863 (1093)
4	No		300	343 (617)	467 (841)	856 (1081)	920 (1197)
5	No		358	353 (635)	487 (877)	904 (1167)	984 (1312)
6	Yes		250	319 (575)	467 (840)	842 (1056)	894 (1150)
7	Yes		307	354 (638)	490 (882)	923 (1202)	974 (1293
8	Yes		356	378 (680)	535 (963)	968 (1283)	1020 (1376)
9	Yes		463	422 (760)	587 (1056)	1073 (1472)	1078 (1480)

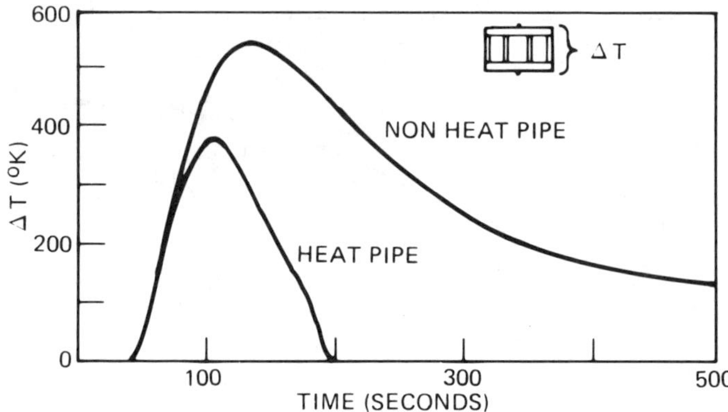

Fig. 17 Comparison of temperature gradients for a heat-pipe and non-heat-pipe sandwich panel.

reaches the back facesheet, the temperature there rises very rapidly as compared to the back facesheet of the non-heat-pipe panel. The temperature at which continuum flow begins and the rate at which the continuum front propagates depend on the working fluid, the temperature, the heat input, and the sonic flow limit of the vapor.

The average reduction in maximum ΔT during startup using a potassium heat-pipe sandwich panel instead of a non-heat-pipe panel is 27%. It is possible that this reduction can be increased by using cesium as the working fluid; this is currently being investigated. Figure 16 gives some idea of the rate of continuum region growth. The results are characteristic of startup of liquid-metal heat pipes as presented in Ref. 9. A typical comparison of temperature gradients through the depth of the honeycomb is shown in Fig. 17. As shown, the non-heat-pipe panel temperature gradient peaks slightly after that of the heat-pipe panel and is 29% higher for this particular run.

Conclusions and Recommendations

Initial studies indicate the heat-pipe honeycomb sandwich panels can be fabricated. The technology and commercial equipment are available to construct all-welded, machine-assembled honeycomb panels. At present, such honeycomb panels are constructed and formed into various shapes for use in airframe structures. Calculations and experiments with subscale test specimens indicate the feasibility of full-scale heat-pipe sandwich structures. Potential applications for heat-pipe sandwich panels include alleviating excessive thermal stresses in jet engines, cooling electronic components and circuit cards, limiting thermal distortions in large structures such as space antennas, and as radiators for space platforms.

Acknowledgments

The authors wish to acknowledge T. R. Lamp, H. Tanzer, and J. T. Burdette of Hughes Aircraft Co., who supported the design, fabrication, processing, and testing of the honeycomb heat-pipe sandwich panels, and T. Bernard of Astech for his support and for the fabrication of honeycomb sandwich segments.

References

[1] Buchmann, O. A., "Thermal-Structural Design Study of an Airframe-Integrated Scramjet," NASA CR 3141, Oct. 1979.

[2] Feldman, K. T. Jr., "Flat Plate Heat Pipe With Structural Wicks," U.S. Patent Application No. 803,582.

[3] Peeples, M. E., Reeder, J. C., and Sontage, K. E., "Thermostructural Applications of Heat Pipes," NASA CR 150906, June 1979.

[4] Fleishman, G. L. and Marcus, B. D., "Flat Plate (Vapor Chamber/Heat Pipes," AIAA Paper 75-7728, 10th Thermophysics Conference, May 1975.

[5] Basiulis, A. and Formiller D. J., "Emerging Heat Pipe Applications," Proceedings of the Third International Heat Pipe Conference, 1978.

[6] Fleishman, G. L., Loose J. P., and Scallon, T. Jr., "Vapor Chambers for Atmospheric Cloud Physics Laboratory," Third International Heat Pipe Conference, 1978.

[7] Dunn, P. and Reay, D. A., Heat Pipes, Pergamon Press, New York, 1976.

[8] Cotter, T. P., "Heat Pipe Startup Dynamics," Heat Pipes: AIAA Selected Reprint Series, Vol. XVI, edited by C-L Tien, AIAA, New York, 1973, pp. 42-45.

[9] Camarda, C. J., "Analysis and Radiant Heating Tests of a Heat-Pipe-Cooled Leading Edge," NASA TN D-8468, Aug. 1977.

Osmotic Pumped Heat Pipes for Large Space Platforms

H. J. Tanzer* and G. L. Fleischman†
Hughes Aircraft Company, Torrance, Calif.
and
D. D. Stalmach‡
Vought Corporation, Dallas, Texas

Abstract

A thermal bus will be required as a thermal control source for future space platforms. The osmotic heat pipe is one candidate device with potential significant payoff toward serving growing thermal management needs. Results of a study evaluating osmotic heat pipes for thermal bus applications are presented. Electrostatic and other techniques are proposed for flow control and solution circulation in zero gravity. Baseline size and performance design parameters of cellulose acetate membrane/sugar-water solution and other combinations were scaled up to predict osmotic pump performance for heat loads and temperatures of 4 to 120°C. A compact hollow-fiber membrane module measuring 20 in. in diameter by 12 in. long and weighing 190 lb is projected for 50-kW heat loads.

Introduction

A thermal bus will be required to provide a centralized thermal utility for multihundred-kilowatt space platforms projected for the 1990's. In previous space systems, thermal management has been achieved either passively or through the use of pumped liquids and electrical heaters. Evolving future space platforms, however, will require a much more significant role of thermal management because of the multiyear mission durations, large quantities of waste heat to be dissipated, long physical

---

Presented as Paper 82-0902 at the 3rd AIAA/ASME Joint Thermophysics, Fluids, Plasma & Heat Transfer Conference, St. Louis, Mo., June 7-11, 1982.

*Project Engineer, Thermal Devices Department, Electron Dynamics Division.

†Project Manager, Thermal Devices Department, Electron Dynamics Division.

‡Lead Thermodynamics Engineer.

distances involved, and variety of payloads and missions that must be accommodated by the platform (Fig. 1).

The idea of a thermal bus has evolved to effectively serve these growing thermal management needs. The function of a thermal bus would be to provide a uniform thermal control source (cooling and heating insensitive to the addition or removal of loads) for space platform electrical, life support, mechanical, scientific, and experimental equipment. The osmotically pumped heat pipe[1] has been identified as one candidate device which

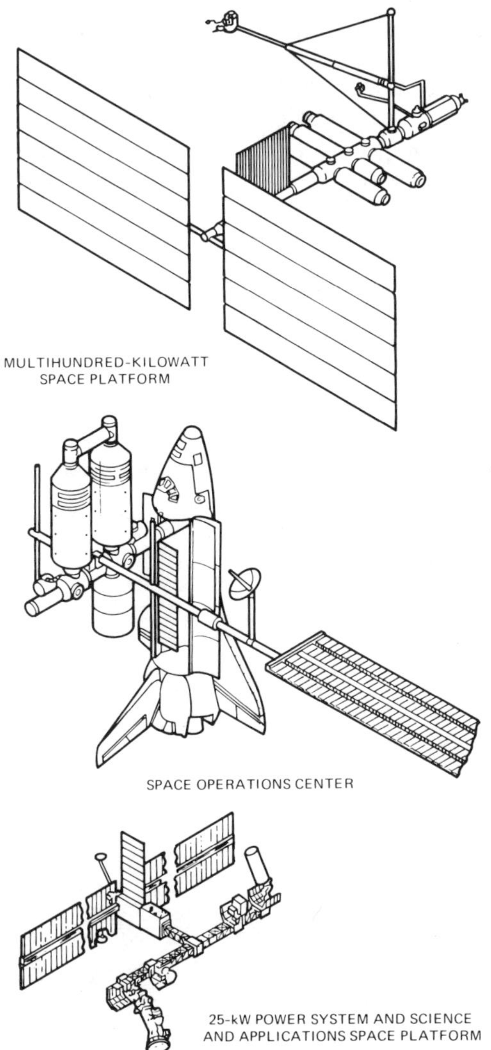

Fig. 1 Future platforms being evaluated by NASA will have increasing thermal management needs.

could provide significant payoff toward making such a system feasible. An investigation to evaluate the osmotic heat pipe for this application can provide information necessary to make concept recommendations and preliminary specifications in determining the direction of further development of a thermal utility system.

The Osmotically Pumped Heat-Pipe Concept

Direct osmosis is employed as the driving force that circulates working fluid inside a closed system. When two fluids of different concentrations are separated by a semipermeable membrane, the solvent will flow into solution to attain equilibrium. The membrane is impermeable to the solute. As long as the concentration gradients are high between solvent and solvent-solute mixture (solution), flow will continue due to direct osmotic forces. The addition of heat to the solution causes solvent to evaporate and flow to the condenser, where it recondenses, giving up heat. The solvent then returns to the membrane. The passage

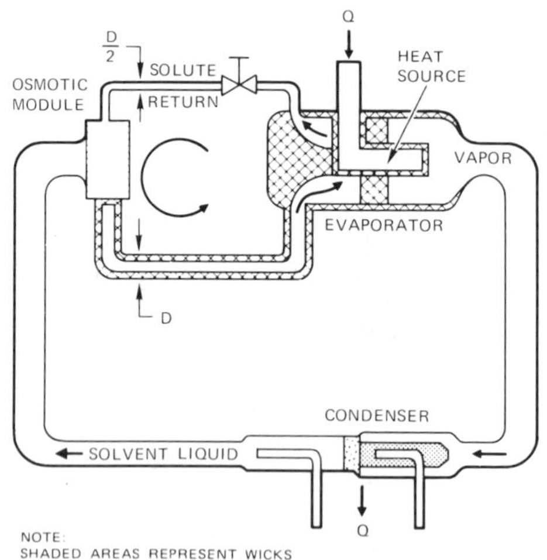

Fig. 2 Conceptual drawing of osmotically pumped module.

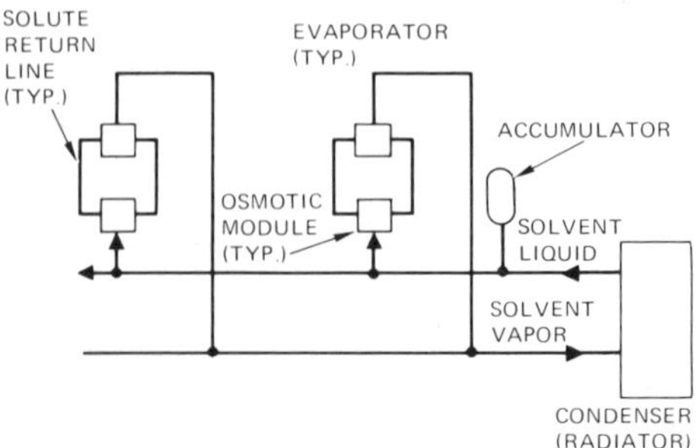

Fig. 3 Osmotic-pump-driven network.

of solvent through an osmotic membrane can create a differential pressure that is orders of magnitude greater than the capillary action created by surface tension in conventional heat-pipe wicks. This osmosis provides the capability to transport large amounts of heat over very long distances with a passive device.

A conceptual osmotic-pump-driven energy transport loop is illustrated in Fig. 2. A wick structure must be employed in the evaporator to prevent solute from being carried over with the solvent vapor and contaminating the system. The pumping force of the osmotic membrane is dependent on the concentration gradient across it. If solute is carried to the solvent side of the membrane, this pumping action will degrade as the concentration gradient is lessened. Eventually the concentration gradient could become so low as to cause cessation of pumping.

An osmotic pump module is located just upstream of each evaporator in an osmotic-pump-driven thermal bus system. Figure 3 shows a network with several osmotic pump components in a parallel flow arrangement. The evaporators represent any of the several types of heat loads at various locations on the space platform.

Design Considerations

Candidate solvent/solute working fluids were evaluated in order to select the best combinations for the specified temperature range of 4 to 120°C. Compatibility with membrane and container, performance, toxicity, flammability, and vapor pressure were taken into account in fluid selection. Membrane materials are chosen based on performance and compatibility with the selected solvent/solute combination.

One of the key requirements before the osmotic heat pipe can be applied to the space environment is a mechanism for return of solute from the evaporator to the membrane surfaces (solution circulation), thereby reducing concentration polarization. In a 1 g field, natural convection buoyancy forces assist in promoting the required solution circulation. However, new techniques are needed for solution circulation as well as flow control in 0 g applications.

In order to reduce weight, it is desirable to reduce the volume of the osmotic membrane modules. This may be accomplished by increasing the ratio of membrane surface area per unit volume. Two types of compact osmotic membrane modules, spiral wound and hollow fiber, have been investigated.

Dimensional details and performance data for state-of-the-art osmotic membranes and their application to heat-pipe closed systems were surveyed. The basic flow rate data obtained from actual tests were used to scale up and predict osmotic pump performance and sizes required for larger heat loads ranging from 1 to 50 kW over a temperature range varying from 4 to 120°C.

Membrane/Fluid Combination

Based on previous experiments and investigations,[2] a near-term and a longer-range advanced technology approach are recommended. The near-term aproach consists of the following membrane and working fluid combination:

Membrane:	Cellulose acetate
Solvent:	Water
Solute:	Sucrose
Temperature range:	
Evaporator:	30 to 100°C
Membrane module:	20 to 75°C

The advantage of this approach is that all materials are readily available so that laboratory and/or flight systems experiments can be performed today. Water has the highest mass flow and separation properties in conjunction with current state-of-the-art membranes such as cellulose acetate. This is largely due to the polar nature of the water molecule and its associated high latent heat. Other, more advanced composite membrane materials have been developed for reverse osmosis, but these have been found to give reduced performance or not work at all in direct osmosis. Sucrose is extremely soluble in water and is essentially impermeable to the membrane in direct osmosis applications. The container should be copper or copper-plated stainless steel for compatibility purposes. This combination, however, is not suited to the entire range of operating temperatures to be addressed in this study. The sucrose will break down and oxidize at temperatures near 100°C. Also, the cellulose acetate membrane cannot withstand sustained operation at temperatures higher than 75°C. Note, however, that the liquid condensate can be subcooled in the condenser region before entering the module. The vapor pressure of water is too low for operation at temperatures below 30°C.

For these reasons, an advanced technology approach is suggested that would extend the operating capability throughout the desired temperature range of 4 to 120°C. The recommended approach consists of the following:

Membrane:	Polybenzimidazole (PBI)
Solute:	Aluminum sulfate
Solvent:	Water (30 to 120°C)
	Ammonia (4 to 50°C)
	Methanol (4 to 120°C)

This approach will require development before it can be implemented. It is felt that methanol and ammonia will be compatible with PBI. However, compatibility and membrane pumping experiments are recommended. Aluminum sulfate was selected primarily because it has good solubility and separation characteristics with water. Other solutes should be investigated for the nonaqueous solutions. In addition, nonmetallic container materials (e.g., glass reinforced polymides) could be developed for weight savings and corrosion resistance.

0 g Osmotic Pumping Techniques

In order to apply osmotic heat-pipe technology to space applications, a means for operating in a 0 g environment must be devised. With the nonexistence of gravitational forces to contain the liquid, the capillary attraction forces of wicking material are required to eliminate any free-floating or unattached liquid. In addition, a means is necessary to actively direct the concentrated solution at the evaporator toward the membrane, where it can mix with dilute solution and increase the concentration in the vicinity of the membrane to promote increased osmotic pressure.

In a 1 g operation, this solute regeneration function is carried out by free-convection currents induced by gravitational forces.[3] Several techniques have been considered to induce return of solute to the osmotic membrane during operation, including the following: 1) electrostatic technique, 2) solvent velocity/displacement, 3) Bernoulli effect, 4) magnetic forces, 5) centrifugal forces, 6) temperature differences.

The electrostatic technique is illustrated in Fig. 4. By use of an electrolytic solute such as aluminum sulfate, a voltage could be imposed across the membrane to maintain the solute concentration in the vicinity of the membrane surface. The required voltages are very small. Table 1 shows the voltage required for several electrolytes. This concept is not a circulation technique but merely a means of holding the solute ions in place. Consequently, it requires little or no power consumption. Any slight current leakage would be due to secondary currents in the electrolyte and could be determined experimentally.

Table 1 Voltage estimated for electrostatic technique

Ion	Voltage[a]
K^+	0.092
Mg^{++}	0.127
Na^+	0.135
Li^+	0.175
SO_4^{--}	0.084
NO_3^-	0.095

[a]Estimated voltage required to counteract solvent velocity, $J_v = 7\times10^{-5}$ cm/s.

If solvent flow can be preferentially channeled (i.e., path of least resistance), then the solvent pumping velocity can be used to drive the solution circulation. As the solvent flows through the membrane into the solution compartment, it carries a portion of the solution with it toward the evaporator. If a return line is provided, new solution will be drawn into the module to maintain continuity. Likewise, when the dilute solution reaches the evaporator, the solvent will evaporate and the resulting concentrated solution will be displaced from the evaporator. Figure 2 illustrates an evaporator concept that will provide for return of the concentrated solution back to the pump module.

A pressure gradient that favors movement of concentrated solution from the evaporator to the membrane region can be created owing to the Bernoulli effect and its associated changes in cross-sectional flow area. Two disadvantages of this effect are that 1) the velocity required to generate a reduced pressure flow section is always upstream of the membrane, and 2) the velocities generated are extremely low. Thus it is difficult, if not impossible, to return the concentrated solution directly to the membrane surface with this approach.

Centrifugal forces, such as those on a spinning spacecraft, could be used to simulate the 1 g acceleration field.

Temperature differences give rise to surface tension gradients in fluids (Marangoni effect) which have been found to dominate in many 0 g materials processing experiments.[4] The surface tension of most fluids increases with decreasing temperature. Since the membrane modules operate at a lower temperature than the evaporator, this effect could be used to advantage. This is an area requiring further investigation by experimentation.

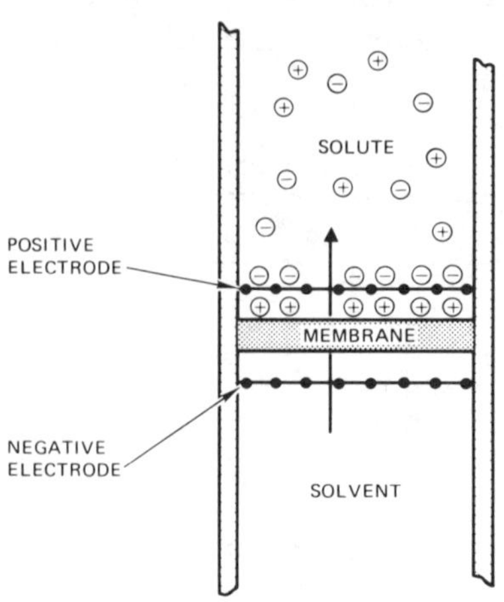

Fig. 4 Electrostatic technique for solute retention.

Another consideration for 0 g operation is the control technique. A flow control technique is necessary, since the osmotic pumping rate is not coupled to the heat input rate, as in a capillary-pumped heat pipe. Without a means of control, the solvent will continue to flow at a constant rate into the solution compartment even after the heat input is reduced or shut off. If this flow continues, the solution will eventually be forced through the wick and into the solvent compartment.

Three possible techniques for 0 g control are solvent depletion, control valve, and potential difference. The most simple and direct approach is to size the solvent condensate volume such that it can be stored in the evaporator or an accumulator on the solution side of the membrane under a no-load condition. Pumping will stop when all of the solvent has been pumped into the solution side. Second, a control valve can be used to regulate the flow rate if the valve is located on the solvent side of the loop. Otherwise the solvent would continue to flow across the membrane until the full osmotic pressure of several hundred psi is reached. This occurrence, known as the "prune effect," could damage the membranes or membrane seals. Finally, if the electrostatic method of solute retention is used, the applied voltage could be varied to control the concentration of ions near the membrane surface and therefore control the flow rate of solvent across the membrane. Although each of these control concepts appears feasible for on/off control, their ability to accurately control flow rates at reduced loads is uncertain and would have to be verified by laboratory experiments.

Note that the maximum length of the solvent return line in an osmotically pumped system will be determined by the flow pressure drop and degree of subcooling. Without subcooling, nucleate boiling would occur in a very short distance, as soon as the liquid pressure dropped below the saturation pressure of the fluid. With an appropriate amount of subcooling, however, the viscous pressure drop of the solvent flow should not exceed the absolute value of the vapor pressure at the condensate/vapor interface. Otherwise the solvent pressure becomes negative, placing the liquid under tension and resulting in separation of the liquid column.[5] This action is similar to the phenomenon that an ordinary suction pump cannot pull a column of water beyond a height of 33 ft, corresponding to a pressure of 1 atm.

Osmotic Pump Performance and Predictions

Experimental data were used to scale-up and predict osmotic pump performance and sizes. Osmotic membranes investigated included flatsheet,[6] shell-and-tube,[3] and spiral-wound

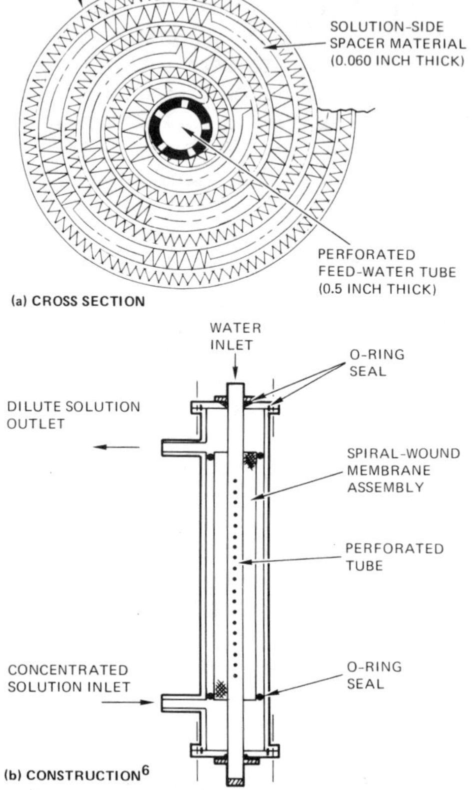

Fig. 5 Spiral-wound module:
a) cross section;
b) construction.[6]

closed-loop[7] heat-pipe systems, as well as spiral-wound and hollow-fiber open-loop systems.[8]

In a steady-state condition, the solvent evaporation rate is balanced by the rate of solvent flow through the membrane. This mass balance can be mathematically expressed in terms of heat input as

$$\dot{Q} = \dot{m}\lambda = J_v A_m \rho \lambda$$

where Q is the evaporator heating rate, W; $\dot{m}$ is the rate of solvent evaporation in evaporator, $g\text{-}s^{-1}$; is the solvent latent heat of vaporization, $W\text{-}s\text{-}g^{-1}$; J_v is the volumetric flow rate (flux) of solvent through the membrane per unit area of membrane surface, $cm\text{-}s^{-1}$; A_m is the membrane area, cm^2; and ρ is the solvent density, $g\text{-}cm^{-3}$.

The basic equation for solvent flux J_{vmax} through an ideal membrane is expressed as[7]

$$J_{vmax} = L_p (\pi - \Delta P)$$

where L_p is the filtration coefficient, cm^3-s^{-1}-$dyne^{-1}$; ΔP is the static pressure difference across membrane, dyne-cm^{-2}, and equals the pressure (solution side) minus the pressure (solvent side); and π is the osmotic pressure, dyne-cm^{-2}.

The percentage improvement in membrane solvent flux J_v has been found to increase exponentially with increasing solution circulation velocity, v' (Ref. 4),

$$V = J_v/J_{vmax} = 1 - e^{-41.5v'}$$

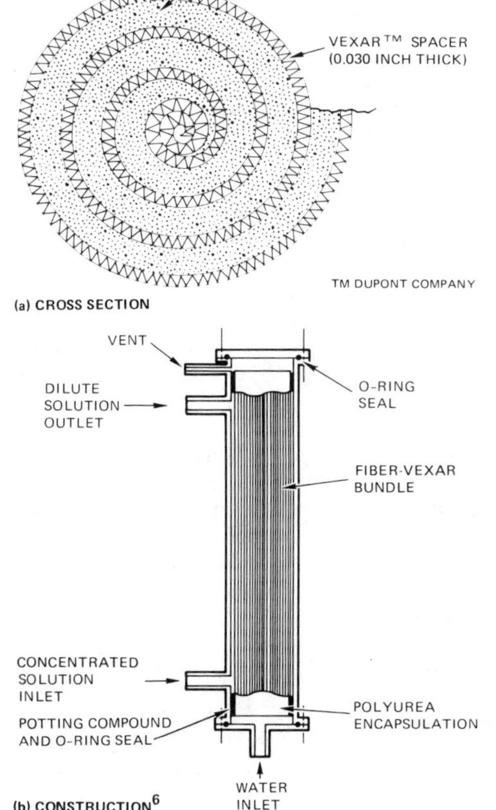

Fig. 6 Hollow-fiber module: a) cross section; b) construction.[6]

Solute return from the evaporator is necessary for continuous membrane pumping, but a point is reached where higher solution circulation velocities provide little additional improvement in performance. This empirical equation represents a percentage improvement, and does not depend on a particular membrane geometry or concentration level.

Rearrangement of the above equations results in the following thermodynamic relationship:

$$\dot{Q}/A_m = \rho \lambda V L p \, (\pi - \Delta P)$$

Solvent flux data taken across cellulose acetate membrane are presented in Table 2. The term J_V is a specific osmotic membrane property that exists for each membrane-solution combination, system pressure, membrane geometry, and solution circulation. Correlation of data in Table 1 results in a design capability for predicting osmotic heat-pipe performance if solvent flux data are known (for equal system pressure drop and solution circulation): Same solvent ,

$$(\lambda, \rho = \text{constant}) \rightarrow \quad \left(\frac{\dot{Q}}{A_m}\right)' = \left(\frac{\dot{Q}}{A_m}\right)_o \left(\frac{J_v'}{J_o}\right)$$

Table 2 Survey of solvent flux across cellulose acetate

Description	Solute	Power Density W/ft^2	J_v cm/s	Reference
Spiral wrap	1 M Sucrose/water	102	4.52×10^{-5}	7
Spiral wrap, open loop	1 M Sucrose/water		7.54×10^{-5}	8
Hollow fiber, open loop	1 M Sucrose/water		5.66×10^{-5}	8
Hollow fiber, open loop	0.79 M KH_2PO_4/water		9.43×10^{-5}	8
Shell and tube	3 M Sucrose/water	220	1×10^{-4}	3
Flat sheet	1 M Sucrose/water	332	1.64×10^{-4}	6
Column rise, open loop	1 M Sucrose/water		6.9×10^{-5}	6
Column rise, open loop	2 M Polyethylene glycol/methanol		6.5×10^{-6}	6

Different solvent,

$$(\lambda, \rho \neq \text{ constant}) \rightarrow \quad \left(\frac{\dot{Q}}{A_m}\right)' = \left(\frac{\dot{Q}}{A_m}\right)_o \left(\frac{J_v' \lambda' \rho'}{J_{v_o} \lambda_o \rho_o}\right)$$

where subscripts o and ' represent original and new condition, respectively.

Baseline design parameters were established from experimental data of several prototype Hughes osmotic heat pipes containing cellulose acetate membrane and a 1 molality (M) water/sucrose working solution. A spiralwound unit containing 5 ft^2 of membrane in a cylindrical module measuring 7.32 cm id and 28.6 cm in length was tested to 500 W steady-state operation in 1 g at temperatures from 40 to 60°C. (The spiral-wound module is shown in Fig. 5.) In addition, dimensional and construction details of a more compact hollow-fiber module (Fig. 6), containing 10 ft^2 of membrane in a volume similar to that in the spiral-wound module, were used in conjunction with measured performance data from the spiral-wound module to obtain a second set of baseline design parameters. These parameters are listed in Table 3.

As the degree of membrane compactness increases, the membrane solvent flux (J_v) decreases, primarily owing to reductions in effective solution circulation that result from the more constricted membrane geometry. In other words, doubling the

Table 3 Design basis for membrane module sizing

Parameter	Spiral wrap	Hollow fiber
Performance achieved, W	500	Not tested
Dimensions, in.	2.88 i.d. x 11.25 length	2.88 i.d. x 11.0 length
Module volume, ft^3	0.041	0.042
Membrane area, ft^2	5	10
Design values		
$\frac{\text{Power, W}}{\text{Unit area, ft}^2}$	100	Assume 100
$\frac{\text{Membrane area, ft}^2}{\text{Unit volume, ft}}$	121.6	241

membrane area does not necessarily mean doubling the total solvent flow, since membrane increase ("compactness") generally carries the penalty of reduced solvent flux. However, for purposes of this study, identical power density of 100 W/ft^2 membrane for both types of modules is assumed. This is based on an average value obtained from ratios of open-loop J_V data, multiplied by the measured spiral-wrap heat-pipe power density:

$$(Q/A)\text{hollow fiber} = (Q/A)\text{spiral wrap}\left(\frac{J_V \text{ hollow fiber}}{J_V \text{ spiral wrap}}\right)$$

$$(Q/A)\text{hollow fiber} = 100 \text{ W/ft}^2\left(\frac{J_V \text{ hollow fiber}}{7.54\text{x}10^{-5} \text{ cm/s}}\right)$$

$$\underset{(1\text{M sucrose})}{5.66\text{x}10^{-5}\text{cm/s}} \leq J_V\text{hollow fiber} \leq \underset{(0.79 \text{ M } KH_2PO_4)}{9.43\text{x}10^{-5}\text{cm/s}}$$

$$\therefore\ 75 \text{ W/ft}^2 \leq (Q/A)\text{hollow fiber} \leq 125 \text{ W/ft}^2$$

$$\text{Assume average: } (Q/A)\text{hollow fiber} = 100 \text{ W/ft}^2$$

Figures 7 and 8 present the results of module sizing for the spiral-wound and hollow-fiber designs, respectively. Scaling-up for heat load increases was accomplished by keeping module length constant and increasing module diameter as necessary.

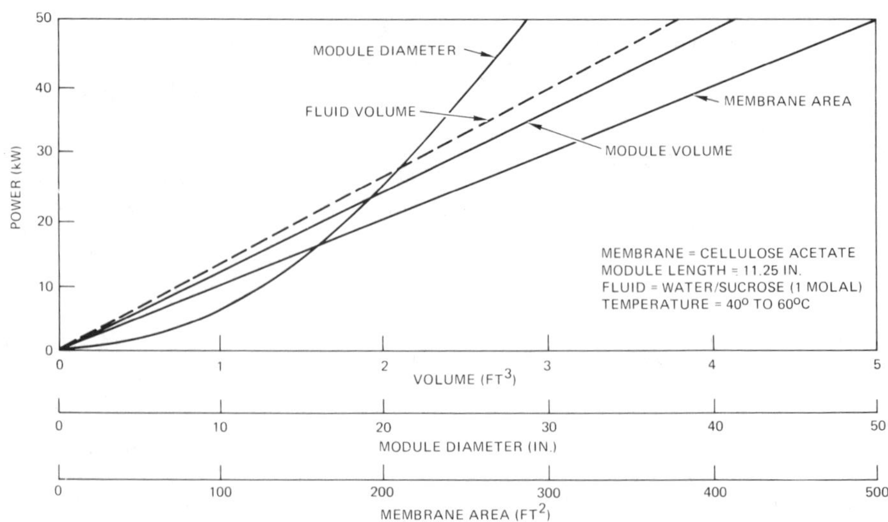

Fig. 7 Spiral-wound module size requirements.

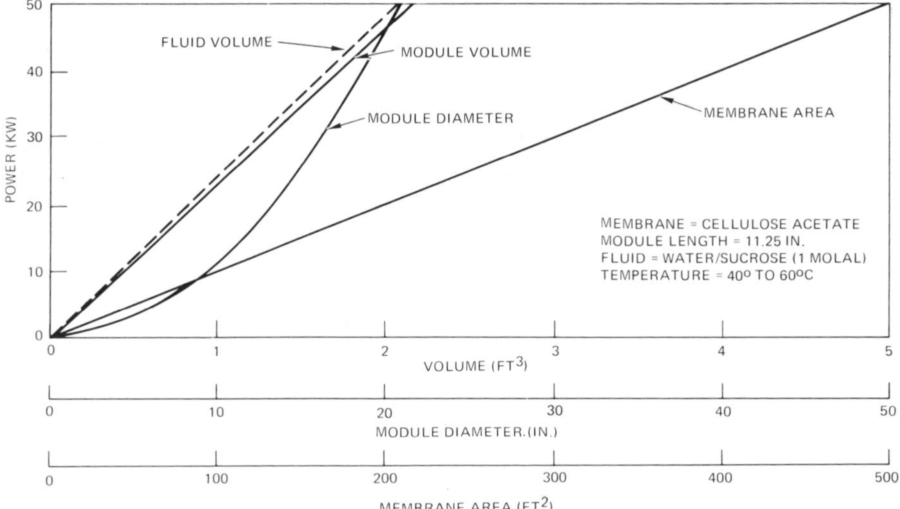

Fig. 8 Hollow-fiber module size requirements.

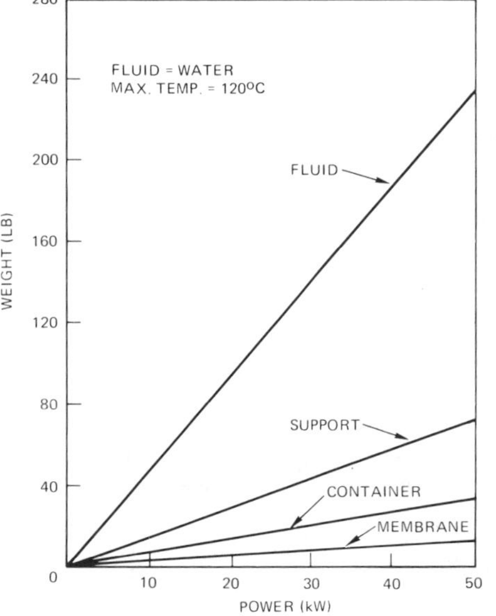

Fig. 9 Spiral-wound module component weights.

Inherent in this approach is the assumption that in 0 g operation, a viable solute circulation technique exists that permits equal pumping rates across a unit membrane corresponding to 1 g. Although optimistic, this approach does portray the advancement expected from future development. Performance predictions at temperatures below 20°C introduce the evaluation of the working fluids (solvents) ammonia and methanol and the corresponding new membrane materials. Sufficient data for osmotic pump performance with these fluids are not available. The expectation of successful future developments justifies the assumption of new membrane/solution combinations performing equivalent to the baseline.

The baseline units were constructed of copper and were not weight optimized. For the purposes of this study, the module

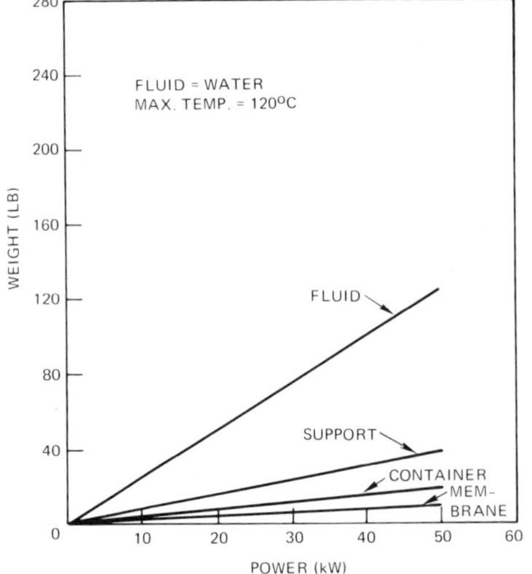

Fig. 10 Hollow-fiber module component weights.

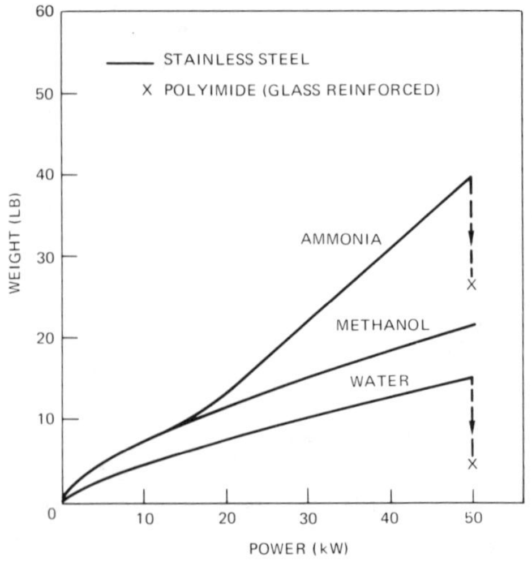

Fig. 11 Hollow-fiber module container weights.

container was assumed to be made of stainless steel, and wall thicknesses were based on containment of working fluid vapor pressure. The basis for weight design is presented in Table 4. Figures 9 and 10 present the module component weights for the spiral-wound and hollow-fiber designs, respectively, with water as the working fluid. The compactness advantage of the hollow-fiber construction is evident in the smaller volume and weight. Hollow-fiber module container weights for the working solvents water, methanol, and ammonia are presented in Fig. 11.

Determination of pressure drops through the modules required modeling of the flow path, considering both solution and solvent sides of the membrane, and application of the appropriate pressure drop equations. As module power requirements increase, cross-sectional flow area increases proportionately to mass flow.

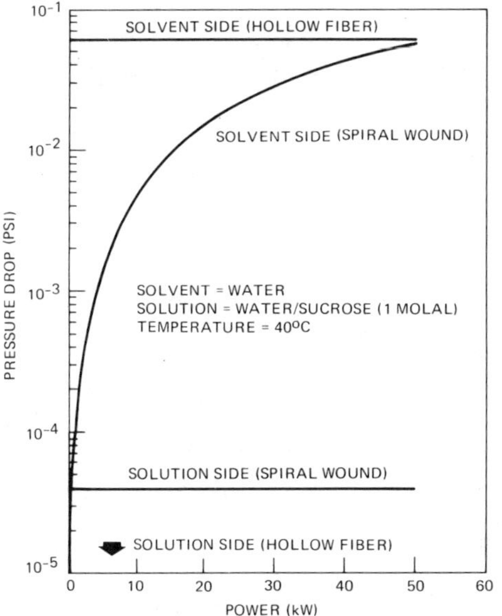

Fig. 12 Module pressure drops.

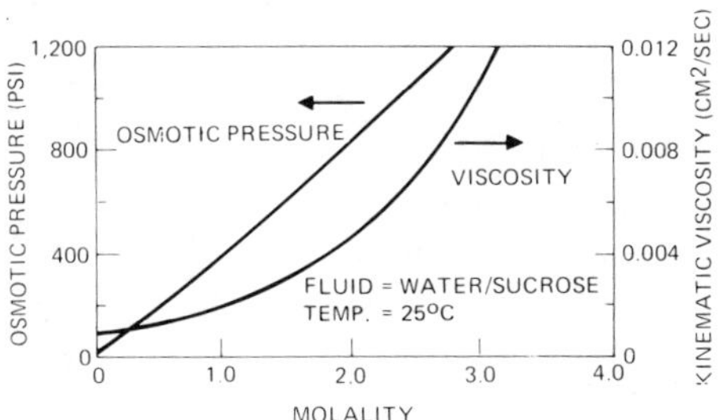

Fig. 13 Osmotic pressure head and viscosity.

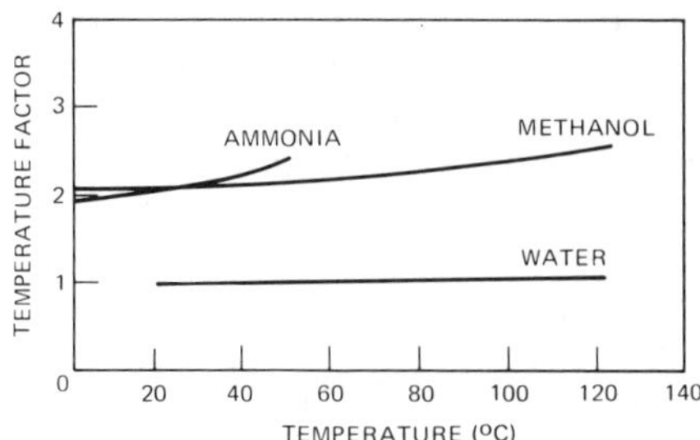

Fig. 14 Temperature factor for module length and mass flow rate.

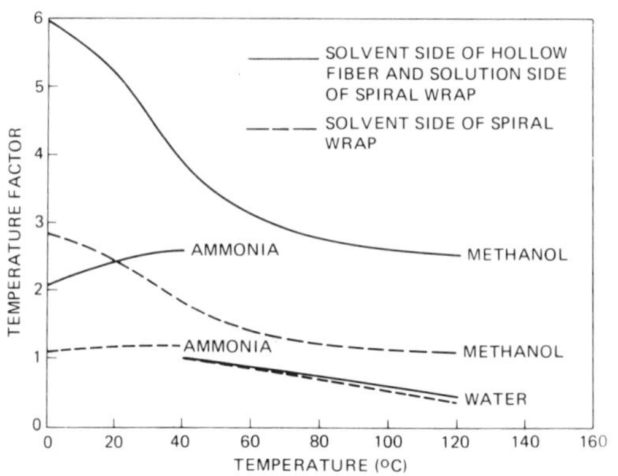

Fig. 15 Temperature factor for pressure drop.

Table 4 Weight design basis

Parameter	Spiral wrap	Hollow fiber
Membrane density	0.034 lb/in.^{3}	0.0013 gr/linear in.
Support vexar (screen) density, lb/in.^{3}	0.010	0.010
Epoxy density, lb/in.^{3}	. . .	0.042
Container (sst): Wall thickness	$t = 4\left(\frac{P_v R}{\sigma \text{ ult}}\right)$	Safety factor = 4
Module volume	Vm = V cylinder + V endcaps	
	$= 2\pi$ RoLt + 4 (π Ro^{2}t)	Safety factor = 4
Fluid (water) volume	Vf = Vm - V support (assume V membrane = 0)	
Weight = volume x density		
Total weight = W membrane + support + W container + W fluid		

Table 5 Module pressure drop

Spiral wrap
Solution side
Darcy's law to calculate pressure drop through wick (support) structure
Flow path along longitudinal axis of module
ΔP = Constant $\neq$ Function (module size/power)
Solvent side
Darcy's law $\Delta P = \left(\frac{\mu \, Leff}{\rho \, Aw \, K}\right) (\dot{m})$
Where Leff = Effective liquid flow path length; Aw = Wick x-sect flow area; K = Wick permeability
Flow path as spiral from center outward
$\Delta P \propto C$ (Leff) as power increases
Hollow fiber
Solvent side
Fanning equation $\Delta P = 4f \frac{L}{Di} \frac{\rho V^2}{2g}$
Parallel flow through fibers
ΔP = constant
Solution side
High-flow cross-sectional area,
ΔP very small, $< 10^{-5}$ psi

Thus pressure drop remains constant for all cases except the solvent side of the spiral wrap, where the spiral flow path distance and thereby the pressure drop increases with power. Module pressure drop details are contained in Table 5. Magnitudes of all pressure drops are quite small owing to low solution/solvent mass flux through the modules (Fig. 12). Pressure rise across the membrane is the osmotic pressure head and is solely dependent on the molality and circulation of the working fluid. Figure 13 shows that the osmotic pressure rise with a cellulose acetate membrane and a 1 M water/sucrose working solution is approximately 2.6×10^6 N/m^2 (400 lb/in.^{2}).

Performance predictions as a function of temperature are accomplished through derived "temperature factors" that are used as multiplication factors applied to performance predictions of the baseline unit H_2O/40°C. Pertinent details are presented in

Table 6 Temperature effects (4, 20, 40, 120°C)

New membrane/solution combinations

Assume solvent mass flux rate (J_v) same

Factors accounted for:

Solvents NH_3 and CH_3OH

Fluid properties change with temperature

Length and mass flux increase by

$$L', \dot{m}' = L_o, \dot{m}_o \left(\frac{\lambda_o H_2O/40^oC}{\lambda' \text{ solvent/temp}} \right)$$

Note: Membrane area increases to let enough mass through to offset Hfg decrease of new solvents

Increase pressure drop by factor

$$\Delta P' = \Delta P_o \left(\frac{\rho_o \mu' \dot{m}'}{\rho' \mu_o \dot{m}_o} \right)$$ for solvent-side spiral wrap

$$\Delta P' = \Delta P_o \left(\frac{\rho_o \mu' \dot{m}' L'}{\rho' \mu_o \dot{m}_o L_o} \right)$$ for solution-side hollow fiber for solvent-side spiral wrap

Where o subscript refers to water solvent/40°C and refers to new solvent/new temperature

Table 6. Temperature factors take into account both the change to solvents other than water which affects membrane area (or module length) and mass flux as shown in Fig. 14, and the variation of fluid properties at temperatures other than 40°C that affect pressure drops through the module (Fig. 15).

Conclusions and Recommendations

The two most promising techniques for solution circulation in a 0 g environment are the solvent velocity/displacement technique and the electrostatic technique. The former can be used with the near-term membrane and fluids combinations with organic solutes. The electrostatic technique is compatible with the long-range approach, where electrolytes are recommended as the working solution. The electrostatic technique also provides a means of control by variation of the voltage potential. Proof-of-principle laboratory experiments are needed to confirm the feasibility of these ideas. The displacement technique can be tested with the apparatus in a horizontal plane.

Another area that should be investigated experimentally is the optimum size of membrane modules. If the modules are made too large, flow distribution problems may detract from the expected performance potential. The number of modules to be used in parallel will depend on the optimum module size determined in this manner. Also, the maximum length of solvent (condensate) return line to prevent separation -- vapor lock -- of the liquid in a 0 g environment should be determined by testing.

In conclusion, it is felt that laboratory, materials, and conceptual studies should be continued for the osmotic pump module. Its promise as a high-performance passive pump makes this concept worth pursuing.

References

[1] Midolo, L. L., "Heat Transfer Apparatus with Osmotic Pumping," U.S. Patent No. 3,677,337, July 18, 1972.

[2] Minning, C. P., Fleischman, G. L., and Giants, T. W., "Development of an Osmotic Heat Pipe," Proceedings of the Third International Heat Pipe Conference, AIAA, New York, 1978, pp. 327-334.

[3] Basiulis, A., Fleischman, G. L., and Minning, C. P., "Design, Development, and Test of a 1000 W Osmotic Heat Pipe," Heat Transfer and Thermal Control: AIAA Progress in Astronautics and Aeronautics, Vol. 78, edited by A. L. Crosbie, AIAA, New York, 1981, pp. 324-344.

[4] Bittence, J. C., "Materials Engineering in the Sky," Materials Engineering, Vol. 92, Oct. 1980, pp. 34-41.

[5] Apfel, R. E., "The Tensile Strength of Liquids," Scientific American, Vol. 227, Dec. 1972, pp. 58-71.

[6] Minning, C. P., Fleischman, G. L., Anderson, R. A., and Perrygo, C. M.. "Design and Test of a Prototype Osmotic Heat Pipe," Heat Transfer, Thermal Control, and Heat Pipes: AIAA Progress in Astronautics and Aeronautics, Vol. 70, edited by W. B. Olstadt, AIAA, New York, 1980, pp. 307-328.

[7] Fleischman, G. L., "Osmotic Heat Pipe," Wright-Patterson Air Force Base, Ohio, Technical Report AFWAL-TR-81-3132, Nov. 1981.

[8] Baker, R. W., Johnson, J. M., and Lee, E. K. L., "Membrane Modules for Osmotic Heat Pipes," Bend Research, Inc., Bend, Ore, Final Report, Feb. 1981.

Development of a Double-Wall Artery High-Capacity Heat Pipe

R. Ponnappan*
Universal Energy Systems, Inc., Dayton, Ohio
and
E.T. Mahefkey†
Wright-Patterson Air Force Base, Dayton, Ohio

Abstract

This paper describes the design and testing of a "double-wall" integral wick-artery heat-pipe concept. The design incorporates a double wall using two concentric tubes. The inner-wall tube has axial arterial channels machined on its outer surface, while the outer wall has fine radial grooves on its inner surface. The inner-annular space is filled with a screen wick which functions as a capillary pump and evaporator reservoir wick. Under the conditions of interest, the three wick structures can be independently optimized to produce high-capacity and high-radial heat flux capability. A proof-of-concept 1.2-m-long copper-water prototype was built and tested, demonstrating an axial capacity of 1600 W-m and evaporator heat fluxes of 16 W/cm^2. The concept offers the potential for higher-capacity and evaporator heat flux.

Nomenclature

A	=	cross-sectional flow area
d_0	=	outer diameter of inner tube
g	=	acceleration due to gravity
H	=	capillary head available for circumferential liquid distribution

Presented as Paper 82-0906 at the 3rd AIAA/ASME Joint Thermophysics, Fluids, Plasma and Heat Transfer Conference, St. Louis, Mo., June 7-11, 1982.

*Thermal Engineer, Research and Development Division.
†Aerospace Engineer, Aerospace Power Division.

K = wick permeability factor (Darcy's law)
L = length of the heat pipe
$\dot{m}$ = mass flow rate
N = mesh number of the wire screen
P_{cm} = maximum capillary pressure
$\dot{Q}$ = heat flow rate
r_c = capillary radius
R_1, R_2 = principal radii of curvature of the liquid meniscus
$\bar{T}$ = temperature (average)
$\Delta P/\Delta x$ = axial pressure gradient
ΔT = temperature difference
δ = depth of artery groove
θ = wetting angle
μ = viscosity
ρ = density
σ = surface tension
ω = width of artery groove

Subscripts

AC = adiabatic and condenser
ADI = adiabatic section
COND = condenser
EA = evaporator and adiabatic
EVAP = evaporator
G = groove
HP = heat pipe
L = liquid
S = screen
V = vapor

Introduction

Advanced spacecraft thermal management requirements motivate the development of improved heat pipes devoid of present technology limitations (capacity, artery gas blockage sensitivity, and evaporator heat flux). There is a need for a high-capacity, high heat flux device which is simple, inexpensive, readily manufactured, and ground testable.

This paper describes a new composite wick-artery heat-pipe concept, its design, and the performance characteristics of a 1.2-m copper-water prototype device.[1] The design goals included a 2000-W-m capacity, 20-W/cm^2 radial evaporator heat flux, and a nominal evaporator to condenser temperature drop of 10°C. The design reported here was not aimed at any specific application, but rather to prove the concept of the new wick-artery design and to evaluate the performance of the integral wick-artery structure under a variety of applied heat fluxes and tilt conditions.

Several composite wick types, such as screen-covered axial grooves, slab wick, pedestal, spiral, and tunnel arteries, have been investigated in the past. Some of the major problems associated with these are summarized in the following[2]:

1) Screen covered axial grooves: evaporator heat transfer is less effective due to poor near-wall thermal conductivity and liquid distribution.
2) Slab wick: low capacity; marginal improvements.
3) Pedestal artery: sensitive to artery priming.
4) Spiral artery: fabrication difficulty.
5) Tunnel artery: gas-bubble blockage in artery.

All these wicks have a radial heat flux limitation (boiling limit) at the evaporator leading to the wick dry-out at heat flux rates greater than 7.5 W/cm^2 for ammonia (30 W/cm^2 for water) under the maximum liquid transport factor conditions. Hence these require a larger evaporator area for a given heat-pipe length and capacity. In other words, heat input fluxes higher than the above rates cannot be handled by those wicks. The exceptions to this are the inverted meniscus wick (circumferential grooves covered with screen mesh) concept described by Saaski,[2] which gave an ammonia evaporator heat flux of 20 W/cm^2 compared to that of an open groove wick value of 4 W/cm^2, and the monogroove concept developed by Grumman Aerospace Corp., which had demonstrated 13.4 W/cm^2 for ammonia.[3,4]

It is generally observed that there are no well-established data on radial heat flux density for a given type of wick and working fluid. Also, it is difficult to evolve a design standard with the limited experimental data available in the literature. Hence the designer must rely on experimental investigations in confirming the performance of of a new wick design

Double-Wall Artery Description

The double-wall artery-wick concept, shown in Fig. 1, is a simple idea of a composite wick-covered artery. The device employs a concentric perforated and externally grooved (axial) inner tube wall, an inner-annuli screen mesh wick, and the outer envelope internally grooved evaporator tube, packaged in a single concentric assembly. The inner tube acts primarily as a multiple artery transport wick and maintains the interannuli evaporator supply screen wick firmly pressed against the outer tube threaded (circumferential) groove wick. The latter feature ensures that the screen wick remains in good contact with the grooved wall evaporator wick throughout its operating life, thereby improving the heat transfer and enhancing the evaporator burn-out limits. The supply wick grooved capillary wall feature has been demonstrated to sustain over 300-W/cm^2 heat fluxes.[5] Suitable vapor cutouts at the evaporator and condenser sections along the axial grooves of the inner tube provide for the vapor injection from the evaporator and the vapor suction into the condenser screen wick, respectively. Likewise, small holes drilled along the grooves on the adiabatic section allow absorption of condensed liquid into the wick during startup condition. The screen wick is formed from spirally wrapped wire mesh.

The double-walled wick concept is characterized by the following additional features:

1) Easy manufacturability - external groove cutting is simpler than internal groove cutting, which is possible only by an extrusion process. The inner tube is easy to clean and inspect and is sinterable, too.

2) No adiabatic section entrainment problem - the vapor and liquid flow paths are not in direct shear.

3) No bubble blockage of arteries - owing to the cutouts and holes provided on the grooves. The arteries are self-venting.

4) No artery priming problem - owing to the easy circumferential liquid distribution by the screen wick. The configuration is testable in a 1 g field. Even if priming or bubble blockage problems occur, only a few grooves would be affected, and the remaining arteries could restore heat-pipe operation at least at reduced capacity.

5) Figure 2 shows the cross-section view of the double-wall concept as compared to a screen over groove design and the relative merits of each.

Liquid and Vapor Transport Models

It was assumed that the liquid transport from the condenser to the evaporator was by parallel flow through the artery channels and the screen annuli through a common pressure drop, according to the general form of Darcy's law,

$$\frac{\Delta P_L}{\Delta x} = - \frac{\dot{m}_L}{AK}\left(\frac{\mu_L}{\rho_L}\right) \tag{1}$$

The total mass flow rate for this parallel flow model is given by the sum of the mass flow rates in the arteries and the screen, or

$$\dot{m}_L = \dot{m}_{L_G} + \dot{m}_{L_S} = -\left(\frac{\Delta P_L}{\Delta x}\right)\frac{\rho_L}{\mu_L}\ (A_G K_G + A_S K_S) \qquad (2)$$

Most of the liquid transport is provided by the high-permeability groove channel artery, while the screen provides the capillary pumping power to prime the arteries in the condenser and supply the wall groove capillaries in the evaporator; under these conditions, the wick priming functions and liquid pumping functions could be independently designed.

In the evaporator section of the heat pipe, a liquid/vapor counterflow model was assumed, with the liquid supply to the capillary wall grooves being radially outward along the screen layer contact points, and the vapor transport from the capillary wall grooves being radially inward (toward the vapor core) through the screen layer voids and the vapor slots. This can be modeled as a simple converging nozzle in the radially inward direction, with the nozzle inlet area being the outer-wall evaporator

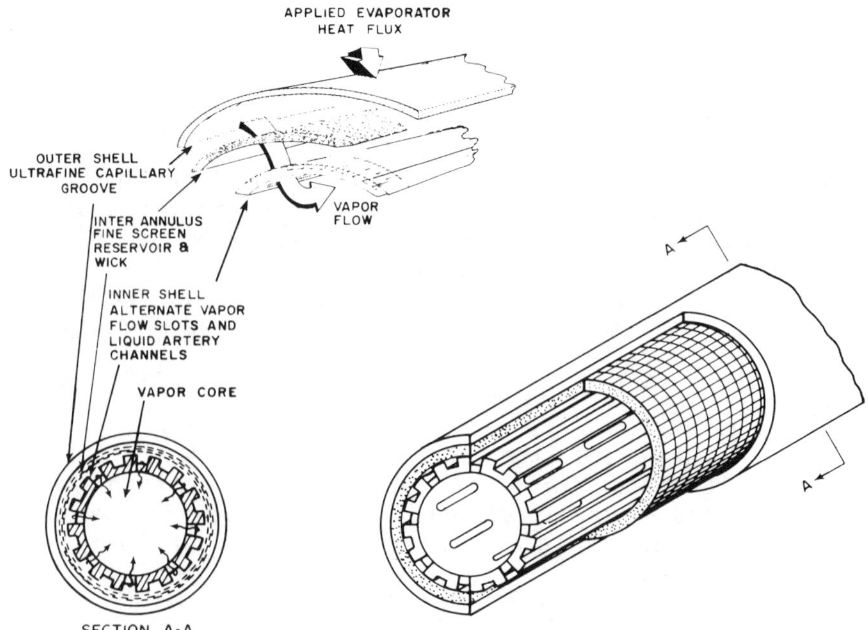

Fig. 1 Double-wall artery heat pipe.

1. WICK GEOMETRY

DOUBLE-WALL ARTERY

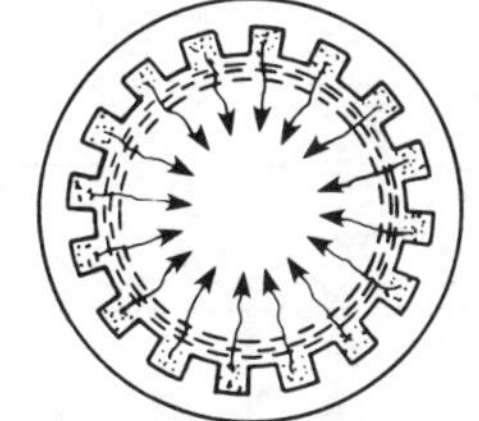

SCREEN COVERED GROOVE

	DOUBLE-WALL ARTERY	SCREEN COVERED GROOVE
2. MECHANICAL DETAILS:	• TWO TUBE CONFIGURATION	• SINGLE TUBE CONFIGURATION
	• SCREEN HELD TIGHTLY	• SCREEN HELD LOOSELY
	• AXIAL GROOVES ON INNER TUBE ; VENTED SUITABLY	• PLAIN AXIAL GROOVES
3. FABRICATION:	• EASY EXTERNAL GROOVE CUTTING	• DIFFICULT INTERNAL GROOVING, NEED TO EXTRUDE
	• SCREEN INSERTION EASY	• EASY, REQUIRES MECHANICAL SCREEN BOND
4. PERFORMANCE:	• VAPOR AND LIQUID FLOWS SEPARATE	• VAPOR AND LIQUID FLOWS INTERACT AND SHEAR
	• NEAR-WALL CONDUCTIVITY IS HIGH; ENHANCED FILM BOILING LIMIT	• POOR NEAR-WALL CONDUCTIVITY LEADS TO FILM BOILING
	• $r_c\|_{screen} < r_c\|_{groove}$ NEAR-WALL BETTER CAPILLARY PUMPING & FLUID DISTRIBUTION	• $r_c\|_{groove} > r_c\|_{screen}$ NEAR-WALL EARLY DRY-OUT AND POOR FLUID DISTRIBUTION
	• CIRCUMFERENTIAL GROOVING OF OUTER WALL POSSIBLE	• NOT POSSIBLE
	• ARTERY RADIALLY SUBCOOLED WITH RESPECT TO VAPOR SLOT AND VAPOR CHANNEL	• ARTERY NOT SUBCOOLED

Fig. 2 Comparison of double-wall artery and screen-over groove wick.

internal surface area and the nozzle outlet area being the evaporator slot area. Entrainment effects and nucleate boiling effects are omitted in this "lean-evaporator" model. Figure 3 shows the schematic model for the double-wall heat-pipe wick-artery configuration under A) saturated wick conditions and B) partially filled wick lean-evaporator conditions. The counterflow liquid/vapor

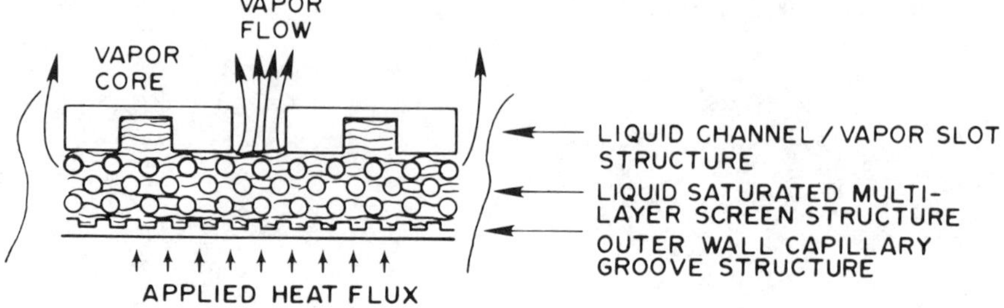

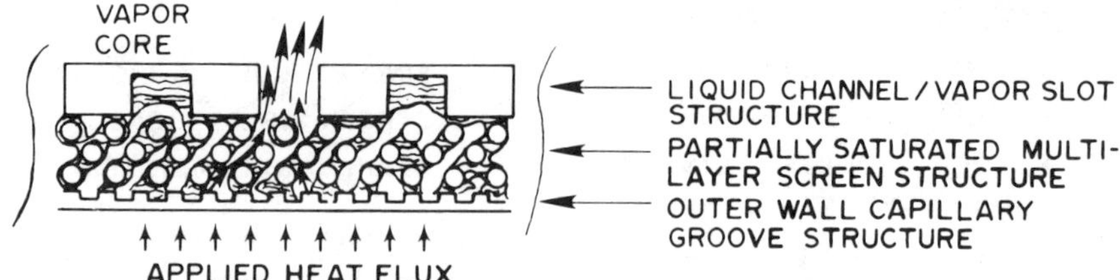

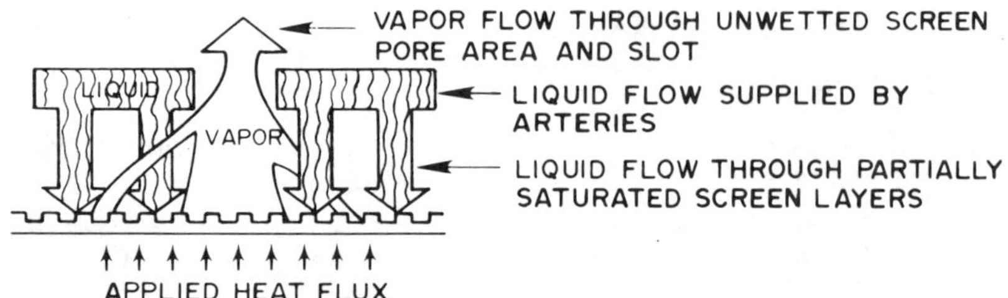

Fig. 3 Evaporator section liquid/vapor flow model.

flow operation of the device is represented in part C of that figure.

Design Details

A copper-water system was chosen to demonstrate the proof of the concept and to evaluate the performance of the new wick design. Water has been selected as the working fluid for the convenience of near-room-temperature operating range (an important temperature range for spacecraft applications) and in view of its highest transport factor (figure of merit), $\sim 10^4$ kW/cm in 25-200°C. Copper, the best compatible envelope and wick material suitable for water, has been chosen. A modest axial heat transport capacity of 2000 W-m with an effective length of 1 m was selected for the present design goal, which is comparable with the established performances of other artery heat pipes (slab: 404 W-m; pedestal: 711 W-m; spiral: 762 W-m; tunnel: 1524 W-m; axial groove - nonconstant groove width: 2040 W-m; and monogroove: 12,232 W-m). These comparative values are based on an equivalent water heat pipe with approximately the same vapor core diameter as this device (1.34 cm).[2,4,6]

A detailed design based on standard design procedure applicable for a nongas filled heat pipe operating in a conventional heat-pipe mode was carried out as outlined in Ref. 7. Using standard available tubing, a suitable heat-pipe diameter and wall thickness were determined which provided a vapor Mach number of 0.2, eliminating vapor compressibility effects, and which could withstand an internal pressure of 47.6 kg/cm^2 (700 psi) at the maximum operating temperature of 250°C. The maximum pumping capacity and the liquid pressure drop effects were decoupled in this composite wick by independently determining the available capillary pumping head of the chosen 100x100-in.$^{-1}$ screen and the laminar flow liquid frictional drop in the grooves.

Artery Groove Dimension

In the conceptual wick design, the screen wick provides the capillary head required for the circumferential liquid distribution which is necessary for priming the axial artery grooves under 1g condition. The capillary head available is calculated from the equation,

$$P_{cm} = 2\sigma_L/r_c = (\rho_L - \rho_v)\, gH \tag{3}$$

where r_c = 1/2 N. For the chosen screen mesh size (100 in.$^{-1}$) and the temperature range of interest, the available head is greater than 6.0 cm, where as the diameter of the heat-pipe envelope is only 2.22 cm. Hence circumferential liquid distribution is no problem even with partial wetting.

For priming considerations, the artery situated at the top of the inner tube is considered as the critical artery. The working fluid has to be pumped to a vertical height equal to the diameter of the inner tube (compared to the bottom artery), since the condensate tends to collect at the bottom during ground testing. Moreover, the arteries

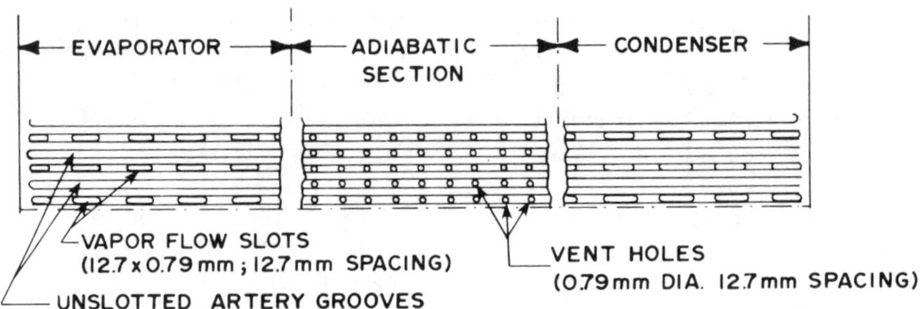

Fig. 4 Groove and vapor-vent details of the inner tube.

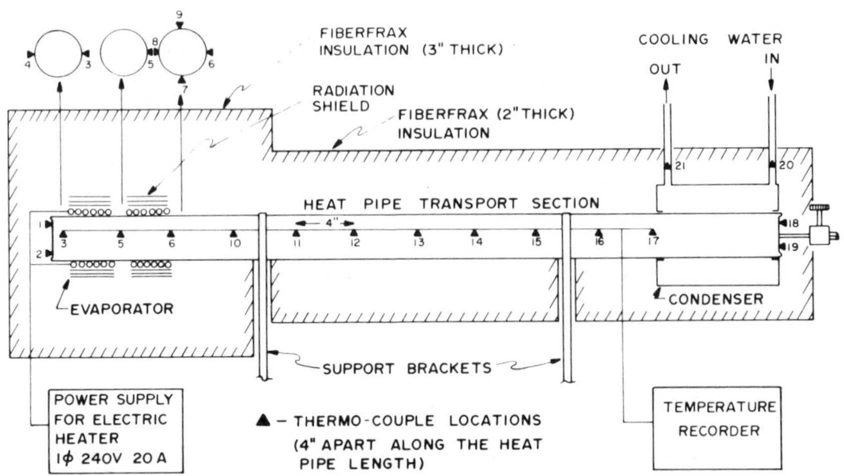

Fig. 5 Schematic of the test setup.

must be able to refill automatically in case they become depleted of the working fluid. The maximum priming height which can be achieved by a capillary for horizontal operation is given by the Laplace-Young equation,

$$\sigma_L \cos\theta(1/R_1 + 1/R_2) = (\rho_L - \rho_v)g(d_o - \delta) \qquad (4)$$

For the purposes of priming the entire length of the artery, the second principal radius of curvature R_2 should be very large or at least equal to the length of the heat pipe. For a rectangular artery, $R_2 \simeq L$ and $R_1 = \omega/2$. Substituting these in Eq. (4) and assuming an aspect ratio of the groove as unity ($\delta/\omega = 1$) for minimum frictional drop, the optimum groove width is given by

$$\omega = \frac{1}{2}\left[d_o - \sqrt{d_o^2 - \frac{8\sigma_L \cos\theta}{g(\rho_L - \rho_V)}}\right] \tag{5}$$

Assuming complete wetting ($\theta = 0$), for the chosen inner tube diameter of $d_o = 1.59$ cm, the optimum artery width has been calculated at various temperatures. The selected artery width, $\omega = 0.794$ mm, corresponding to the maximum liquid transport factor conditions (127°C) has been used. A summary of the design parameters of the prototype double-wall artery heat pipe is given in Table 1. It has been determined that the heat pipe built according to this design would have the transport capacity limited by its capillary limitation in the entire working temperature range. Other limitations, such as viscous, sonic, entrainment, and boiling limits, are very large.

Fabrication, Assembly, and Test Procedure

Fabrication

Type K hard copper standard tube stocks were used to make the 2.22-cm (7/8-in.) o.d. outer and 1.59-cm (5/8-in.) o.d. inner tubes. Circumferential threads (100 threads per inch) were machined on the inside of the outer tube over the 20-cm evaporator length. Twenty-five longitudinal grooves were cut on the inner tube on a horizontal milling machine using a precision saw-cutter of 0.79 mm (1/32 in.) width. Rectangular vent-slots of 0.79x12.7 mm were made (as seen in Fig. 4) for vapor transport by ELOX electrodischarge machining. Only 12 out of 25 grooves were slotted (vented) alternately, leaving 13 grooves unslotted in order to derive good liquid distribution and heat-transfer characteristics at the evaporator and condenser sections. All the grooves were drilled with 0.79-mm (1/32-in.) diameter holes 12.7 mm (1/2 in.) apart on the adiabatic section in order for the arteries to vent. One rectangular piece of 32.77x121.9-cm copper screen (100x100 in.$^{-1}$) was carefully trimmed at the edges for making the wick. All the parts were thoroughly vapor degreased, chemically cleaned, and dried. The mesh screen was wrapped over the inner tube and carefully inserted into the outer tube using simple ring fixtures. The end-caps and fill-tube were TIG (tungsten inert gas) welded to the outer tube ends inside a dry glove-box in an argon inert gas environ-

Table 1 Double-wall artery heat-pipe design summary

Material	
Outer tube, end-caps, and fill-tube	Copper
Inner tube	Copper
Wick	Copper screen mesh
Working fluid	Water
Overall details	
Temperature range	25-250°C
Heat transport capacity	2163 W-m at 127°C (capillary limited)
Effective length	1.0 m
Total length	1.2 m (4 ft)
Evaporator length	0.2 m
Adiabatic section	0.8 m
Condenser length	0.2 m
Vapor core diameter	1.339×10^{-2} m (0.527 in.)
Overall diameter	2.22×10^{-2} m (0.875 in.)
Wick	
Type	Double-wall artery wick (multiple groove arteries and screen mesh)
Screen size	100x100 $in.^{-1}$
Number of wraps	6
Artery grooves	25 square grooves; cut on the outer surface of the inner tube
Groove width	0.79 mm (1/32 in.)
Groove depth	0.79 mm (1/32 in.)

(Table continued on next page)

Table 1 (cont.) Double-wall artery heat-pipe design summary

Heat transport limits (horizontal mode of operation)	
Maximum capillary pumping head	841.9 N/m^2 at 127°C
Maximum available pumping head	668.0 N/m^2 at 127°C
Liquid pressure drop in artery channels	0.309 N/m^2 per W-m at 127°C
Vapor pressure drop in vapor core	0.008 N/m^2 per W-m at 127°C
Capillary limitation	2163 W at 127°C
Entrainment limitation	>10,000 W in 20-300°C
Sonic limitation	3433 W at 40°C
Viscous limitation	3501 W at 21°C
Boiling limitation	4225 W at 200°C
Temperature drop across the saturated wick at the evaporator	~30°C
Temperature drop in vapor along the vapor flow passage	<1°C
Fluid inventory	
Quality	Demineralized deionized water (18 MΩ-cm resistance)
Quantity	91 cm^3 filled at 20°C

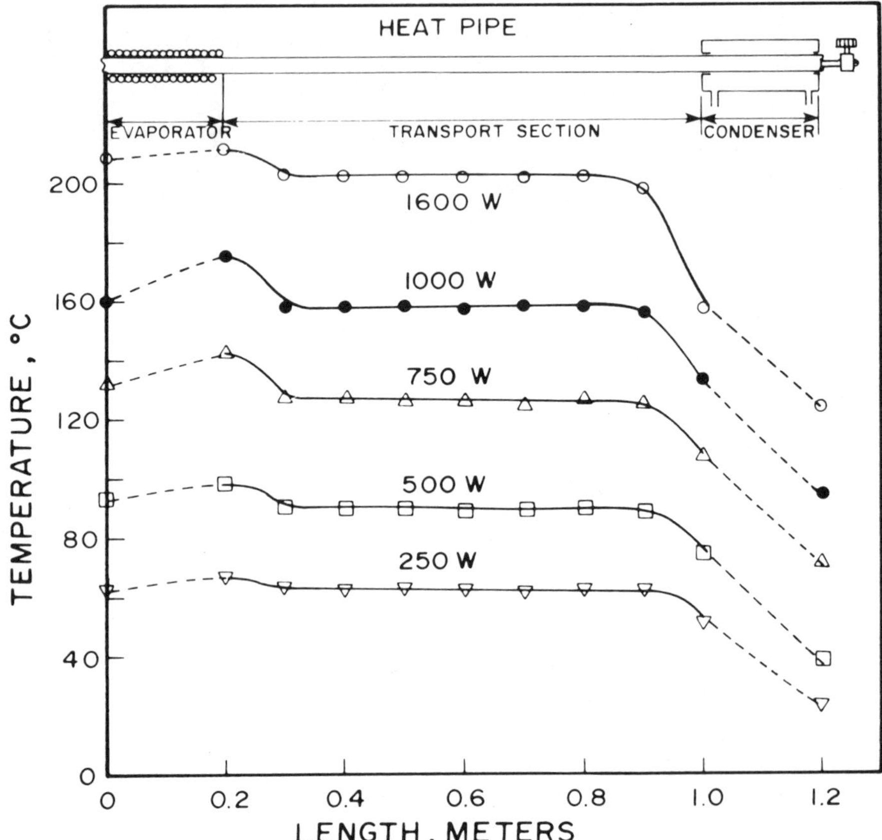

Fig. 6 Axial temperature profile (water-circulated heat exchanger cooling system).

ment. A bellows-type valve attached to the fill-tube facilitates sealing, purging, and changing of fluid inventory inside the heat pipe.

The heat pipe was filled with the required (91 cc) amount of degassed, demineralized, deionized water after evacuating the pipe using a specially built heat-pipe filling station which has a calibrated burette and nitrogen overpressure line.

Test Setup and Procedure

The schematic of the test setup is shown in Fig. 5. An electric resistance heating coil closely wound over the 20 cm length of the evaporator provided the heat input to the heat pipe. Cooling at the condenser was accomplished by two separate methods, by a 20-cm-long water-circulated heat exchanger and droplet-fed evaporative heat exchanger. Copper-constantan thermocouples (21 locations) were mounted along the length of the heat pipe to monitor the temperature profile. A turbine-type flowmeter was installed to monitor the cooling water flow rate. The whole length of the heat-pipe assembly was insulated with (3-6)-in.-thick "Fiberfrax" insulating material. The tilt of the heat pipe could be adjusted using a set of leveling screws on the support bracket.

Tests were performed to establish the maximum heat transport capacity of the heat pipe before wick dry-out at positive, negative, and zero tilt angles. The criterion used was to ensure near-isothermality of the heat pipe under steady-state conditions and the absence of large-temperature difference (ΔT_{EA}) between the evaporator and adiabatic sections. Evaporator dry-out condition was defined when $\Delta T_{EA} > 10^\circ C$.

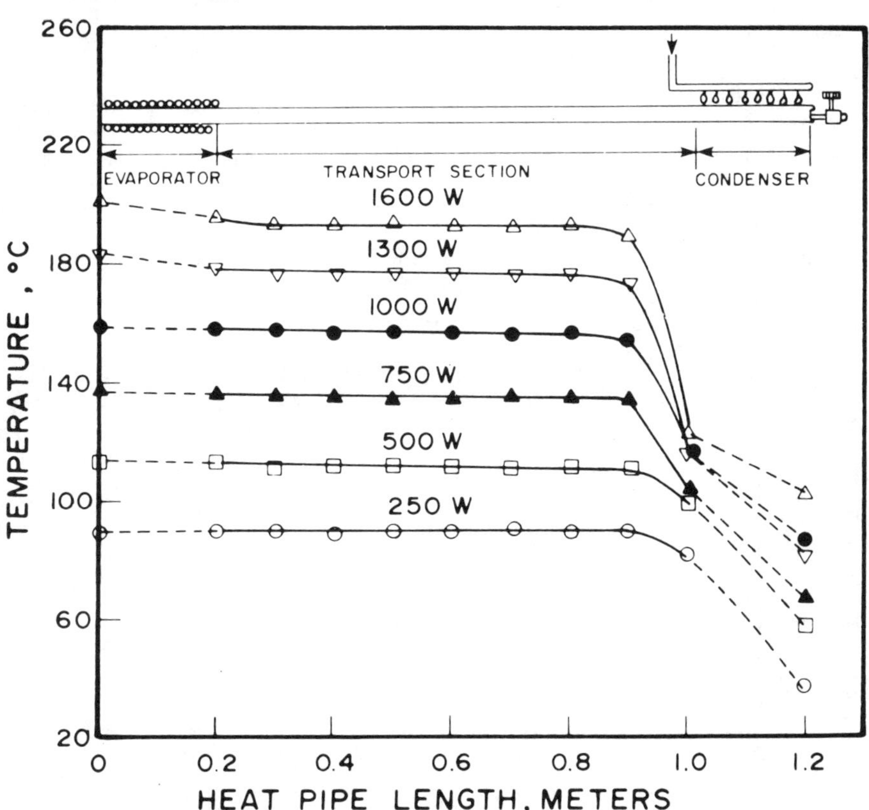

Fig. 7 Axial temperature profile (water-droplet/forced-air cooling).

Preliminary tests were conducted with a preheated-water-circulated condenser heat exchanger to evaluate the performance with a quenched condenser ($T_{COND} << T_{ADI}$). To obtain high mean condenser temperatures and to evaluate the performance of the device at higher condenser mean temperatures, the droplet evaporative cooling system was used.

Results and Discussion

Calorimetric Accuracy

The insulation and the instrumentation systems provided a reasonably good calorimetry. Heat losses as a function of temperature were estimated by comparing the electric power input and the sensible heat increase of the external flow. Twenty to thirty equilibrium tests at various heat input conditions under zero tilt showed an average heat loss of less than 12%.

Overall Performance

Figures 6 and 7 show the axial temperature distribution of the heat pipe for various heat loads at zero tilt with water-circulated heat exchanger cooling and droplet/

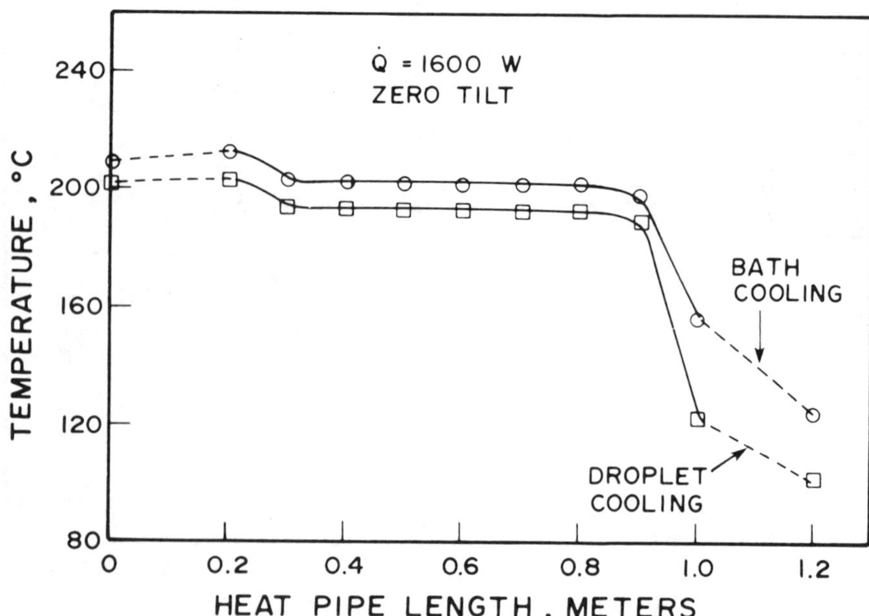

Fig. 8 Effect of condenser cooling on the temperature profile.

forced-air evaporative cooling systems, respectively. In both the cases the adiabatic section temperature profiles are isothermal within 1°C. In the droplet cooling system, the cleaned forced-air jet was used to cool the steam generated from the external wick, thereby improving the convective heat-transfer coefficient. This produced lower adiabatic temperatures at higher heat loads compared to the bath cooling techniques. At lower heat loads (<1000 W), higher adiabatic temperatures were observed due to the maximum cooling capacity of the evaporative cooling at 100°C. Figure 8 compares the effect of the cooling method on the temperature distribution at 1600 W.

Figure 9 shows the heat transport rate as a function of the average adiabatic section temperature. Experimental results show the same trend as that of the predicted performance. The reason for the right-shift of the experimental results is attributable to the idealization of the theoretical model in which the heat pipe has been assumed to be isothermal at the average adiabatic temperature at which the heat-pipe fluid properties are evaluated. In other words, the evaporator and condenser temperature differences are not accounted for in plotting the theoretical curve $\dot{Q}$ vs $\overline{T}$. Figure 10 shows the overall

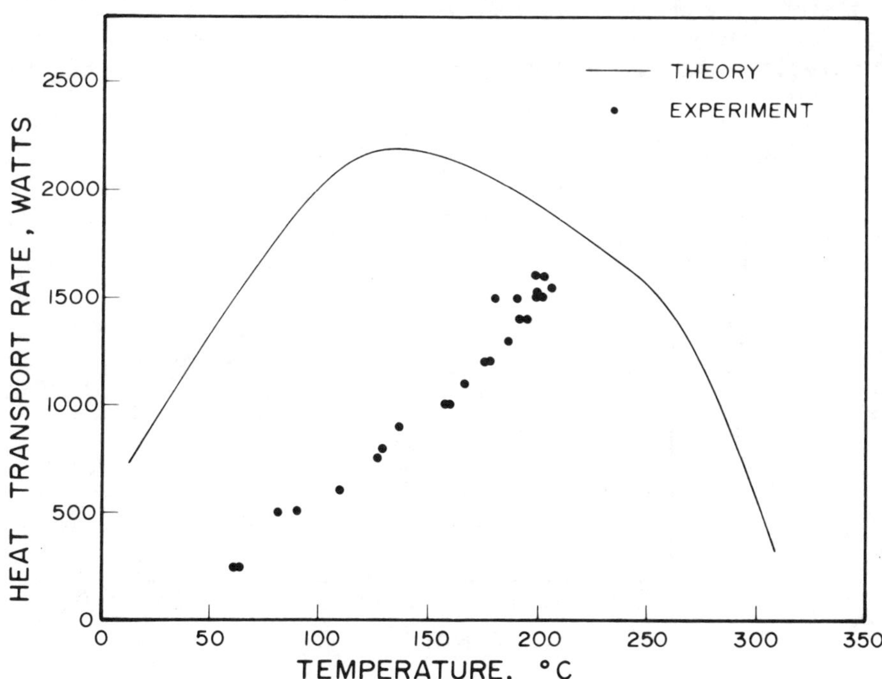

Fig. 9 Heat transport capacity vs temperature.

temperature differences of the heat pipe as a function of the heat transport rate. It is seen that the condenser ΔT is larger than the evaporator ΔT, leading to a lower heat-transfer coefficient at the condenser. This situation is indicative of poor pressure recovery, possibly due to non-condensible gas accumulation at the condenser. After several hundred hours of performance tests, the heat pipe began to show gradual evidence of gas blockage of the condenser. Gas was vented from the heat pipe when it was operating at 100°C, and it was observed that the ΔT_{HP} returned to original conditions instantly.

Maximum heat transport capacity achieved was 1600 W-m as against the design value of 2000 W-m.

Condenser Performance

The exact vapor condensation mechanisms within the double-walled heat-pipe condenser are uncertain. It is thought that the outer-wall, inner-wall, and the inner-screen layers all provide subcooled surface area for condensation. The poor condenser heat transfer coefficient ($h_C \simeq 0.36$ W/cm^2°C) compared to that of the evaporator ($h_e \simeq 1.57$ W/cm^2°C) is apparent due to the larger ΔT_{AC} than ΔT_{EA} for any heat load as shown in Fig. 10.

In order to eliminate the condenser limitation, two solutions had been considered. One was to increase the

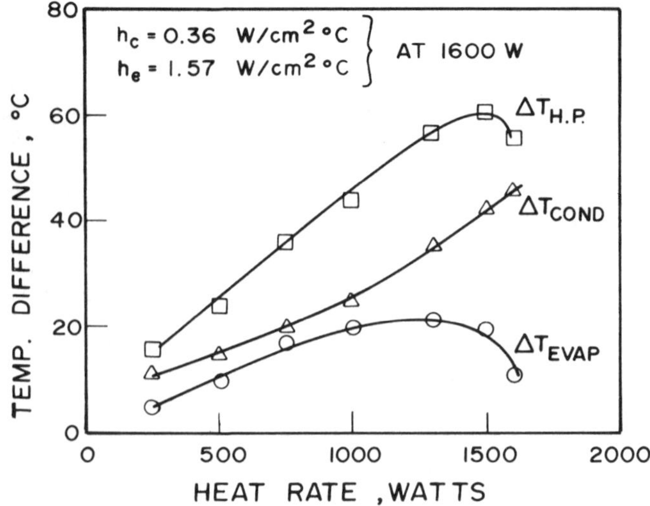

Fig. 10 Temperature difference vs heat transport rate.

condenser external heat-transfer coefficients by evaporative cooling, as mentioned earlier under test procedure. This had been done and only marginal improvements were noticed. The other possibility was to increase the condenser length and area by opening more vapor slots on the inner tube. A second heat-pipe unit, geometrically identical to the present unit and having approximately twice the condenser vapor slot area, was fabricated and tested.[8] The results of the additional tests on both units showed that the condenser temperature drop (ΔT_{AC}) was significantly reduced (approx. 20°C) by increasing the vapor flow slot area (for a fixed condenser length) or by increasing the external surface area of the condenser (for a fixed vapor slot area geometry). Another factor affecting the condenser heat transfer is wick saturation/flooding at the condenser. If the vent slots are filled with the condensate puddle blocking the subcooled surface areas, resistance to heat transfer increases. Obviously, this situation would also call for increased condenser length and area as discussed earlier.

Further improvements in condenser performance may be attainable by heat-transfer augmentation techniques utiliz-

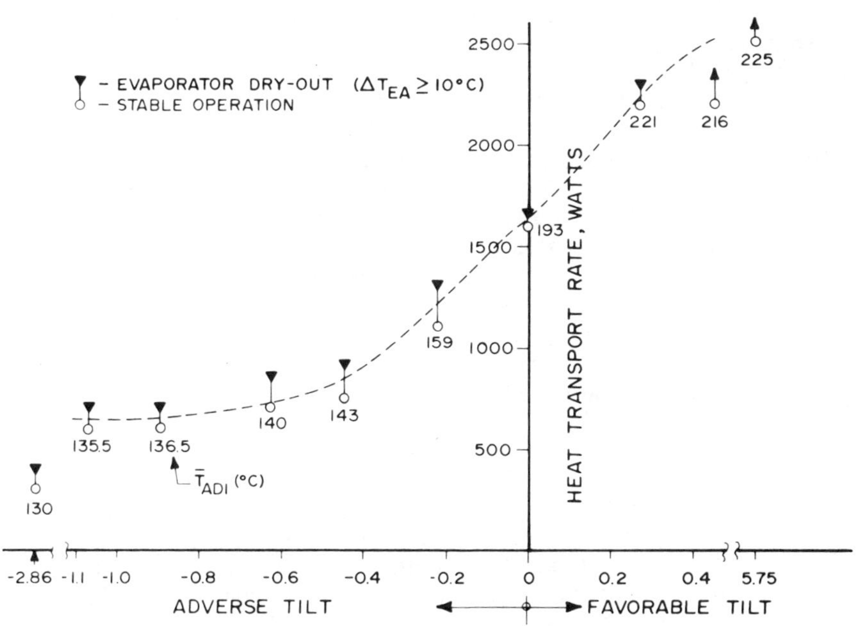

Fig. 11 Heat transport rate vs tilt angle.

ing twisted tape vapor flow swirlers to enhance turbulence or tapered plug to enhance vapor velocity. Larger condensation surface areas for fixed condenser length can be achieved by radially expanding the condenser.[8]

Evaporator Performance

The evaporator has performed very well near the design expectations. Also, it is observed that the double-wall wick has no inherent limit on the evaporative heat flux and that the evaporator operates on "lean-evaporator" conditions. Experimental ΔT_{EA} has been less than 30°C always, and this confirms the design value.

In order to determine the radial heat flux handling capacity of the evaporator, positive tilt tests were done and a maximum flux of 25 W/cm^2 within allowable ΔT_{EA} (10°C) has been obtained. The tests had to be limited to 2500 W owing to heater capacity limits and operational safety. It is expected that the evaporator can handle heat fluxes larger than 25 W/cm^2 without dry-out if a matching condenser section is used. From the tests and results reported here, it has been observed that problems such as film or nucleate boiling at the evaporator leading to quick dry-out or gas-bubble blockage of artery grooves leading to sudden depriming of arteries were not encountered.

Adverse tilt tests were done at several tilt angles, and the results are plotted in Fig. 11. The change of slope in the curve at 900 W and -0.4 deg should be noted. This explains the fact of composite wicking arrangement that has different capillary pumping capabilities for the screen and groove (artery) under adverse tilt conditions. At the transition point, the artery deprimes and the screen becomes the primary transport wick. Since the screen wick has a larger pumping head, even at a large adverse tilt (-2.86-deg) angle, the heat pipe is able to perform as a heat pipe with lower capacity.

Conclusions

A new composite wick design called the "double-wall" artery has been developed and a 1.2-m-long 22.2-mm-diam (o.d.) proof-of-the-concept copper-water heat pipe has been fabricated and tested successfully. A nominal axial heat transport capacity of 1600 W-m and a radial flux density of 16 W/cm^2 were achieved under zero tilt conditions. This device was simpler to fabricate and had superior performance compared to several other conventional artery configurations (such as spiral and tunnel arteries). The evaporator withstood 25 W/cm^2 of radial heat flux with a temperature difference (ΔT_{EA}) of <10°C without dry-out under positive tilt (5.75 deg).

The condenser needs futher improvement to match the evaporator performance. Minor modifications, such as increasing vapor slot area or condenser length, reduced condenser ΔT substantially. Further performance improvements appear attainable through design changes. Several design variations are possible with this concept and it appears feasible to use this configuration for other temperature regimes, particularly liquid-metal and cryogenic heat pipes.

Acknowledgments

The work described in this paper was conducted at the Air Force Wright Aeronautical Laboratories under the sponsorship of the Air Force Office of Scientific Research. The technical assistance provided by Charles Hall (AFWAL) and John Tennant (UES) throughout this project is acknowledged.

References

[1] Mahefkey, T. and Ponnappan, R., "Double Walled Artery Wick for a High Performance Heat Pipe," Disclosure and record of invention; application submitted to the U.S. Air Force, Nov. 20, 1981.

[2] Grumman Aerospace Corporation, "Prototype Space Constructable Long-Life Radiator System - Program Review," presented at NASA Johnson Space Center, Dec. 18, 1980.

[3] Saaski, E. W., "Investigation of an Inverted Meniscus Heat Pipe Wick Concept," NASA CR-137, 724, Aug. 1975.

[4] Alario, J., Haslett, R., and Kosson, R., "The Monogroove High-Performance Heat Pipe," AIAA Progress in Astronautics and Aeronautics Series: Spacecraft Radiative Transfer and Temperature Control, Vol. 83, edited by T. E. Horton, AIAA, New York, 1982, pp. 305-324.

[5] Jacobson, D. L., Bickford, W., Kidd, J., Barthelemy, R., and Bloomer, R. H., "Analysis and Testing of a Heat Pipe Mirror for Lasers," Journal of Energy, Vol. 1, Sept.-Oct. 1977, pp. 306-311.

[6] Schlitt, K. R., "Development of an Axially Grooved Heat Pipe with Non-Constant Groove Widths," AIAA Paper 78-375, 3rd International Heat Pipe Conference, Palo Alto, Calif., May 1978.

[7] Chi, S. W., Heat Pipe Theory and Practice, McGraw-Hill Book Company, New York, 1976.

[8] Ponnappan, R. and Mahefkey, T., "Performance Characteristics of the Double-Wall Artery High Capacity Heat Pipe," AIAA Paper 83-0318, 21st Aerospace Sciences Meeting, Reno, Nev., Jan. 1983.

The Marangoni Effect in Axially Grooved Variable-Conductance Heat Pipes

R. L. Kosson* and W. Harwell†
Grumman Aerospace Corporation, Bethpage, N.Y.

Abstract

An analysis is developed for the Marangoni effect, due to surface tension variation with temperature, in axially grooved variable-conductance heat pipes (VCHP) and compared with test data. The axial momentum equation is solved, accounting for the axial variation in surface tension forces due to changes in meniscus length, as well as changes in surface tension and liquid-channel wall friction due to the recirculating Marangoni flow in the gas-blocked region. The NASA groove analysis program has been modified to include these changes. Results indicate very substantial reductions in transport capacity for VCHPs compared with original isothermal values.

Nomenclature

A = cross-sectional area, m^2
F = force in axial direction, N
k = nondimensional constant, Eq. (13)
ℓ_b = diffusion zone length, m
ℓ_m = total meniscus length in cross section, m
m = nondimensional summation parameter, Eq. (14)
N = number of axial grooves
n = nondimensional summation integer, Eq. (14)

Presented as Paper 82-0904 at the 3rd AIAA/ASME Joint Thermophysics Fluids, Plasma & Heat Transfer Conference, St. Louis, Mo., June 7-11, 1982.

*Technical Specialist, Thermodynamics.

†Senior Engineer, Thermal Systems.

P = pressure, N/m^2
R_x = meniscus radius at axial station x, m
R_t = fin tip radius, m
t = equivalent rectangular groove depth, m
w = equivalent rectangular groove width, m
x = axial distance from evaporator end, m
α = angle complement of θ
δ = half-angle subtended by fin tip flat
η = groove taper half-angle
θ = fin tip contact point angle
μ = angle of meniscus radius at fin contact point
σ = surface tension, N/m
ϕ = half-angle to center of fin tip radius (also groove aspect ratio parameter)

Subscripts

b = blocked (or body force)
l = liquid
m = meniscus
s = single groove value
v = vapor

Introduction

The thermal canister built by Grumman for NASA Goddard Space Flight Center (GSFC) contains 20 axially grooved variable-conductance heat pipes (VCHP) for thermal control. When these pipes operate in a partially blocked mode the blocked region drops in temperature, causing an increase in surface tension which acts to resist the flow of liquid from the condenser to the evaporator, reducing the heat pipe thermal transport capacity. This has been termed the "Marangoni effect" and has been discussed in Ref. 1 for idealized rectangular-groove heat pipes.

The thermal canister heat pipes make use of a NASA GSFC-developed extrusion with a somewhat more complicated geometry. Performance calculations had been performed for these pipes using the NASA groove analysis program (GAP), but this computer code did not account for the Marangoni effect. To provide an estimate of Marangoni effects for the thermal canister VCHPs, a more general analysis was derived which incorporates the concept developed by Eninger and Marcus (Ref. 1) into a modified version of the GAP code (MGAP).

The new analysis is based on use of the axial momentum equation and accounts for the axial surface tension force gradient due to changes in meniscus length, meniscus contact position on the fin tips, and surface tension. Gas pressure acting on the meniscus surface is also taken into account. These terms are self-cancelling, except for the terms due to surface tension variation which can act to significantly reduce the calculated transport capacity of the pipe.

This paper presents the analysis, some calculated results, and a comparison of calculated results with some experimental data.

Axial Momentum Equation for Liquid Flow

Consider a control surface about a differential axial length of liquid, as shown in Fig. 1. For steady liquid flow with negligible change in liquid momentum, the sum of the forces acting on the liquid must be equal to zero, i.e.,

$$[P_1A_1 + (P_v + \frac{dP_v}{2})\, dA_1 - (P_1 + dP_1)(A_1 + dA_1)]$$

$$+ [(F_m + dF_m) + dF_x - F_m] + dF_b + dF_w + dF_{v,1} = 0$$

where the first bracketed expression accounts for pressure forces, the second for meniscus surface tension forces, dF_b for body forces, dF_w for wall shear forces, and $dF_{v,1}$ for vapor-liquid shear forces on the meniscus surface.

Simplifying and neglecting second-order differential terms gives

$$-A_1dP_1 + (P_v - P_1)A_1 + dF_m + dF_x + dF_b + dF_w + dF_{v,1} = 0 \quad (1)$$

For axial-grooved heat pipes, the geometry may be described as in the NASA GAP, a detail of which is shown in Fig. 2. At any x location

$$P_v - P_1 = \sigma/R_x \quad (2)$$

where σ is the surface tension, which may vary with x, and R_x is the meniscus radius of curvature. Differentiating,

$$-dP_\ell = d\sigma/R_x - (\sigma/R_x^2)\, dR_x - dP_v \quad (3)$$

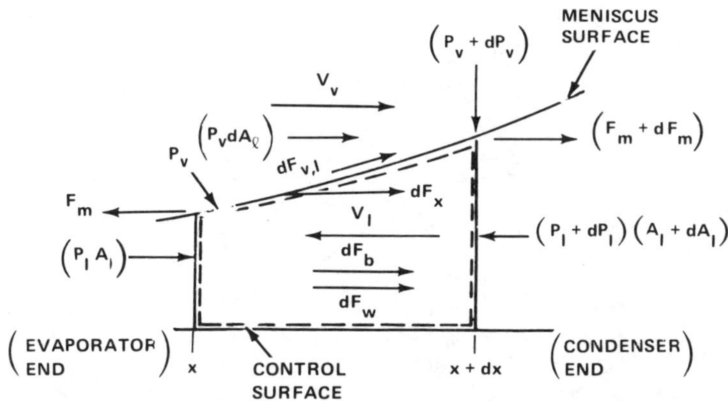

Fig. 1 Control volume schematic for liquid axial momentum.

Referring to Fig. 2, the differential change in area for a single groove is given by

$$\tfrac{1}{2}\, dA_{\ell,s} = (56875) = (1341) + (4574) - (3683) - (1561)$$

or

$$dA_{1,s} = (R_x + R_t)^2\, d\alpha + R_x^2\, \mu - (R_x^2 + 2R_x\, dR_x)\mu - R_x^2\, d\mu - R_t^2\, d\alpha$$

where $\mu = \theta + \phi$.

Setting $d\mu = -d\alpha$ and $dA_1 = N\, dA_{1,s}$ where N is the number of grooves, gives

$$dA_1 = -2NR_x\, [(R_x + R_t)\, d\theta + (\theta + \phi)\, dR_x] \tag{4}$$

The meniscus axial force is given by

$$F_m = 2N\, R_x\, (\theta + \phi)\, \sigma$$

from which, differentiating

$$dF_m = 2N\, R_x\, (\theta + \phi)\, d\sigma + 2N\, R_x\, \sigma\, d\theta + 2N\, (\theta + \phi)\, \sigma\, dR_x \tag{5}$$

The change in meniscus contact position on the fin tip (from point 5 to 6 in Fig. 2) provides a projected length of the contact line on a plane normal to the x axis, and a force

$$dF_x = 2N\, R_t\, \sigma\, d\,\theta \tag{6}$$

This force term is in the -x direction (since $d\theta$ is negative), assisting condensate return to the evaporator.

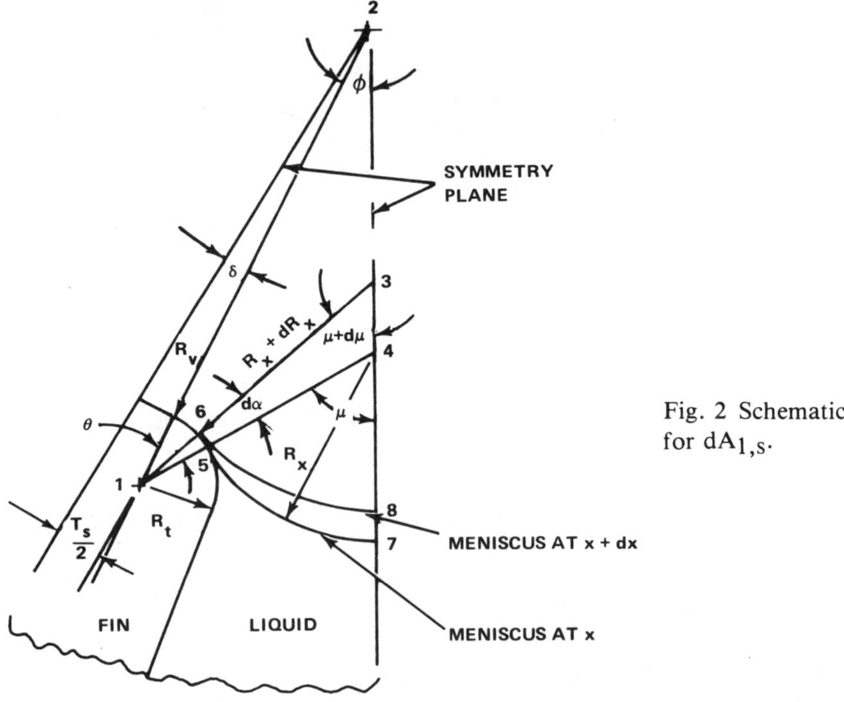

Fig. 2 Schematic for $dA_{l,s}$.

Combining Eqs. (2) and (4-6), the $d\theta$ and dR_x terms cancel, leaving

$$(P_v - P_l)\, dA_l + dF_m + dF_x = 2N\, R_x\, (\theta + \phi)\, d\sigma \tag{7}$$

Equations (1), (3), and (7) can then be combined to give

$$\frac{\sigma}{R_x^{\,2}} \frac{dR_x}{dx} = \left(\frac{1}{R_x} + \frac{\ell_m}{A_l}\right)\frac{d\sigma}{dx} - \frac{dP_v}{dx} + \frac{1}{A_l}\left(\frac{dF_b}{dx} + \frac{dF_w}{dx} + \frac{dF_{v,l}}{dx}\right) \tag{8}$$

where ℓ_m is the meniscus length defined by

$$\ell_m = 2N\, R_x\, (\theta + \phi) \tag{9}$$

and

$$(\theta + \phi) = \sin^{-1}\left[\frac{(R_v + R_t)\sin\phi}{(R_x + R_t)}\right] \tag{10}$$

Equation (8), for the variation in R_X with x, differs from that in GAP[2,3] only by the addition of the $d\sigma/dx$ term. The vapor temperature and hence the meniscus surface temperature and surface tension are assumed constant in the operating portion of the pipe making $d\sigma/dx = 0$. In the blocked region of the pipe, a diffusion zone length ℓ_b and blocked zone temperature T_b must be specified, from which

$$\frac{d\sigma}{dx} = \frac{\sigma_b - \sigma_o}{\ell_b} \tag{11}$$

where σ_b and σ_o are the surface tensions in the blocked and operating portions of the pipe, respectively.

Within the blocked region of the pipe there is no net liquid flow or net vapor flow. GAP therefore calculates

$$\frac{dP_v}{dx} = \frac{dF_w}{dx} = \frac{dF_{v,1}}{dx} = 0 \tag{12}$$

Because of the surface tension variation, which acts as a shear stress term at the meniscus surface, there is a recirculating flow within the liquid channels, with flow toward the condenser end of the pipe near the meniscus surface and flow toward the evaporator end of the pipe near the base of the grooves. Following the work of Eninger and Marcus,[1] the wall friction force associated with this recirculating flow can be written

$$\frac{dF_W}{dx} = k\ell_m \frac{d\sigma}{dx} \tag{13}$$

resisting the return of condensate to the condenser. The coefficient k is obtained from the analysis by Hufschmidt[4] of flow in rectangular channels subject to surface shear. For zero net liquid flow, Hufschmidt provides a solution for k as a function of the width-to-height (w/t) ratio of the groove, which can be rearranged to the form,

$$k = \left[1\text{-}6 \sum_{n=0}^{\infty} \frac{\text{sech}\,(m/\phi)}{m^4}\right] \Big/ \left[1\text{-}6\phi \sum_{n=0}^{\infty} \frac{\tanh(m/\phi)}{m^5}\right]^{-1} \tag{14a}$$

where $m \equiv [(2n+1)/2]\pi$ and $\phi \equiv w/2t$.

For the NASA grooved heat pipe geometry,

$$\phi \cong \ell_m{}^2/2N\,A_1 \tag{14b}$$

The series expressions for k in Eq. (14a) are rapidly convergent, and may be truncated at n = 5.

Considering Eqs. (12) and (3), the momentum equation (8) can be rewritten as

$$\frac{\sigma}{R_X^2}\frac{dR_X}{dx} = \left[\frac{1}{R_X} + \frac{\ell_m}{A_l}(1+k)\right]\frac{d\sigma}{dx} - \frac{dP_v}{dx} + \frac{1}{A_l}\left[\frac{dF_b}{dx} + \frac{dF_w}{dx} + \frac{dF_{v,l}}{dx}\right] \tag{15a}$$

where dF_w/dx is understood to be only that portion of the liquid channel wall force which is due to a net liquid flow. In the operating portion of the pipe, Eq. (15a) reduces to

$$\frac{\sigma}{R_X^2}\frac{dR_X}{dx} = -\frac{dP_v}{dx} + \frac{1}{A_l}\left[\frac{dF_b}{dx} + \frac{dF_w}{dx} + \frac{dF_{v,l}}{dx}\right] \tag{15b}$$

and, in the blocked portion of the pipe, Eq. (15a) reduces to

$$\frac{\sigma}{R_X^2}\frac{dR_X}{dx} = \left[\frac{1}{R_X} + \frac{\ell_m}{A_l}(1+k)\right]\frac{d\sigma}{dx} + \frac{1}{A_l}\frac{dF_b}{dx} \tag{15c}$$

Results

The modified GAP program (called MGAP) was run for VCHPs similar to those in the thermal canister. Typical input values, taken directly from the computer program output listing, are shown in Table 1, which also contains the remainder of the output for one operating condition.

The distribution of the pressure loss is shown in the output from the pressure ratio subroutine (PRERAT), see Table 1. In the blocked region of the pipe, the induced liquid channel friction loss associated with Marangoni recirculation is shown under the liquid viscous loss column, and the direct meniscus loss is shown under the meniscus loss column. The sum of these two losses is the loss due to surface tension gradient shown in Eq. (15c). For the case shown, with the VCHP operating at adverse tilt, these losses exceed the gravity loss in the blocked region and account for approximately 18% of the overall pressure loss.

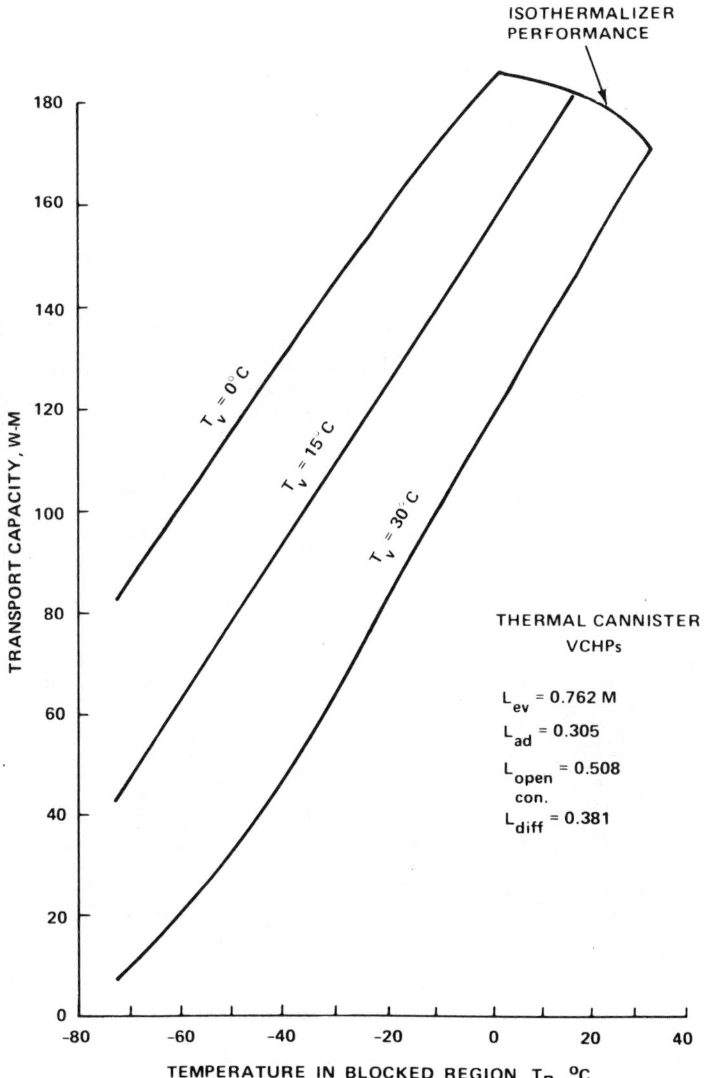

Fig. 3 Transport capacity vs blocked zone temperature, thermal canister VCHPs.

The program was run for heat pipes similar to those used in the thermal canister at operating temperatures of 30, 15, and 0°C. The temperature in the blocked zone (T_b) was varied from operating temperature to -73°C. Results, in terms of transport capacity vs blocked zone temperature, are shown in Fig. 3. These same results, normalized to isothermalizer transport capacity, are given in Fig. 4.

Table 1 Modified groove analysis program (MGAP)

A. REPRINT OF INPUT VALUES

WORKING FLUID PROPERTIES	
WORKING FLUID	AMMONIA
OPERATING TEMPERATURE (K)	306.00
LIQUID DENSITY (KG/M/M/M)	0.591010E+03
LIQUID VISCOSITY (N-S/M/M)	0.122800E-03
LIQUID TRANSPORT FACTOR	0.989153E+11
WETTING ANGLE (RADIAN)	0.0
VAPOR DENSITY (KG/M/M/M)	0.986641E+01
VAPOR VISCOSITY (N-S/M/M)	0.103100E-04
SURFACE TENSION (N/M)	0.181400E-01
HEAT OF VAPORIZATION (J/KG)	0.113300E+07
WICKING HEIGHT FACTOR (M-M)	0.313196E-05
MOLECULAR WEIGHT (KG/MOLE)	0.170320E+02
RATIO OF SPECIFIC HEATS	1.34
GROOVE CHARACTERISTICS	
GROOVE CROSS-SECTION	DIVERGENT
NUMBER OF GROOVES	27
GROOVE TIP RADIUS (VAPOR CORE RADIUS) (M)	0.429300E-02
GROOVE ROOT RADIUS (M)	0.534800E-02
GROOVE TIP CORNER RADIUS (M)	0.198000E-03
GROOVE ROOT CORNER RADIUS (M)	0.127000E-03
GROOVE WIDTH (M)	0.646744E-03
GROOVE DEPTH (M)	0.105500E-02
ASPECT RATIO (GROOVE DEPTH /GROOVE WIDTH)	1.63125
TIP LAND THICKNESS (M)	0.396000E-03
INTEGRATION OF KA/E/R FROM RAIN TO RV	0.110814E-08
GROOVE LAND TAPER ANGLE (RADIAN)	0.760000E-01
MENISCUS ANGLE (THETA+PHI), DEGREES	0.897443E+02
MINIMUM MENISCUS RADIUS (M)	0.323372E-03
HEAT PIPE GEOMETRY	
EVAPORATOR LENGTH (M)	0.304800E+00
TRANSPORT LENGTH (M)	0.152400E+00
CONDENSER LENGTH (M)	0.254000E+00
HEAT PIPE TOTAL LENGTH (M)	0.101600E+01
HEAT PIPE OUTER DIAMETER (M)	0.139200E-01
HEAT PIPE INNER DIAMETER (M)	0.106960E-01
WALL THICKNESS (M)	0.161200E-02
VCHP PARAMETERS	
DIFFUSION ZONE LENGTH (M)	0.304800E+00
BLOCKED ZONE TEMPERATURE (K)	0.286000E+03
BLOCKED ZONE SURFACE TENSION (N/M)	0.228500E-01

(Table 1 continued on next page)

Table 1 (cont.) Modified groove and analysis program

B. OUTPUT

AXIAL HEAT FLOW LIMITS

LIMITING AXIAL HEAT FLOW AT SONIC CONDITION (W)	0.133771E+06
LIMITING AXIAL HEAT FLOW FOR ENTRAINMENT (W)	0.109127E+04

CAPILLARY PUMPING LIMIT FOR NOMINAL CHARGE CONDITION

HEAT PIPE ELEVATION (EVAPORATOR HIGHER) (M)	0.483000E-02
GRAVITATIONAL ACCELERATION (M/SEC/SEC)	0.980000E+01
MAXIMUM HEAT FLOW FOR THESE CONDITIONS (W)	0.104466E+03
EQUIVALENT 1-G HEAT TRANSPORT CAPABILITY (W-M)	0.451085E+02
CORRESPONDING FLUID INVENTORY (KG)	0.105110E-01
MAXIMUM AXIAL REYNOLDS NUMBER (VAPOR, 1-G)	0.132620E+04
GROOVE STATIC HEIGHT (M)	0.895576E-02
0-G MAXIMUM HEAT FLOW (W)	0.208739E+03
EQUIVALENT 0-G HEAT TRANSPORT CAPABILITY (W-M)	0.901333E+02

AXIAL GROOVE LIQUID FLOW CHARACTERISTICS AND PRESSURE LOSSES IN 1-G ENVIRONMENT:

AXIAL LOCATION (M)	MENISCUS RADIUS (M)	PERMEABILITY (M-M)	GROOVE AREA (M-M)	GRAVITY LOSS (%)	LIQ-VISC LOSS (%)	LIQ-VAPOR SHR LOSS (%)	VAP-VISCOUS LOSS (%)	MENISCUS LOSS (%)	TOTAL-PRESSURE GRAD (N/M/M/M)	CUM % LOSS
0.0	0.3234E-03	0.1769E-07	0.1166E-04	0.0	0.0	0.0	0.0	0.0	0.0	0.0
0.5080E-02	0.3240E-03	0.1861E-07	0.1205E-04	93.23	6.51	0.21	0.06	0.0	0.21877E+02	0.207
0.6096E-01	0.3335E-03	0.2120E-07	0.1317E-04	58.71	39.48	1.39	0.42	0.0	0.34740E+02	3.293
0.1270E+00	0.3509E-03	0.2293E-07	0.1397E-04	43.82	53.53	2.00	0.65	0.0	0.46545E+02	8.493
0.1880E+00	0.3735E-03	0.2416E-07	0.1458E-04	36.62	60.23	2.35	0.80	0.0	0.55699E+02	14.514
0.2489E+00	0.4038E-03	0.2519E-07	0.1512E-04	32.01	64.46	2.61	0.93	0.0	0.63723E+02	21.541
0.3048E+00	0.4406E-03	0.2603E-07	0.1559E-04	29.01	67.16	2.80	1.03	0.0	0.70304E+02	28.766
0.3099E+00	0.4444E-03	0.2610E-07	0.1563E-04	29.12	67.05	2.80	1.03	0.0	0.70044E+02	29.453
0.3404E+00	0.4687E-03	0.2651E-07	0.1587E-04	29.73	66.40	2.82	1.05	0.0	0.68605E+02	33.525
0.3708E+00	0.4952E-03	0.2688E-07	0.1609E-04	30.29	65.81	2.83	1.07	0.0	0.67335E+02	37.519
0.4013E+00	0.5243E-03	0.2722E-07	0.1630E-04	30.81	65.25	2.84	1.09	0.0	0.66195E+02	41.441
0.4318E+00	0.5565E-03	0.2753E-07	0.1650E-04	31.30	64.73	2.86	1.11	0.0	0.65156E+02	45.300
0.4572E+00	0.5860E-03	0.2778E-07	0.1666E-04	31.69	64.32	2.87	1.12	0.0	0.64351E+02	48.470
0.4623E+00	0.5922E-03	0.2783E-07	0.1669E-04	32.21	63.82	2.85	1.12	0.0	0.63322E+02	49.096
0.5131E+00	0.6557E-03	0.2827E-07	0.1698E-04	38.12	58.19	2.64	1.05	0.0	0.53510E+02	54.810
0.5639E+00	0.7204E-03	0.2862E-07	0.1721E-04	45.93	50.79	2.34	0.94	0.0	0.44405E+02	59.599
0.6147E+00	0.7837E-03	0.2891E-07	0.1741E-04	56.96	40.40	1.88	0.77	0.0	0.35809E+02	63.524
0.6655E+00	0.8423E-03	0.2913E-07	0.1756E-04	73.95	24.43	1.15	0.47	0.0	0.27582E+02	66.625
0.7112E+00	0.8874E-03	0.2929E-07	0.1767E-04	100.00	0.00	0.00	0.00	0.0	0.20396E+02	68.739
0.7163E+00	0.9005E-03	0.2933E-07	0.1770E-04	41.48	15.83	0.00	0.00	42.69	0.49175E+02	69.313
0.7772E+00	0.1112E-02	0.2987E-07	0.1808E-04	41.39	15.98	0.00	0.00	42.63	0.49281E+02	76.698
0.8382E+00	0.1414E-02	0.3038E-07	0.1843E-04	41.23	16.16	0.00	0.00	42.61	0.49472E+02	83.407
0.8992E+00	0.1879E-02	0.3087E-07	0.1878E-04	41.02	16.37	0.00	0.00	42.61	0.49722E+02	89.539
0.9576E+00	0.2649E-02	0.3133E-07	0.1910E-04	40.79	16.58	0.00	0.00	42.63	0.50003E+02	94.945
0.1016E+01	0.4267E-02	0.3180E-07	0.1942E-04	0.0	0.0	0.0	0.0	0.0	0.0	99.951

0-G BLOCK LENGTH PER % OF OVERCHARGE (M)	0.307169E-02

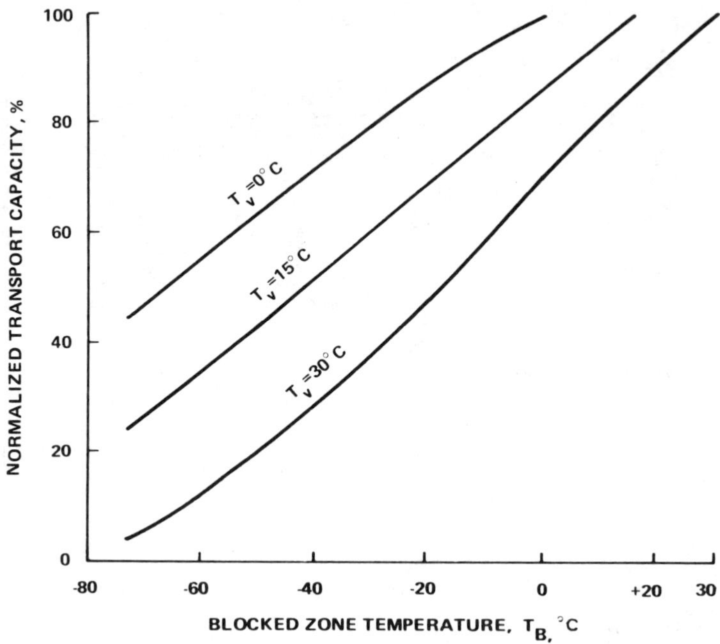

Fig. 4 Normalized transport capacity vs blocked zone temperature, thermal canister VCHPs.

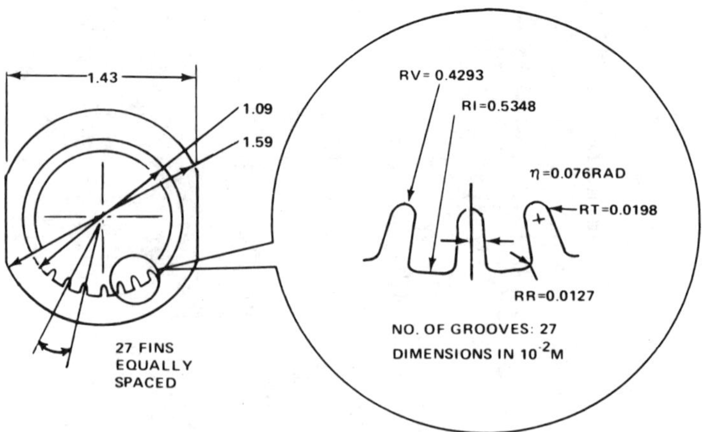

Fig. 5 Longitudinally grooved aluminum extrusion.

Experimental Data

Almost all of the VCHP test data in the heat pipe literature is concerned with thermal control aspects of performance when the pipes are operating in the partially blocked mode. The only data reported on maximum transport capacity for grooved heat pipe VCHPs operating in the partially blocked mode appear to be those of Eninger and Marcus. This VCHP was made from the aluminum extrusion developed by NASA GSFC as a backup for the heat pipes used on the ATS-6 satellite.

Additional data on maximum transport capacity was obtained at Grumman with a VCHP fabricated from what is believed to be the same batch of aluminum extrusion used by Eninger and Marcus. The Grumman VCHP was an early developmental test article for the Deployable Heat Pipe Radiator Project.[5] The pipe was 1.37 m (54 in.) in length, exclusive of the reservoir which was simply a 500 mℓ metal charge bottle attached to the heat pipe charge valve, providing a reservoir-to-condenser vapor volume ratio of 8.6. Since the bottle lacked an internal capillary structure, it was positioned vertically to permit any condensate to drain back to the pipe.

A schematic of the pipe cross section, including dimensions as required by the GAP program, is shown in Fig. 5. A schematic diagram of the test setup, including details of the thermocouple (Cu/Con) installation, is given in Fig. 6. The setup included a single-sided heat input to the evaporator, a short transport section, and a long (mostly single-sided) condenser. A layer of H-film was wrapped around the entire condenser section to spoil the effectiveness of the water spray bath, which was used as the heat sink. This made it easier to track the movement of the inert gas interface since it created a greater temperature difference between the active and inactive portions of the condenser. Otherwise, it would be difficult to determine which part of the condenser was blocked since a good sink coupling would force the condenser closer to the sink temperature.

Test Description

The gas reservoir was first charged with inert dry nitrogen gas to 134 psig (at room temperature) and then attached to the pipe's charge valve. A tee fitting with vacuum pump was used to evacuate the connecting line to preclude air entrapment. The reservoir

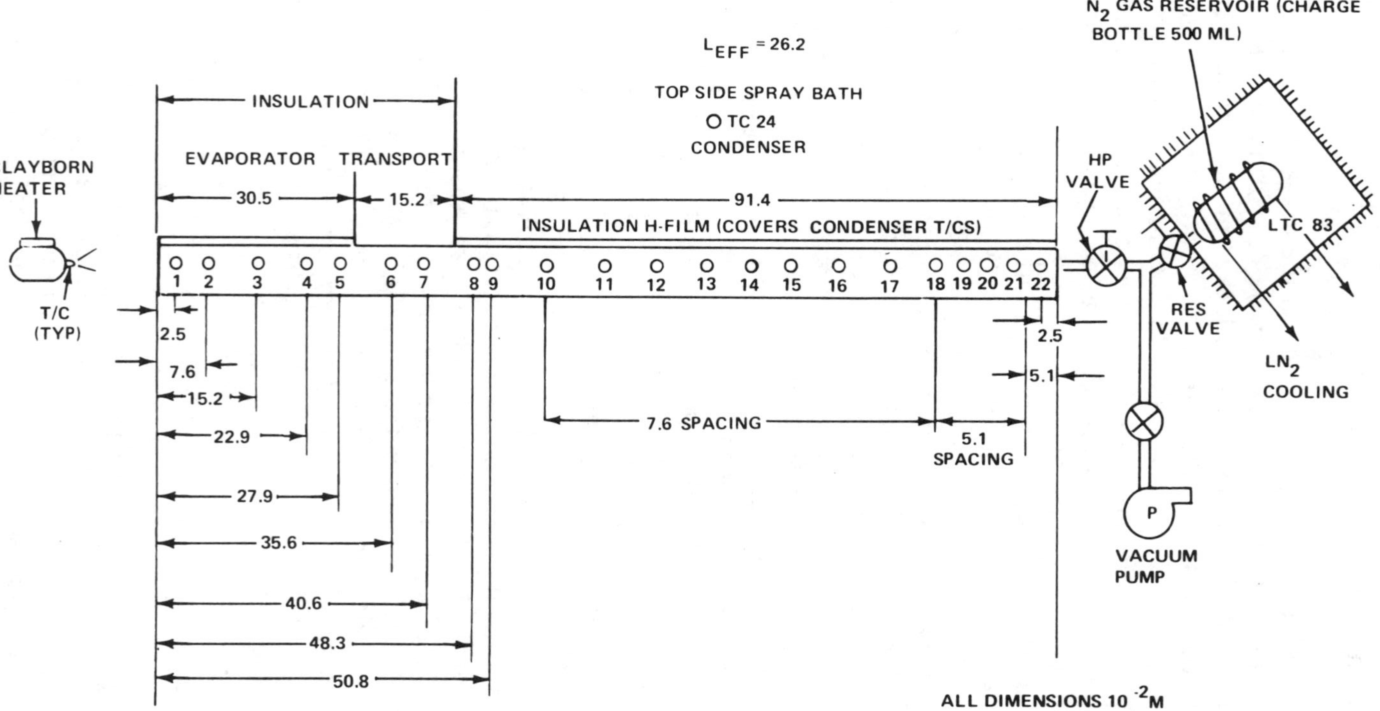

Fig. 6 Gas-loaded grooved heat pipe feasibility demonstration.

pressure corresponds to a conveniently chosen test condition requiring a full-open condenser at a vapor temperature of 75 °F and a reservoir temperature of -20°F. The reservior temperature was controlled by using LN_2 cooling coils which were fitted around the charge bottle. An insulated container, packed with crushed ice, surrounded the charge bottle and coils. The ice served to fill in the voids between the copper coils and the reservoir and provided an improved conductive coupling, albeit still small.

The heat pipe, with its existing 17.4 g ammonia charge, was first tested as a single-fluid device to obtain a baseline performance map of Q_{MAX} vs adverse tilt. All testing was done at a 55°F spray bath temperature. At each specified tilt value the heat load was increased in 20 W increments until an evaporator burnout was witnessed, as indicated by runaway temperatures. The heat load which immediately preceded the burnout value was designated as Q_{MAX}, the maximum functional power level of the pipe.

After completion of the single-fluid tests, the pressurized nitrogen gas reservoir was attached to the condenser and the heat pipe tested as a VCHP to determine if the presence of nitrogen gas degraded

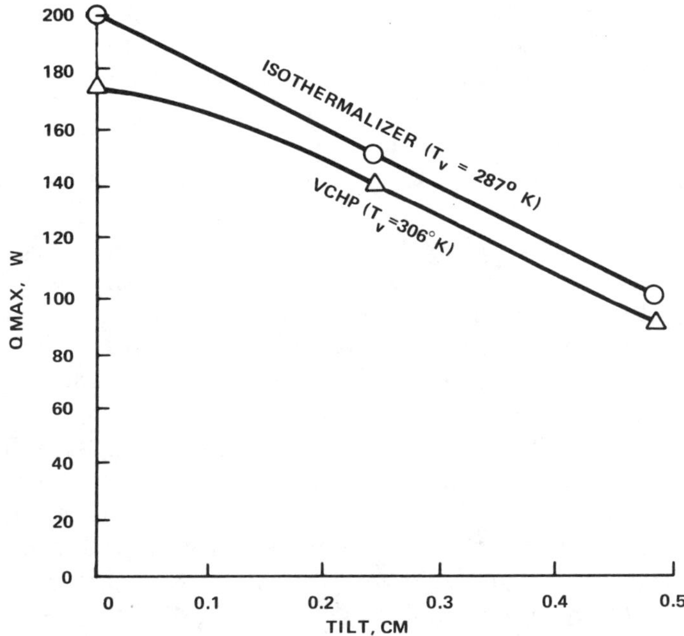

Fig. 7 Q_{max} vs tilt test results.

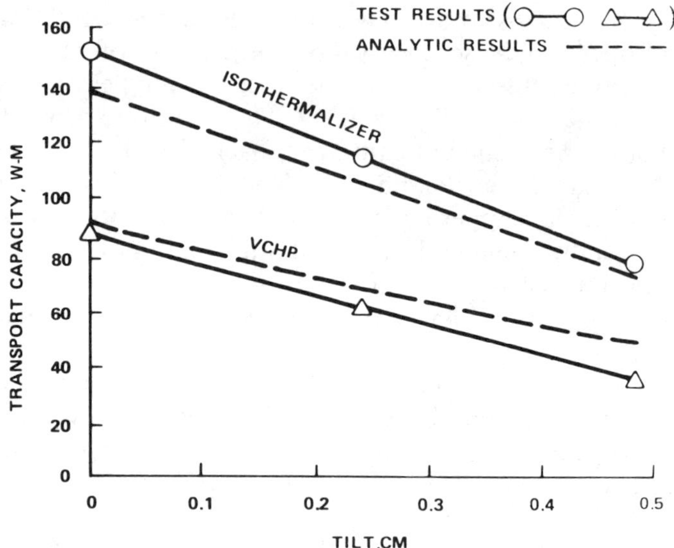

Fig. 8 Transport capacity vs tilt.

the transport capacity of the pipe. For the data of interest in this paper, the reservoir was kept at about 7 °C. This positioned the upstream end of the diffusion zone to provide an active condenser length of 0.18-0.38 m. The diffusion zone length in this pipe was about 0.23 m.

Test Results

Maximum measured values of heat-transfer and heat-transport capacity as a function of adverse tilt are given in Figs. 7 and 8 for both the isothermalizer and VCHP operating modes. The vapor temperature in the VCHP mode was higher than that of the isothermalizer mode because a common condenser sink temperature (13 °C) was used.

Figure 8 also shows predicted values of transport capacity for both operating modes. Predicted values were about 9% low for the isothermalizer mode and about 13% high for the VCHP mode. These test data indicate a somewhat greater drop in performance for the VCHP mode compared with isothermalizer values (~45%) than that predicted by the analysis (~35%). Part of the discrepancy between test and predictions may be due to the lack of a capillary structure between the reservoir and pipe in this particular

Table 2 Eninger and Marcus VCHP test

Working fluid		Ammonia
Pipe lengths:	Evaporator	0.38 m
	Transport section	0.10 m
	Condenser	0.76 m
Effective length, isothermalizer		0.67 m
Effective length, VCHP		0.32 m
Vapor temperature		294 K
Blocked Zone Temperature		222 K
Pipe tilt (evaporator raised)		0.0038 m

laboratory setup. There was a smaller charge in the pipe in the VCHP mode due to vapor diffusion into the reservoir.

The analysis was also applied to the data of Eninger and Marcus.[1] Some pertinent test parameters are given in Table 2. Reference 1 reports experimental Q_{MAX} values of 90 and 125 W for the VCHP and isothermalizer modes, respectively. Adjusting for differences in effective length, the corresponding thermal transport capacity values were 29 and 84 W·m, respectively, or a ratio of 0.35. Using the same groove geometry as for the Grumman test article, the analysis presented in this paper predicts thermal transport capacity values of 19 and 82 W·m, respectively, or a ratio of 0.24. Here, the predicted loss in performance is somewhat greater than that measured experimentally. For this same set of data, the method of Ref. 1 predicts a ratio of 0.56.

Conclusions

1) The modified NASA groove analysis program (MGAP) provides a reasonable estimate of the reduction in VCHP thermal transport capacity due to the Marangoni effect.

2) The Marangoni effect is substantial, and should be considered in grooved VCHP designs.

3) Additional data on VCHP thermal transport capacity is desirable to validate the analysis over a wider range of test conditions.

References

[1]Eninger, J. E. and Marcus, B. D., "Marangoni Effect and Capacity Degradation in Axially Grooved Heat Pipes," *AIAA Journal*, Vol. 17, July 1979, pp. 797-799.

[2]Jen, H. F., "User's Manual for Groove Analysis Program," BK012-1007 (NASA Contract NAS5-22562), June 1976.

[3]Jen, H. F., "Summary Report for Axially Grooved Heat Pipe Study," BK012-1009 (NASA Contract NAS5-22562), July 1977.

[4]Hufschmidt, W. et al., "Der Einfluss der Scherwirkung des Dampfstromes auf den laminaren Flussigkeitstrom in Kapillaren von Warmerohren (Liquid Vapor Interaction in Heat Pipes)," *Warme-und Stoffubertragung*, Vol. 2, 1969, pp. 222-239.

[5]Edelstein, F., "A 2.2 Sq M (24 Sq Ft) Self-Controlled Deployable Heat Pipe Radiator," ASME Paper 75-ENAs-43 (NASA Contract 8-29905), July 1975.

Chapter IV. Material Properties and Interfaces

Some Thermophysical Properties of Paraffin Wax as a Thermal Storage Medium

A. Haji-Sheikh,* J. Eftekhar,† and D. Y. S. Lou‡
The University of Texas at Arlington, Arlington, Texas

Abstract

An experimental study is conducted to determine the suitability of paraffin wax SUNTECH P116 as a phase change material for storage of thermal energy. Certain temperature-dependent thermophysical properties in the neighborhood of the melting point useful for this study, but not adequately available in the literature, are measured. They include thermal conductivity, density, thermal expansion coefficient, and viscosity. It is observed that the thermal conductivity of paraffin wax, in solid phase, is not a monotonic function of temperature as reported in the literature. Other thermophysical properties of the liquid phase measured vary monotonically with temperature.

Nomenclature

C_D	= drag coefficient
C_F	= calibration factor
C_{pl}	= specific heat of liquid
C_{ps}	= specific heat of solid
d	= diameter = 2r
g	= gravitational acceleration
k	= thermal conductivity
r	= radius of spheres
Re	= Reynolds number = Ud/ν

Presented as Paper 82-0846 at the 3rd AIAA/ASME Joint Thermophysics, Fluids, Plasma & Heat Transfer Conference, St. Louis, Mo., June 7-11, 1982.

*Professor, Mechanical Engineering Department.

†Graduate Student, Mechanical Engineering Department.

‡Chairman, Mechanical Engineering Department.

T_c = temperature of cold plate
T_f = temperature of Teflon surface
T_h = temperature of hot plate
U = terminal velocity
v = specific volume
β = thermal expansion coefficient
ρ_ℓ = density of liquid
ρ_s = density of solid spheres

Introduction

Storage of thermal energy in the form of latent heat of fusion in lieu of sensible heat has been studied during the past several years for industrial and environmental applications. Among the many possible thermal storage materials, paraffin wax (SUNTECH P116) appears to have some attractive features as a medium for storage of thermal energy, especially in solar energy applications. Its melting point in the neighborhood of 44°C is sufficiently high to provide heated air for introduction to a conditioned environment; yet, it is still low enough to permit absorption of solar insolation using flat plate collectors, or other low-temperature thermal energy sources such as waste heat.

Some advantages of this material are low cost, chemical stability in thermal cycling, and high value of heat of fusion, etc. The major disadvantages are its tendency to react with copper and its unusual values of thermal conductivity in the temperature range of 30 to 44°C.

Paraffin wax is a mixture of several long-chain hydrocarbons, see Appendix. It is a by-product of the oil refining process. In addition to the melting point, the following thermophysical properties are cited in the literature[1,2] or furnished by suppliers,[3] Table 1. The ongoing research on the utilization of paraffin wax (P116) not reported in this paper reveals that there is a significant inconsistency between the rate of the melting (or freezing) in rapid heating (or cooling) with that predicted analytically. In addition, there is an onset of natural convection which enhances heat transfer across the liquid layer. In order to study the source of the inconsistency, it is necessary to measure certain temperature-dependent thermophysical properties of the material. The thermal conductivity of paraffin is measured by using a specially designed thermal conductivity cell. In order to determine natural convection as a possible means to enhance heat transfer to and from the material, a knowledge of the thermal expansion coefficient, density, and viscosity of the liquid wax is necessary for determin-

Table 1 Some thermophysical properties of paraffin wax

Property	Value
Specific heat of solid	2.95 kJ/kg-°C
Specific heat of liquid	2.51 kJ/kg-°C
Density of solid	818 kg/m
Density of liquid	760 kg/m
Thermal conductivity of solid	0.24 W/m-°C
Thermal conductivity of liquid	0.24 W/m-°C
Heat of fusion	266 kJ/kg

ing the Rayleigh number. They are measured using standard laboratory techniques.

Measurement Techniques

Different thermophysical properties require different measurement techniques. Each is described briefly in this section.

Thermal Conductivity

The thermal conductivity cell as shown in Fig. 1 contains a thin, vertical, cylindrical bowl machined from Teflon. It is placed on a 6.5-mm-thick copper disk cooled from below via a cooling chamber. A constant temperature bath supplies water at a preset temperature to the cooling chamber. Another 6.5-mm copper disk is placed on top of the Teflon bowl and kept at a constant temperature by resistance heating using direct current. The one-dimensionality of heat conduction in the cell is assured by a properly placed and controlled guard heater ring. The thermal conductivity of Teflon and paraffin wax is nearly the same, therefore it is not expected to encounter a measurable two-dimensionality in temperature distribution.

Four thermocouples are imbedded in each copper disk and four more are placed flush with the surface of the Teflon bowl (see Fig. 1). As a preliminary to thermal conductivity measurement, the cell is calibrated using distilled and deionized water whose thermal conductivity is well documented in the literature.[4] The thermal conductivity cell is placed horizontally so that heat flux is in the direction of the gravitational field. Since heat flows through the liquid layer heated from the top, it minimizes the contribution from natural convection. In order to assure one-dimensional heat transfer in the cell, the guard heater is controlled so temperatures indicated by the four thermocouples imbedded in

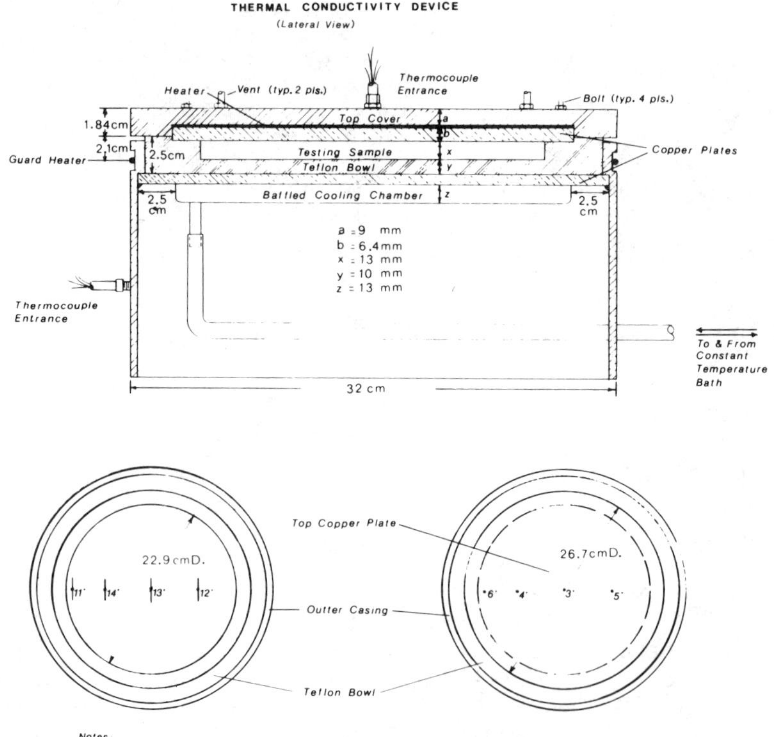

Fig. 1 Thermal conductivity cell and location of thermocouples.

the Teflon dish are nearly identical. Two vents placed at the top of the device are used to add liquid wax to the Teflon bowl to insure that the cell is fully filled during the measurement. If the temperature of the hot disk is T_h, Teflon surface T_f, and cold disk T_c, the calibration factor based on a one-dimensional conduction analysis is

$$C_F = k(T_h - T_f)/(T_f - T_c) \tag{1}$$

The value of C_F, which is proportional to the thermal conductivity of teflon, is found to have a near linear dependence on temperature. The largest deviation from linear regression observed is less than 6%. Once the device is calibrated, the thermal conductivity of a sample can be calculated from the following equation:

$$k = C_F(T_f - T_c)/(T_h - T_f) \tag{2}$$

This cell calibration is also verified by measuring the thermal conductivity of ethylene glycol. At a mean temperature of 36.8°C, it is measured to be 0.260 W/m-K, which agrees favorably with the value of 0.255 W/m-K reported in Ref. 4. A duration of 10 to 24 h is required to attain a steady-state condition for a single data point, depending on the initial temperature distribution in the cell and across the fluid layer. Twenty-eight tests were conducted and the thermal conductivity results are plotted as a function of temperature as shown in Fig. 2. The experimental data indicate that at temperatures below 25°C, the thermal conductivity of paraffin wax is the same as that reported in current literature. However, it decreases initially upon heating from the solid phase. It reaches a minimum value at 34°C, and then increases to an asymptotic value for the liquid phase. Hence the material in its solid phase exhibits a tendency to insulate the liquid wax during both melting and solidification processes. This is not a desirable effect since it impedes the heat transfer to or from the storage system.

A temperature difference of approximately 5°C across the sample is maintained during all tests except in the vicinity of 34°C, where a 4°C temperature difference is used. Inasmuch as it is difficult to predict this temperature difference apriori, slight deviations from the aforementioned numbers are expected.

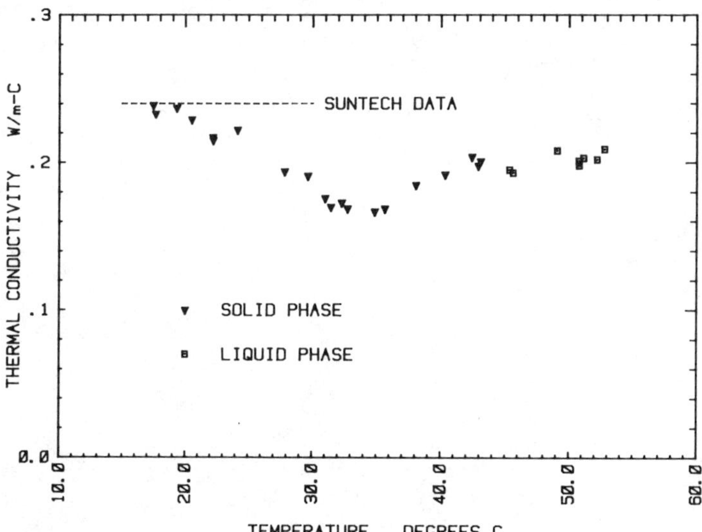

Fig. 2 Thermal conductivity of SUNTECH P116 paraffin wax in the neighborhood of melting point.

The reproducibility of the experimental data reported here is satisfactory. In fact, each set of data is collected under different laboratory conditions. For instance, the thermal conductivity data are measured consecutively as the mean temperature of the sample is changed discretely. The test started when the temperature T_f was just over 44°C. After completion of the first test, the mean temperature is discretely increased for subsequent data. This guarantees that there is a good contact between liquid and the upper copper disk. In fact, a considerable loss of liquid due to its volumetric expansion is experienced. This process is repeated in descending order. Consequently, it is necessary to replenish the liquid for each data point. A similar procedure is repeated for measurement of thermal conductivity in the solid phase starting at a solid mean temperature below 44°C. After the thermal conductivity at the lowest mean sample temperature is measured, the process is repeated in ascending order. Care should be exercised to ascertain that there are no voids in the solid due to 7% shrinkage of volume following solidification. The upper plate is heated to slightly above 44°C while cold water is passed through the lower plate cooling chamber. A thin melted layer then forms at the top layer of the wax which permits the liquid to be added in order to fill up the voids without removing the top cover of the thermal conductivity cell.

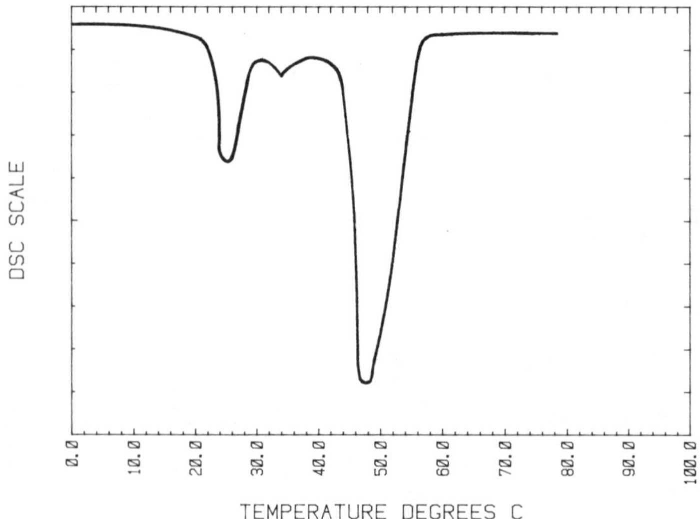

Fig. 3 DSC thermograph for SUNTECH P116.

A differential scanning calorimeter trace[3] of the paraffin wax is presented in Fig. 3. The thermograph clearly shows that there are crystal modifications occurring at 25 and 34°C. The influence of the crystal modification, immediately before melting or following the freezing process, can be understood by examining a simple cooling curve of the material.

A beaker containing 500 ml of liquid paraffin wax placed in the laboratory away from any heat source is used to obtain such a curve. A thermocouple is inserted centrally in the wax to monitor the temperature in the neighborhood of the hot spot. The data are stored on a magnetic tape for a suitable data reduction scheme. The time trace of temperature and the time derivative, dT/dt, as functions of time and temperature are plotted in Figs. 4a and 4b to explain this peculiar physical behavior. From Fig. 4a, one observes that -dT/dt stops above 0 in the melting process. If the phase change process completes itself at the melting point of 44°C, it will follow a path shown by the dashed line in Fig. 4b. The extrapolated dashed line in the liquid phase has a higher value of -dT/dt at 44°C than in the solid phase because $\rho_s C_{ps}$ is less than $\rho_\ell C_{p\ell}$. The latent heat is released over a range of temperatures between 44 and 34°C which coincides with the peculiar behavior of its measured thermal conductivity as shown in Fig. 2.

Thermal Expansion Coefficient

The thermal expansion coefficient of the wax is measured by placing 20 ml of paraffin wax in a precision-bored pyrex tube which has a negligible thermal expansion coefficient in comparison with paraffin wax. The pyrex tube containing paraffin wax is placed in a constant temperature bath which has stability and uniformity within 0.01°C. A micrometer is used to measure the location of the liquid surface as a function of temperature after allowing sufficient time to establish thermal equilibrium. The data are then reduced, and the thermal expansion coefficient is calculated as a function of temperature,

$$\beta = 0.00273/(1.146 + 0.00273T) \qquad (3)$$

The results reveal a slight linear decrease in the thermal expansion coefficient between temperatures of 44 and 60°C as indicated in Fig. 5. The value of the thermal expansion coefficient for the liquid paraffin wax is obtained from the empirical equation for specific volume in the inset of Fig.

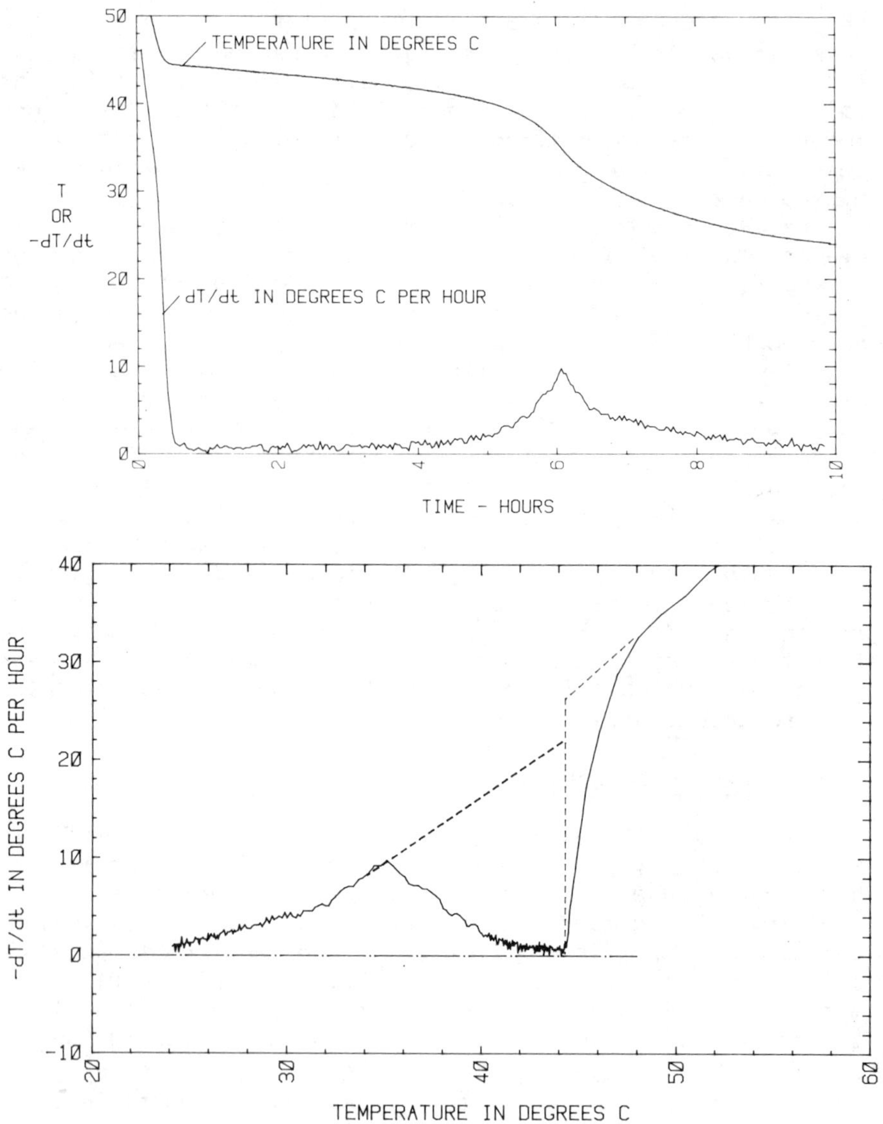

Fig. 4 Typical cooling curve for paraffin wax.

5. The specific volume of the paraffin wax is also plotted in the same figure.

Viscosity Coefficient

Initially, an attempt was made to utilize a Saybolt viscosimeter to measure the viscosity coefficient. The reason for this choice was the convenience of maintaining liquid paraffin at a constant temperature prior to and during the measurement. The experimental results obtained were below 100 universal Saybolt seconds; a range not recommended for this device.

The next attempt consists of a series of drag measurements on spherical bodies. High-density polyethylene spheres of 3.15 ± 0.005 mm in diameter are selected. The density of these spheres measured at 0.957 g/cm^3. Paraffin is placed in a tube with a 16 mm i.d. The tube is placed in a constant temperature bath until thermal equilibrium is attained. At the time of each test, the tube is removed and placed in front of a calibrated strobe light and camera. A polyethylene sphere is dropped in the liquid and the strobe light identifies the trajectory of the sphere at 0.325-s intervals. Figure 6 is a typical photograph taken during this experiment. The photograph shows that the sphere achieves its terminal velocity during free fall within a very short distance. The distance between successive images is very uniform except, as expected, at the beginning and near the end of the trajectory. The former is due to acceleration of the sphere and the latter is caused by the effect of the bottom of the tube. Enlarged photographs are used for calculation of the distance traveled by the spheres

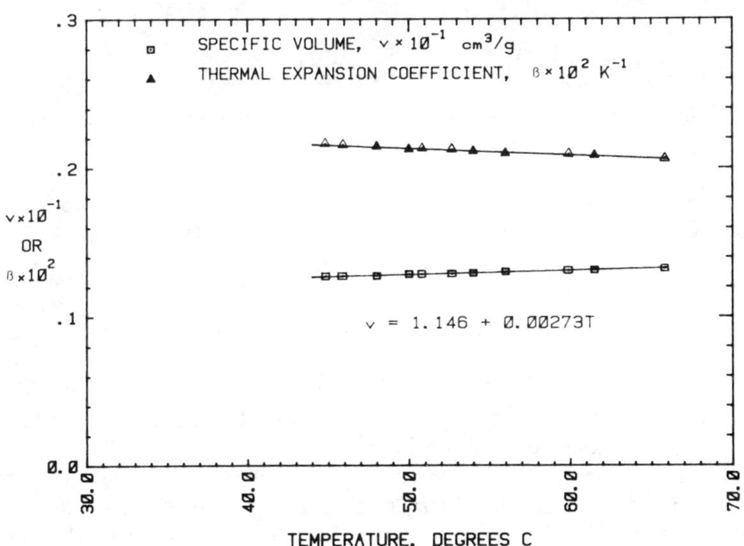

Fig. 5 Variation of specific volume and thermal expansion coefficient with temperature.

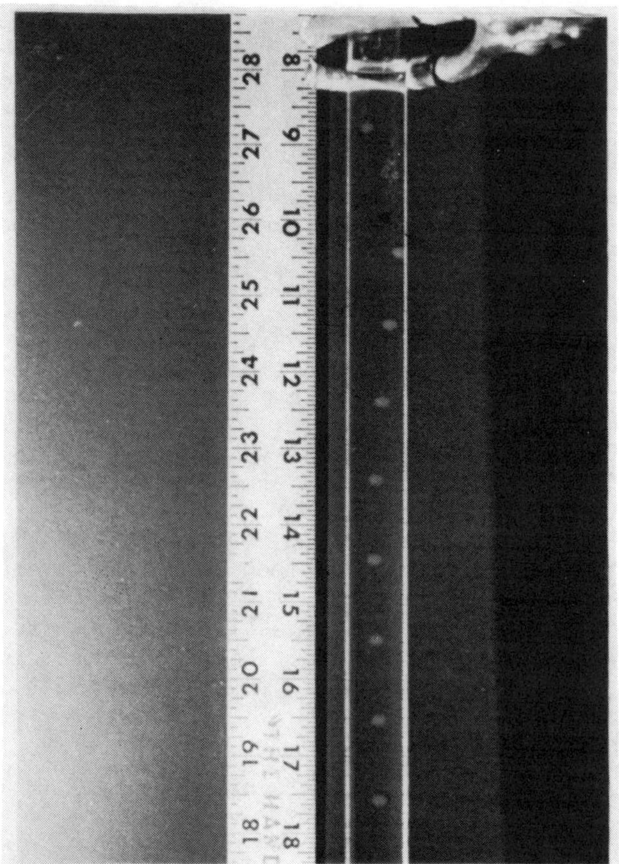

Fig. 6 Location of sphere as it travels through liquid.

using the scale depicted in the background of the photographs. The drag coefficient is then calculated by the equation

$$C_D = 8rg(\rho_s/\rho_\ell - 1)/3U^2 \quad (4)$$

The relation between the drag coefficient and Reynolds number is well known.[5,6] The value of the drag coefficient measured varies between 2 and 12.

The next step is to determine the influence of the tube walls on the drag coefficient measured. This is done by using the same apparatus but with water as the working fluid to determine experimentally the influence over the same

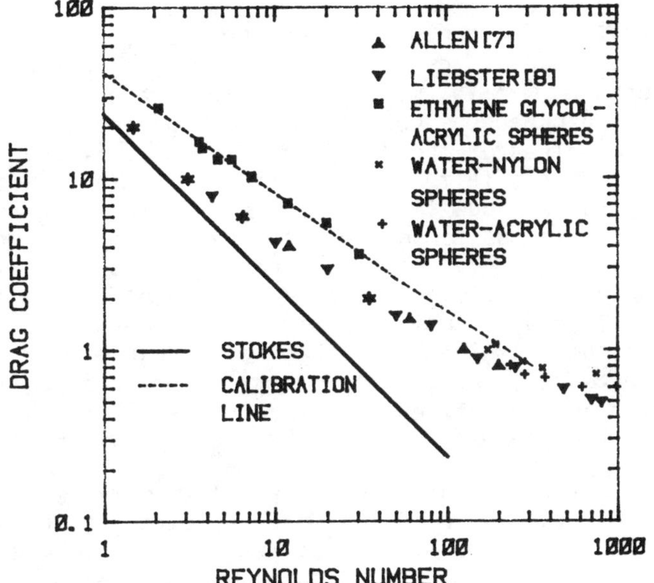

Fig. 7 Drag coefficient as a function of Reynolds number.

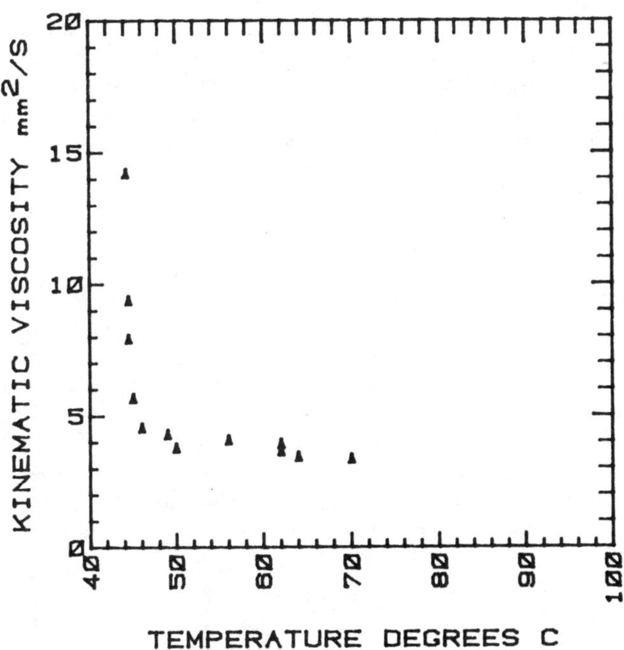

Fig. 8 Variation of viscosity of paraffin wax with temperature.

range of Reynolds numbers. The viscosity of water is well documented[4] with far greater accuracy than that needed for paraffin wax. Acrylic spheres, 3.30 mm in diameter, and nylon spheres, 3.15 mm in diameter, are selected. In addition, ethylene glycol and acrylic spheres are used to obtain the values of the drag coefficient at lower Reynolds numbers. In this latter experiment the terminal velocity is extremely slow and it is measured using a stopwatch. The drag coefficient as a function of the Reynolds number is plotted in Fig. 7. The uncorrected data indicate a 13% higher drag coefficient than the experimental data,[7,8] available in the literature, at higher Reynolds numbers. The magnitude of deviation becomes larger as the Reynolds number decreases. Deviations of this type are common and are caused by the influence of the tube walls.[6] The data in Fig. 7 that appear to be asterisks are in fact two triangles superimposed.

In light of the aforementioned information, the drag coefficient is measured and the Reynolds number is calculated using the calibration line designated by the dashed line in Fig. 7. For convenience of computation, a portion of the dashed line between the Reynolds numbers of 2 and 30 is curve-fitted using a least-squares technique. Following the computation of Reynolds numbers, the values of the kinematic viscosity ν for liquid paraffin wax are calculated and plotted in Fig. 8. The experimental data extend from near the melting point to 70°C. The kinematic viscosity decreases slowly as temperature increases above 50°C. However, it increases rapidly as the temperature decreases toward 44°C.

Remarks

The high value of the thermal expansion coefficient and reasonably low kinematic viscosity of the paraffin wax can cause natural convection in a sufficiently thin liquid layer. A 10°C temperature difference across a layer of liquid wax, a few millimeters thick, will result in a buoyancy force sufficiently high to cause natural convection to begin. Therefore any prediction of thermal energy stored in paraffin wax should include the possibility of heat transfer across the liquid layer by natural convection.

Acknowledgments

Partial funding by the Organized Research Fund at UTA and the assistance of John Thomson, Michael Stevenson, and

Girishkumar C. Patel, former students at UTA, are greatly appreciated. Also, we wish to thank R. J. Stenger, manager of Wax Technology at SUNTECH, for providing us with material and technical assistance.

Appendix

The chemical composition of P116 paraffin wax as provided by the supplier[3] is

$C_{20}H_{42}$ = 2.0%
$C_{21}H_{44}$ = 5.5%
$C_{22}H_{46}$ = 14.0%
$C_{23}H_{48}$ = 23.0%
$C_{24}H_{50}$ = 22.0%
$C_{25}H_{52}$ = 14.0%
$C_{26}H_{54}$ = 6.5%
$C_{27}H_{56}$ = 3.0%
$C_{28}H_{58}$ = 2.5%
$C_{29}H_{60}$ = 2.0%
$C_{30}H_{62}$ = 1.7%
$C_{31}H_{64}$ = 1.5%
$C_{32}H_{66}$ = 1.3%

The hydrocarbons with carbon chain lengths of 20 to 32 constitute 99% of the mixture. The hydrocarbons with carbon chain length longer than 32 constitute the remaining 1% with exponentially reducing fraction. The composition of this substance may change due to manufacturing variance.

References

[1] Bundy, F.P., Herrick, C. S., and Kosky, P. G., "The Status of Thermal Energy Storage," General Electric Company, Report 76 CRD 041, 1976.

[2] Lunde P. J., Solar Thermal Energy, Wiley & Sons, New York, 1980, Chap.13.

[3] Stenger, R. J., private communication, SUNTECH Group, Marcus Hook, Pa, 1980.

[4] Eckert, E. R. G. and Drake, R. M. Jr., Analysis of Heat and Mass Transfer, McGraw Hill, New York, 1972.

[5] Schlichting, H., Boundary Layer Theory, 4th ed., McGraw Hill, New York, 1960.

[6] Prandtl, L. and Tietjens, O. G., Applied Hydro- and Aerodynamics, Dover Publications, New York, 1957.

[7] Allen, H. S., "The Motion of a Sphere in a Viscous Fluid," Philosophical Magazine, Vol. 50, 1900, p. 323.

[8] Liebster, H., "On the Resistance of Sphere, "Ann. Physik, Vol. 82, 1927, p. 541.

Ultraviolet and Electron Irradiation of DC-704 Siloxane Oil on Zinc Orthotitanate Paint

D.L. Mossman,* M.K. Barsh,† and S.A. Greenberg‡
Aerojet Electrosystems Company, Azusa, Calif.

Abstract

Discrepancies exist between accelerated laboratory simulation and geosynchronous orbit flight data for zinc orthotitanate (ZOT) paint degradation. The effects of ultraviolet and electron irradiation on ZOT contaminated with DC-704 silicone oil are reported. In situ solar absorptance (α_S) and emittance changes for contaminated and clean specimens are discussed with reference to post-test surface morphology, determined by scanning electron microscope analysis. Features of the contaminated ZOT degradation kinetics correlate with orbital performance.

Background

Requirements exist for stable thermal control surfaces with low solar absorptance (α_S) and high emittance. Second surface mirrors (silvered fused silica) have been utilized successfully.[1] However, mirrors are not readily amenable to application on curved surfaces, are costly, and exhibit reduced emittance at cryogenic temperatures. Silvered FEP Teflon (Ag/FEP) has also been employed as a low α_S/ε surface. Flight experience and laboratory simulation data[2] have demonstrated that Ag/FEP exhibits significant degradation in the geosynchronous orbit environment.

Paint systems, consisting of metal oxide pigments such as zinc oxide and titanium dioxide in silicone binders have

---

Presented as Paper 82-0865 at the 3rd AIAA/ASME Joint Thermophysics, Fluids, Plasma & Heat Transfer Conference, St. Louis, Mo., June 7-11, 1982.

*Supervisor, Materials and Processes.

†Senior Specialist, Technical Staff, Mechanical Engineering Department.

‡Manager, Mechanical Engineering Department.

unstable optical properties when subjected to the electromagnetic and particulate radiation components of the space environment.[3] Development of zinc oxide pigments with potassium silicate binders (Z-93) by the Illinois Institute of Technology Research Institute (IITRI) resulted in a coating with good ultraviolet radiation resistance. However, electron and proton radiation degrades the solar absorptance of Z-93. Intensive research in the area of more stable thermal control paints led to the investigation of zinc orthotitanate pigments by IITRI.[4] More than a decade of development effort has resulted in the formulation of a zinc orthotitanate pigment in potassium silicate binder paint system (ZOT). This coating exhibits low α_S and good radiation stability. In addition, the high emittance properties are maintained at cryogenic temperatures.

The promising optical properties of ZOT, together with the preliminary radiation test results, have indicated that this coating might provide reliable long-term thermal control for spacecraft operating in geosynchronous orbits. Consequently, ZOT coatings on aluminum substrates were included in a geosynchronous orbit flight experiment, together with second surface mirror reference specimens. Solar absorptance properties of the materials were determined calorimetrically over a three-year period.

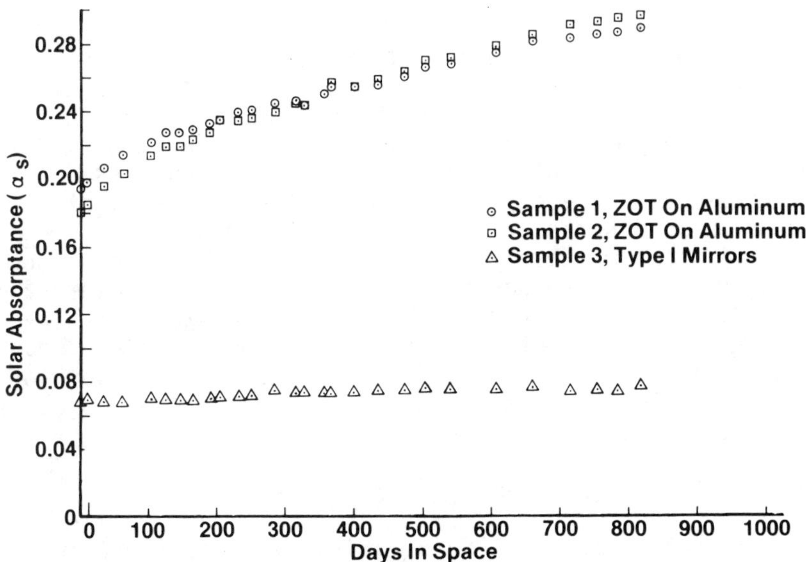

Fig. 1 Degradation of ZOT in geosynchronous orbit relative to second surface mirrors.

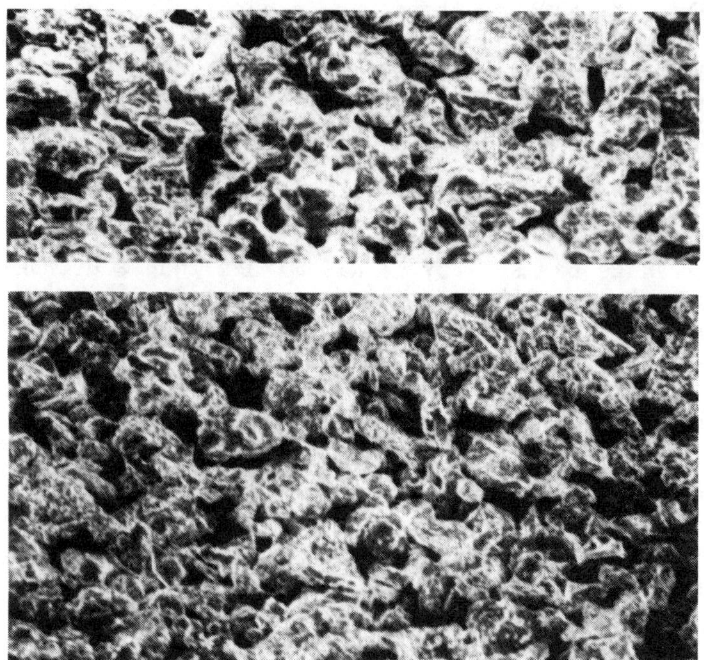

Fig. 2 SEM of the two ZOT test specimens in as-received condition (magnification: 400X).

The degradation kinetics of the ZOT specimens during the first year of exposure were consistent with laboratory predictions. However, the continued increase of α_S into the third year was not fully anticipated. Contamination of the calorimeter materials from spacecraft outgassing is considered to be unlikely owing to the constancy of the adjacent second surface mirror solar absorptances (Fig. 1).

A possible cause of continuing ZOT degradation is contamination during prelaunch handling and testing. Whereas the second surface mirrors are cleaned prior to launch, the porous character of ZOT precludes effective cleaning. In order to verify the contamination hypothesis, laboratory space simulation tests were conducted with ZOT specimens contaminated with DC-704 vacuum pump oil.

Experimental Procedures

Aluminum plates (0.115 in. thick) were coated with ZOT paint (approximately 0.009 to 0.012 in. thick) by Y. Harada of IITRI. These plates were prepared in the identical manner as the samples used for the flight calorimeter experiments. When the coated plates were received, strict precautions were taken to avoid contamination by handling. One-inch-diameter disks were punched from the plates and used as experimental samples.

Prior to mounting the specimens for irradiation, scanning electron microscope (SEM) pictures were taken in order to choose two nearly identical ZOT morphologies (similar porosity) and obtain pretest characterization (Fig. 2). Subsequently, the two selected specimens were each mounted in the radiation facility on separate calorimeters (Fig. 3) on a turntable in a high-vacuum, LN_2 cooled-shrouded, black, high-ε chamber. This chamber is equipped with a window for solar simulator input, two ultraviolet (uv) sources, and a radioactive beta-electron source. The calorimeters were isolated from each other by LN_2 cooled walls. The turntable allowed position control (Fig. 4). This configuration permitted independent uv and electron irradiation of each sample as well as calorimetric measurement of α_s and ε at designated intervals.

The uv sources consisted of a deuterium lamp with output in the 180-450 nm spectral region and a Lyman-α (121.6/123.6 nm) lamp providing a flux of 10^{11} photons/cm^2/s corresponding to 0.25 equivalent Lyman-α suns. In situ measurement of the Lyman-α source was performed with a nitric oxide detector mounted in the sample plane. A Keithley Model 602 electrometer was employed to monitor the nitric oxide cell output during exposure.

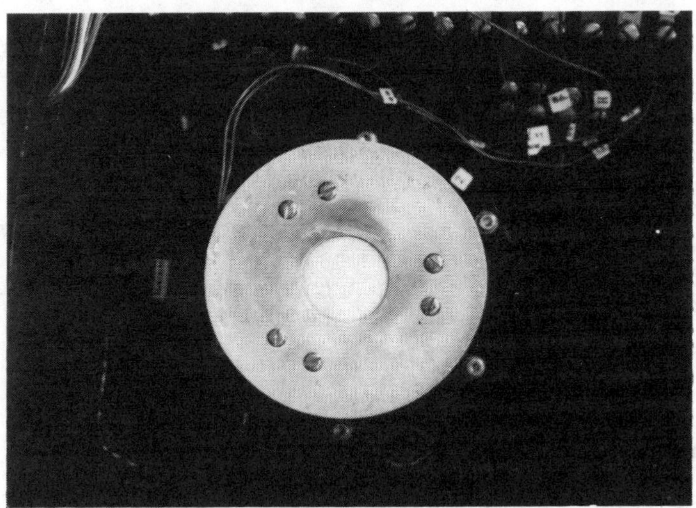

Fig. 3 Specimen mounted in laboratory calorimeter.

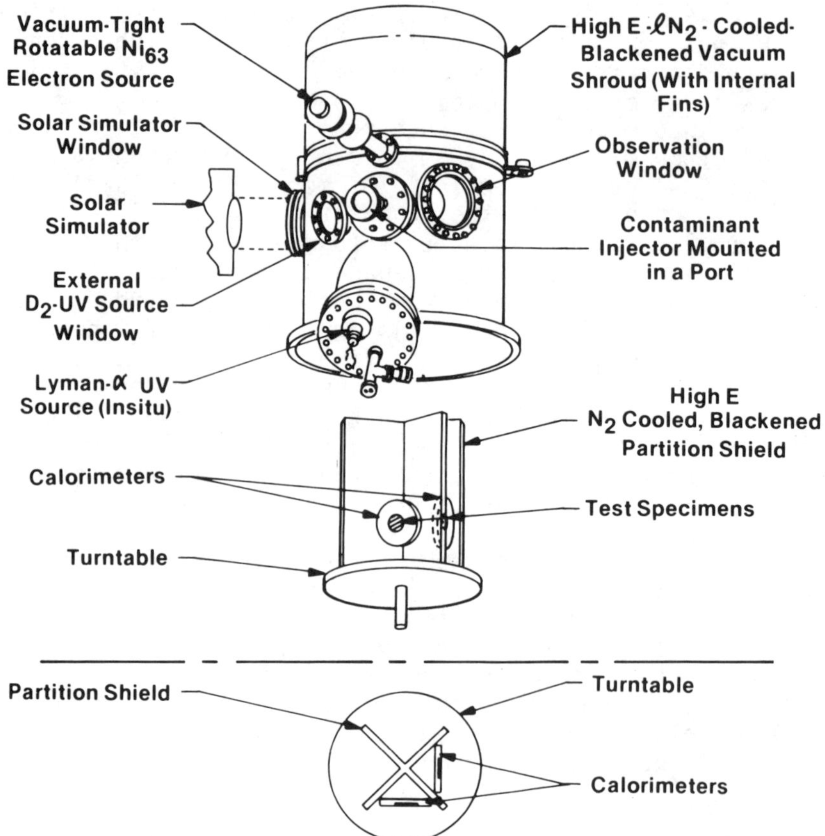

Fig. 4 Schematic representation of laboratory test facility.

The electron source consisted of a movable disk of radioactive nickel (Ni_{63}), providing a nominal flux of 1.6×10^{10} electrons/cm^2/s with an energy of 0.67 MeV. This electron flux corresponds approximately to a threefold acceleration relative to the geosynchronous electron environment. However, the energy deposition rate is approximately 50 times the orbital rate when considerations of electron energy and deposition profile are applied.

The system was evacuated to 10^{-7} Torr (sorption and ion pumping), and the reference values of α_s and ε for both of the specimens were determined with the aid of the solar simulator. One of the ZOT specimens was then deliberately contaminated by spraying the surface with DC-704 vacuum-pump oil (tetramethyl tetraphenyl trisiloxane) using a syringe-type injector. When the chamber pressure stabil-

Table 1 Effects of laboratory exposure on the absorptance and emittance properties of ZOT

	Clean		Contaminated	
Time,days	α	ε	α	ε
Reference	0.210[a]	0.84	0.211[a]	0.84
0	0.279[b]	0.84	0.373	
1			0.385	
2			0.384	0.83
3			0.392	
4	0.286	0.84	0.397	0.83
5				
6				
7	0.285			
9				
10				
11	0.288	0.84	0.405	0.825
12			0.400	
13				
14	0.287			
16				
18				
20			0.40	
22				
24				
25	0.285	0.84		
26			0.41	0.83

[a]At time zero, reference measurement with both samples uncontaminated.
[b]Sample after one month in vacuum environment.

ized (in approximately 1 h), the αs measurement of the contaminated specimen was taken. The specimen was then rotated to the position for uv and electron irradiation. During the first two days, frequent α_S measurements were made. Thereafter, the radiation proceeded for 25 days with periodic measurements of α_S and ε.

At the conclusion of the contaminated ZOT specimen exposure, irradiation of an uncontaminated specimen was performed. The clean specimen, which was previously isolated from the contamination source and radiation by baffles, was rotated into the test position. Measurements of α_S and ε were performed and then immediately rotated to the irradiation position. Periodic optical property measurements were performed during the exposure sequence.

SEM analyses were performed at the conclusion of the tests. Magnifications of 400X and 2000X were used to in-

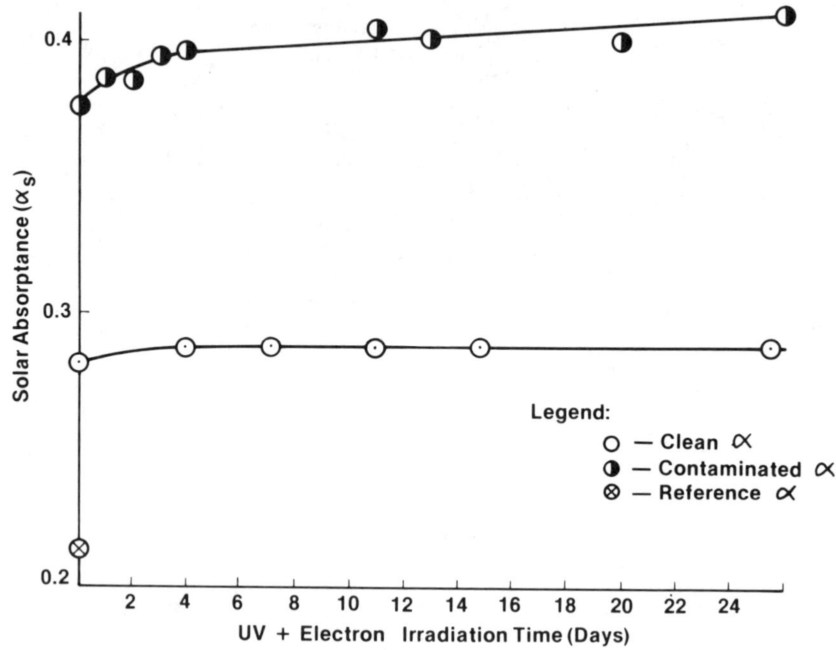

Fig. 5 Effects of radiation on the solar absorptance of clean and contaminated ZOT.

vestigate surface modifications of ZOT resulting from contamination and irradiation processes.

Results

The measured values of α and ε for clean and contaminated ZOT specimens, as a function of the test parameters, are presented in Table 1.

Figures 5 and 6 show the trends of α and ε, respectively, for clean and contaminated ZOT specimens. Since the acceleration parameters for uv and electrons differ significantly, the exposure scale is presented in real-time exposure days rather than equivalent orbital exposure units.

The major observations derived from this study are summarized as follows:

1) Immediately following contamination of the ZOT surface, the α expectedly increased from the reference value of 0.211 to 0.375. The of the clean ZOT surface, which was in the vacuum chamber for approximately one month before irradiation was initiated, only increased from 0.210 to 0.285.

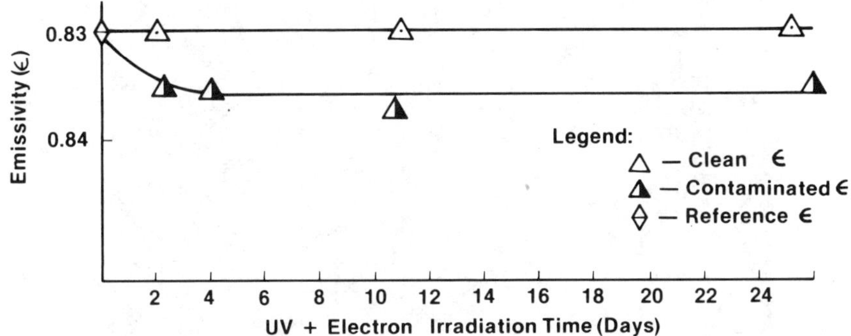

Fig. 6 Effects of radiation on the emittance of clean and contaminated ZOT.

2) The initial reference ϵ values of the contaminated ZOT surface decreased from 0.84 to 0.825 and remained relatively constant thereafter. The clean ZOT remained at 0.84 throughout the test.

3) During the first four days of exposure to radiation, the contaminated ZOT specimen exhibited more rapid increase of α_s then the clean specimen.

4) After four days, the α_s values of the contaminated specimen appeared to increase at approximately a constant rate ($6.0x10^{-4}$/day).

5) After four days, the α values of the clean specimen showed no significant increase.

6) The SEM photos taken of the "as received samples" before testing (Fig. 2) show ZOT to be a very porous structure with individual jagged particles. Two of the samples appeared to have similar porosity in numerous SEM-analyzed areas and thus were used in the experiment obtaining similar initial optical properties.

7) After irradiation, the contaminated sample SEM (Fig. 7) showed a marked loss of porosity and loss of granular particles. High magnification (2000X) shows a fibrous structure filling the crevices which is a product of the polymerized DC-704 silicone oil.

8) Both the irradiated clean specimen (Fig. 8) and the specimen maintained in vacuum for two months (Fig. 9) appear to have a slightly collapsed structure compared to pretest material. The particle surfaces are less irregular and more nearly spherical.

Discussion

Analysis of the ZOT degradation kinetics from the flight data (Fig. 1) indicates that there are three dis-

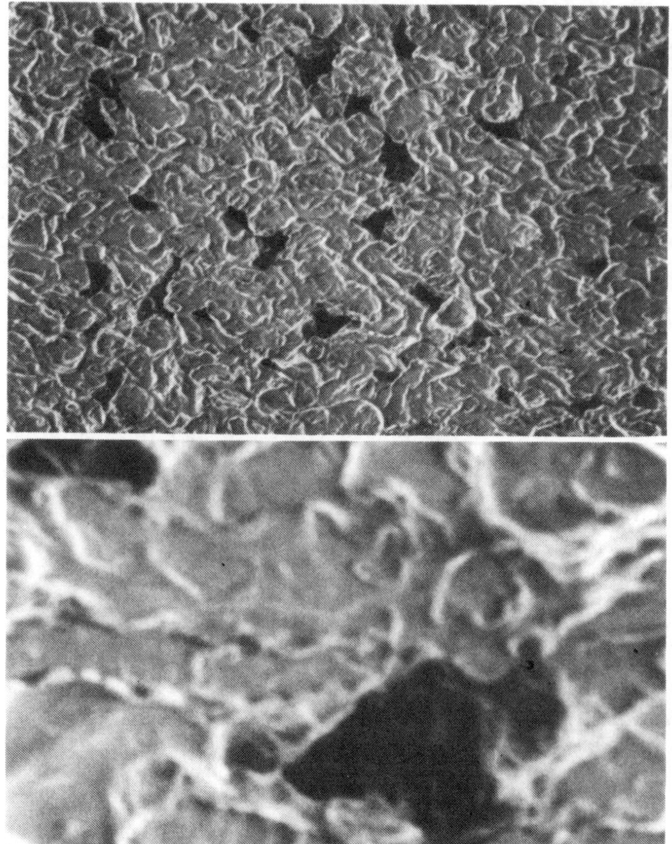

Fig. 7 SEM of irradiated contaminated ZOT (magnification: 400X and 2000X).

tinct stages. During the initial rapid degradation (first 200 days), α_S increases at the rate of 2.5×10^{-4}/day. The second regime occurs in the 200-700 day period where the rate decreases to 1.1×10^{-4}/day and finally after 200 days, the $\Delta\alpha_S$ rate is equivalent to 6.5×10^{-5}/day. The laboratory simulation data show three similar degradation regions which can be correlated with the flight kinetics. Table 2 shows the relationship of flight and laboratory degradation rates for contaminated ZOT.

During the first two stages of ZOT degradation, the ratio of laboratory to flight α_S increase rate is nearly constant (45-48). This acceleration rate is equivalent to the nominal 50X acceleration of laboratory electron irradiation relative to the space dose rate. This correla-

Table 2 Laboratory and flight test degradation kinetics

Orbital exposure, days	Orbit degradation rate, $\Delta\alpha$/day	Laboratory exposure, days	Laboratory degradation rate,[a] $\Delta\alpha$/day	Laboratory/orbit rate
0-200	2.5×10^{-4}	0-1	120.0×10^{-4}	48
200-700	1.1×10^{-4}	1-4	50.4×10^{-4}	45
>700	6.5×10^{-5}	4-26	60.0×10^{-5}	9

[a]Contaminated ZOT specimen.

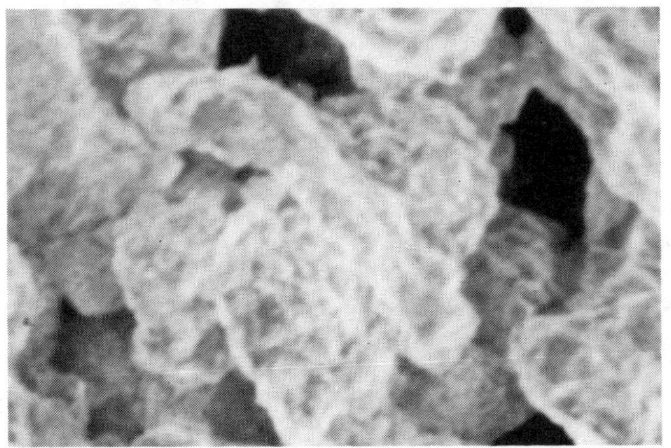

Fig. 8 SEM of irradiated clean ZOT (magnification: 2000X).

tion suggests that electron damage may be dominant relative to the uv effects, which are undersimulated in the laboratory. During the last part of the exposure, the acceleration rate falls to 9X, indicating that the degradation is approaching saturation and the mechanism may be changing. The close relationships between the degradation kinetics of the contaminated ZOT and the flight data strongly suggest the possibility that the flight specimens may have become contaminated prior to launch.

The small changes in α_s observed for the clean ZOT specimen ($\Delta\alpha_s < 0.01$) can be attributed to intrinsic damage from the uv and electron irradiation. It is unlikely that unintentional contamination of these samples occurred, since even small traces of DC-704 result in measurable increases of α_s upon uv irradiation.[5] SEM analysis revealed no indication of foreign surface material after exposure.

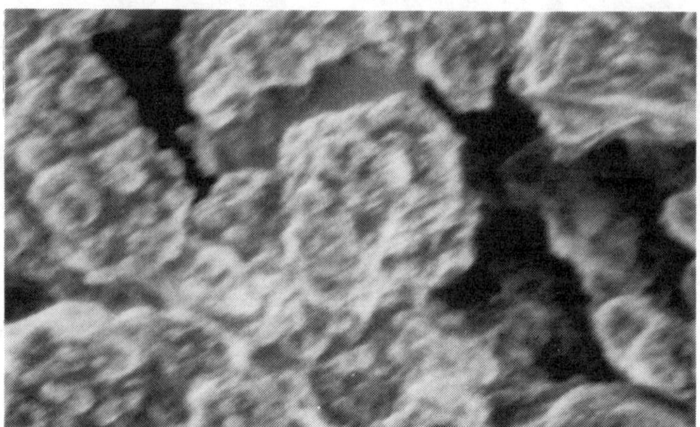

Fig. 9 SEM of vacuum conditioned ZOT (magnification: 2000X).

In contrast, the contaminated ZOT specimens, following exposure, showed substantial polymerized material which tended to fill the voids in the porous structure. The small decrease in emittance during the early stages of irradiation is likely to be the result of decreased surface roughness.

Comparison of the optical property stability of the clean and contaminated specimens clearly demonstrates the susceptibility of ZOT to degradation enhancement by contaminants. The remarkable correlations between the contaminated ZOT and flight data, despite the inadequacies of the space environmental simulation, suggest that the flight data may not accurately represent the true degradation of clean ZOT paint.

Notwithstanding the uncertainties related to the effects of contaminants on the space environmental stability of ZOT, the optical and physical properties of this thermal control coating make it a viable candidate for spacecraft thermal control. Values of α_s in the range of 0.3 after three years in the geosynchronous environment, is a characteristic possessed by few, if any other, paint systems.

References

[1]Greenberg, S. A., Vance, D. A., and Streed, E. R., "Low Solar Absorptance Surfaces with Controlled Emittance," Thermophysics of Spacecraft and Planetary Bodies: AIAA Progress in Astronautics and Aeronautics, Vol. 20, edited by Gerhard B. Heller, Academic Press, New York, 1967, pp. 297-314.

[2]Fogdall, L. B. and Cannaday, S. S., "Effects of High Energy Simulated Space Radiation on Polymeric Second Surface Mirrors," Boeing Aerospace, Seattle, Wash., Final Report NAS 1-13530, CR-132725, Oct. 1975.

[3]MacMillan, H. F., Sklensky, A. F., and McKellar, L. A., "Apparatus for Spectral Bidirectional Reflectance Measurements During Ultraviolet Irradiation in Vacuum," Thermophysics and Temperature Control of Spacecraft and Entry Vehicles: AIAA Progress in Astronautics and Aeronautics, Vol. 18, edited by Gerherd B. Heller, Academic Press, New York, 1966, pp. 129-140.

[4]Zerlaut, G. A., Gilligan, J. E., and Ashford, N. A., "Space Radiation Environmental Effects in Reactively Encapsulated Zinc Orthotitanates and Their Paints," AIAA Paper 71-449, 6th Thermophysics Conference, Tullahoma, Tenn., April 1971.

[5]Mangold, V. L. "The Origin of Deposits Formed on the Surface of Thermal Control Materials by the Action of Extreme Ultraviolet Radiation," U.S. Air Force Flight Dynamics Laboratory, Wright-Patterson Air Force Base, Ohio, AFFDL-TR-68-155, Feb. 1969.

The Thermal Contact Conductance of Dissimilar Metals

David L. Padgett*
Bell Telephone Laboratories, Whippany, N.J.
and
Leroy S. Fletcher†
Texas A&M University, College Station, Texas

Abstract

The design, development, and verification of a new test facility to measure the thermal conductance of dissimilar metals is described. The design permits a reversal of the heat flow direction without removing the samples or disturbing the contact. Experimental thermal conductance data were obtained for aluminum/stainless steel interfaces for apparent contact pressures from 68.9×10^3 N/m^2 to 13.3×10^6 N/m^2, with mean junction temperatures up to 195°C for a selected range of surface conditions. Data are presented which indicate thermal rectification effects at higher apparent contact pressures. The results of this investigation will be useful in the analysis of energy transfer problems involving dissimilar metal junctions.

Nomenclature

FD =flatness deviation, m
h_c =contact conductance, kJ/ms^2°C

Presented as Paper 82-0885 at the 3rd AIAA/ASME Joint Thermophysics, Fluids, Plasma & Heat Transfer Conference, St. Louis, Mo., June 7-11, 1982.

*Member Technical Staff, Interconnection Design Department.

†Professor and Associate Dean, College of Engineering.

P_o =apparent pressure, MN/m^2
q =heat flux, kJ/m^2s
RD =roughness deviation, m
T_m =mean interface temperature, °C
ΔT =temperature drop across interface, °C

Introduction

Because of concern about the possible depletion of current energy resources, efforts are being made to improve the efficiency and reliability of existing energy-related technologies. These efforts normally involve some type of engineering analysis of inefficiencies, frequently related to energy transfer and utilization. In the determination of energy-use characteristics, an analysis of the heat transfer throughout the system is required. For high-performance equipment, it is important to consider every contribution to the total resistance to heat flow. When composite structures consisting of layers of different materials, or junctions of dissimilar materials are considered, the overall heat flow is influenced not only by the individual material properties, but also by the interfaces between the materials. The thermal resistance resulting from these interfaces may significantly influence the overall thermal efficiency of the system.

There are many situations in which the thermal resistance of dissimilar metal interfaces may be important: in heat exchangers when a fin is press-fit onto a tube, in solar collectors when a tube is press-fit into a plate or fins are press-fit onto a tube, in nuclear fuel rods, in electronic circuit boards, in electrical machines, and in cryogenic systems. In most heat-transfer analyses of such systems, the thermal contact resistance is estimated very roughly, determined experimentally for each case, or ignored. For situations in which the thermal resistance must be reliably known, it would be advantageous to be able to predict such resistance without experimentally measuring a value for every anticipated operating condition.

There are several prediction techniques for dissimilar metal interfaces, as noted by Somers et al.[1] These techniques, however, usually apply only to specific situations. Several experimental investigations of dissimilar metallic interfaces also have been conducted, as reviewed by Padgett and Fletcher.[2] Yet these experimental data cover only a portion of the range of parameters necessary for a thorough evaluation of the thermal resistance of dissimilar metal contacts. Further, the reported data do not provide a sufficient range of test parameters to evaluate the phenomenon of thermal rectification. (The phenomenon of thermal rectification occurs when an interface composed of dissimilar metals results in a variation of the thermal resistance depending upon the direction of heat flow.)

The limited availability of design information for heat-transfer calculations involving dissimilar metal contacts suggests the need for additional studies. The increased emphasis on improved efficiency of energy-related systems, and associated refinements in analysis techniques, suggest the need for more complete experimental data over a wider range of test parameters. In order to meet these needs, this paper describes the design, development, and verification of a new test facility to measure the thermal conductance of dissimilar metals and provides thermal conductance data for aluminum/stainless steel interfaces.

Experimental Test Facility

The apparatus was designed to permit accurate measurement of the thermal resistance between similar or dissimilar metals. Particular emphasis was placed on the investigation of dissimilar metals because the inconsistency of existing data suggests the need for more work in this area. To permit a more thorough investigation of the fundamental behavior of thermal rectification, the facility allows testing at higher temperatures and pressures than previous facilities. In addition, the design permits a reversal of the heat flow direction without removing the test samples or disturbing the contact. This is very important because disturbing the contact can both rearrange the relative positions of surface irregularities and alter the surface characteristics, either of which may change the nature of the contact and invalidate any measurement of directional bias. In order to provide flexibility in the use of the apparatus for future investigations, a broad range of test parameter capabilities was incorporated in the facility.

The experimental test apparatus was enclosed in a high vacuum facility consisting of a two-stage rotary vacuum pump and a 15.24-cm oil diffusion pump. Thermocouple and ionization gages measured the vacuum level, and all data were obtained at a test chamber ambient pressure of approximately 10^{-5} Torr. The test apparatus, shown in Fig. 1, was located in a 45.7-cm by 76.2-cm pyrex bell jar. The test samples of 2024-T4 aluminum and 304 stainless steel were 2.54 cm in diameter. The load was applied to the two test interfaces by means of a compressed air bellows system and measured with a 0 to 9000-N compression load cell and strain indicator system. The heat sources were three 200-W cartridge heaters, providing a heat source which could be varied from 0 to 600-W. Water was used as the heat sink coolant. The test specimens were instrumented with Chromel-Alumel thermocouples, permitting the measurement of temperatures up to 1000°C. Radial heat losses were minimized to insure that all heat passed through the interface.

The heat flux through the interface was determined by measuring the axial temperatures of the specimens. Three thermocouples were placed at 1.27 cm intervals. The nearest was 1.27 cm away from the interface. The interfacial temperature drop was found by extrapolation. Axial losses were determined by monitoring the surface temperatures of the samples. Pressure was applied with a high pressure bellows system. The magnitude of the pressure was sensed with a load cell. Details on the test facility and instrumentation have been reported by Padgett and Fletcher.[2]

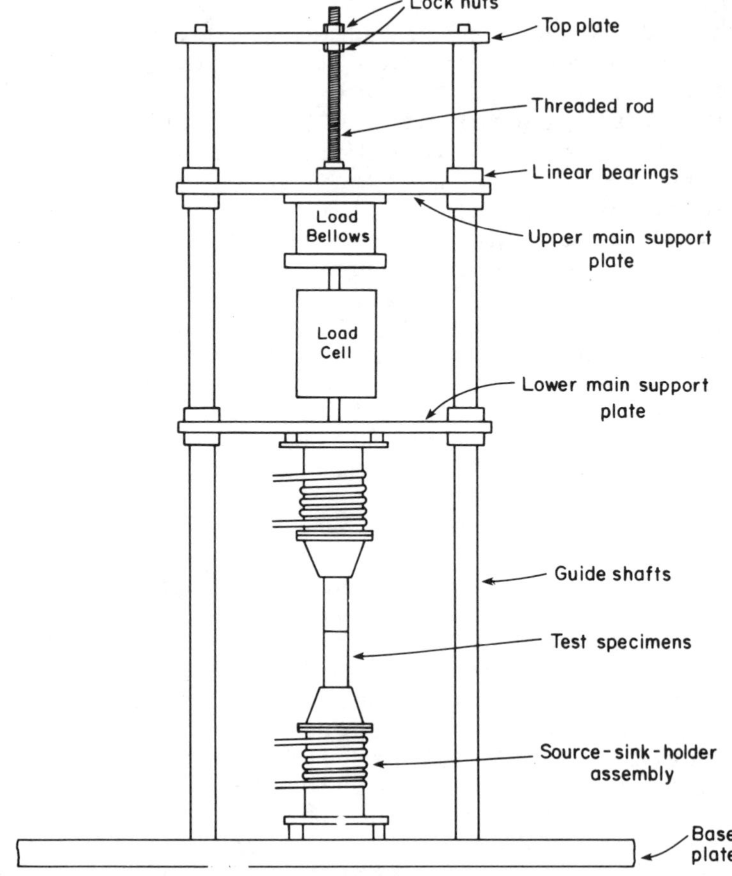

Fig. 1 Schematic of the thermal contact conductance apparatus.

Table 1 Results of heat flow direction reversal tests aluminum 2024-T4

Test	P_a, MN/m	T_m, °C	q, kJ/m²s	T, °C k	h_c, kJ/m²sK	Percent change
1	1.16	57	86.7	10.5	8.25	5.7
1R	1.15	60	85.2	9.8	8.75	...
2	4.17	117	206.1	3.4	60.1	1.0
2R	4.14	114	218.8	3.7	59.5	...
3	6.47	58	98.1	2.3	42.9	5.9
3R	6.49	60	97.9	2.2	45.6	...
4	10.16	117	221.4	1.7	133.2	17.3
4R	10.45	128	243.8	1.5	161.1	...

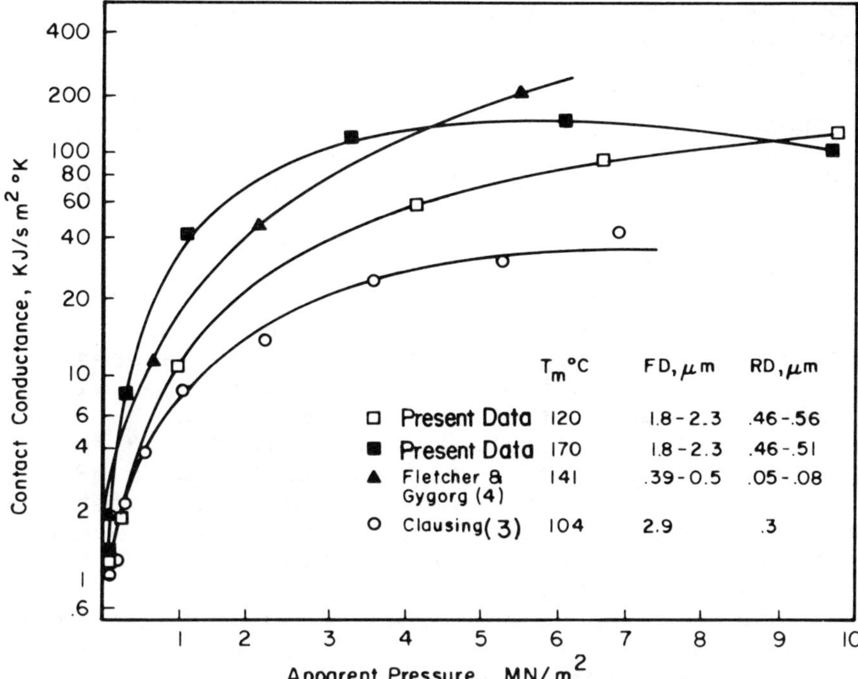

Fig. 2 Variation of contact conductance with apparent interface pressure for aluminum 2024-T4, smooth surface.

Results and Discussion

Preliminary experiments were conducted to verify the performance of the apparatus. These tests were performed with samples made from annealed 2024-T4 aluminum, in order to permit comparison with other published data. Results of these tests are shown in Table 1 for both directions of heat flow. It is interesting to note that the change in thermal contact conductance as a function of direction of heat flow was insignificant when the variations for temperature and pressure were considered. Since the thermal rectification effects should not be present for similar metal contacts, these results were considered quite good. The last data set in Table 1 exhibits the greatest change in conductance value. This difference is reasonable because of the large differences in temperature and pressure. As expected, the variation in contact pressure has a greater influence on the conductance value than the variation in temperature.

These preliminary conductance data were compared with other published data obtained at similar test conditions, as shown in Fig. 2. One set of the present data were obtained at an average mean interface temperature of 123°C and the other set of data at a mean interface temperature of 177 °C. These data compare favorably with the data of Clausing and Chao[3] and Fletcher and Gyorog,[4] allowing for the differences in surface finish and mean interface temperatures.

Experimental tests also were performed with samples of larger flatness and roughness deviations. These data are shown in Fig. 3 for comparison with the work of Clausing and Chao[3] and Fried.[5] The results of the present investigation compare favorably with the published data, although there appears to be some disagreement at the lower contact pressures.

Based on these and other comparisons, it was ascertained that the test facility is suitable for the investigation of thermal contact conductance over a broader range of test conditions than previously possible. Specific details of the tests, along with associated test parameters, have been reported by Padgett and Fletcher.[2] Experimental tests of 2024-T4 aluminum/304 stainless steel interfaces were conducted to provide test data for analysis of dissimilar metal interfaces and the associated thermal rectification. These data are compared with tests for thermal rectification from other sources to establish a more thorough understanding of the phenomenon. The test parameters given in Table 2 are shown for comparison in Fig. 4.

The data presented in Fig. 4 show a wide variation in both magnitude and trend. The difference in magnitude is partially explained by observing Table 2 and noting the wide variation in surface characteristics. Curves 2 and 3 have smoother and flatter surfaces than some of the other tests, and they show the expected high values of h_c. Curves 5-7 have rougher and less flat surfaces and exhibit the expected lower values of h_c.

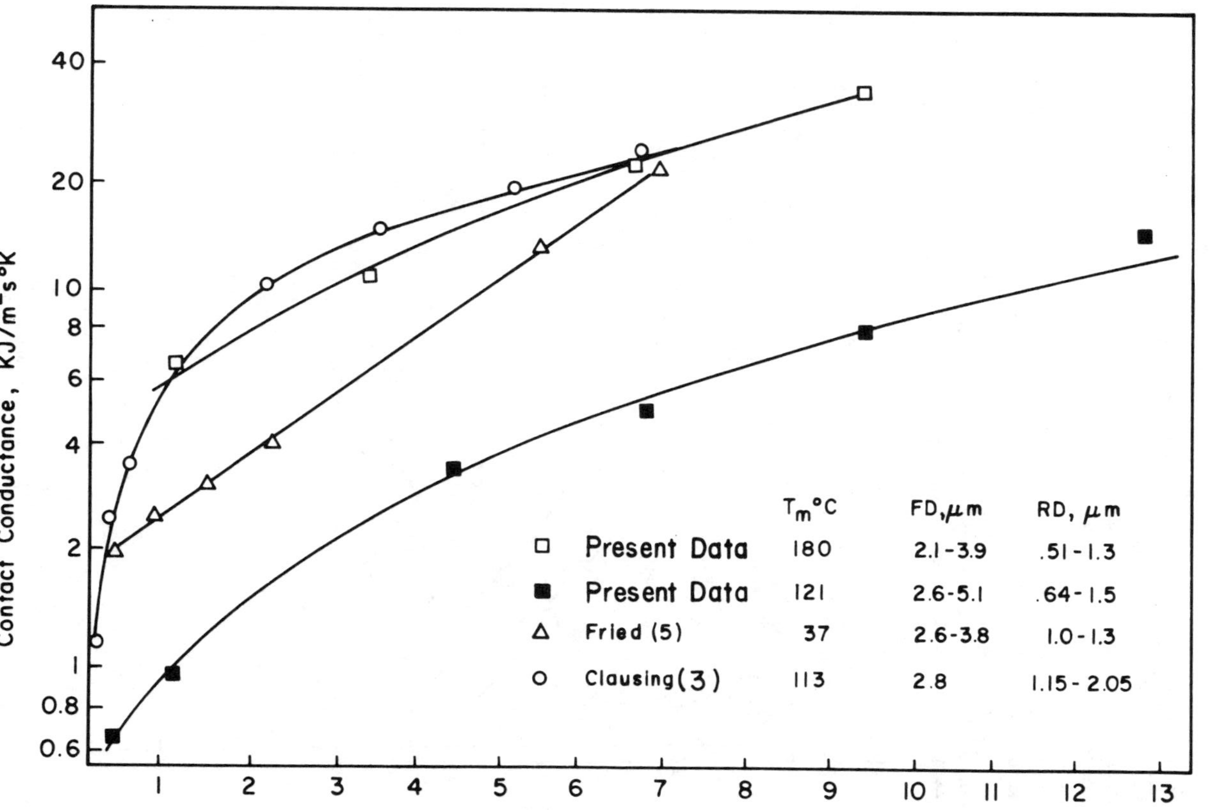

Fig. 3 Variation of contact conductance with apparent interface pressure for aluminum 2024-T4, medium surfaces.

Table 2 Parameters of samples used in tests shown in Fig. 4

Curve[a]	Investigator	Flatness, μm Al	Flatness, μm SS	Roughness, μm Al		Roughness, μm SS	Mean interface temperature, °C SS → Al		Mean interface temperature, °C Al → SS
1	Present data	2.1	2.2	0.513		0.256	78		195
2	Barzelay et al.[6]	5.1	5.1	0.25	to	3.1	90	to	200
3	Lewis and Perkins[7]	0.81	0.61	0.81		0.61	56		113
4	Somers et al.[8]	79.9	95.9	0.165		0.064	20		90
5	Clausing[9]	79.9	95.9	0.165		0.064	20		90
6	Barber[10]	79.9	95.9	0.165		0.064	20		90

[a]Curves 5 and 6 are semiempirical correlations. Flatness, roughness, and temperature values used as input.

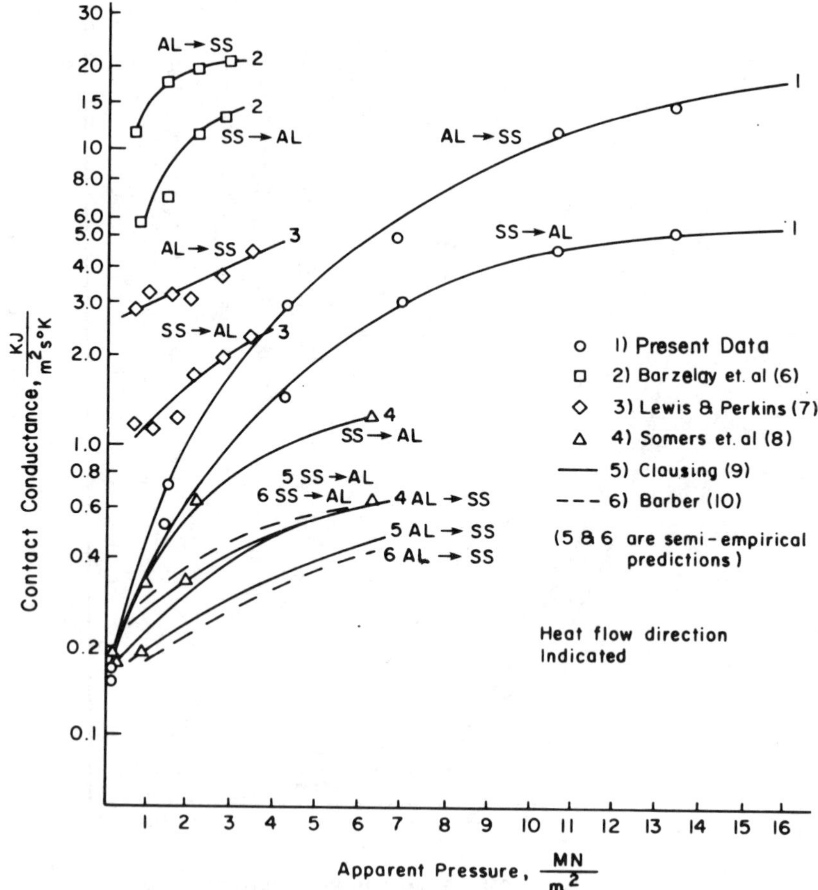

Fig. 4 Variation of contact conductance with apparent interface pressure from aluminum 2024-T4/stainless steel 304 dissimilar metal surfaces at different mean junction temperatures.

Examination of Fig. 4 shows that although the magnitudes of the thermal conductance are reasonable, there are discrepancies regarding both the direction of heat flow yielding the higher value for thermal conductance and the trend of the rectification effect at higher pressures. Curve 1 shows the thermal conductance higher for heat flowing from aluminum to stainless steel (Al → SS) and the rectification effect becoming increasingly more pronounced at higher pressures. Curves 3 and 4 also show the thermal conductance higher for heat flow from aluminum to stainless steel (Al → SS), but they show the rectification effect decreasing as the pressure increases. Curves 4-6 show an increasing rectification effect as the pressure increases, but they show the thermal conductance higher for heat flowing from stainless steel to aluminum (SS → Al). Some of these apparent discrepancies may be explained in terms of the relative magnitudes of flatness and roughness deviations, but discussion of this is beyond the scope of this work.[7,9]

The apparatus designed for this report was built after an exhaustive study of other units reported in the literature. It was constructed to provide reliable data at high pressures and temperatures, and it incorporates shielding and measurement techniques which help to assure accurate data. The consistency of the data shown in curve 1, even in the high contact pressure region, indicates that these values are reliable. The increasing rectification effect at these pressures suggests that it may be worthwhile to investigate the effect of parameter changes at these pressures to more precisely determine their effect. More experimental work is needed to resolve the apparent inconsistencies in the data for stainless steel and aluminum contacts.

Summary and Conclusions

An apparatus was developed to investigate thermal contact conductance and thermal rectification. The apparatus was verified with tests on 2024-T4 aluminum samples. The results of these tests indicated that the facility was suitable for producing reliable and accurate data. Tests were then performed on 304 SS and 2024-T4 AL samples to investigate the thermal rectification effect at high pressures. These tests yielded very consistent data and indicated an increasing rectification effect with increasing pressure. These data also indicated a higher conductance for heat flowing from aluminum to stainless steel (Al → SS). Comparison with data from the literature revealed a consistency in magnitude based on surface parameters. Variations in conductance occur when considering the effect of the direction of heat flow yielding the higher conductance value and the behavior of the directional effect at higher pressures. The current data are considered accurate, and indicate that studying the thermal rectification effect at higher pressures may be worthwhile because the relatively large magnitude will make subtle changes more easily observable. Much more data on dissimilar metal contacts, derived from experimental facilities designed explicitly for that purpose, are needed to define the true nature of these contacts and the associated thermal rectification.

This paper, then, reports a heretofore unpublished investigation of thermal contact conductance involving 1) the design and development of an experimental test facility for dissimilar metals and thermal rectification studies over a wide range of test parameters, and 2) an experimental investigation of 2024-T4 aluminum/304 stainless steel interfaces. Experimental data were obtained for apparent contact pressures from 68.9×10^3 N/m^2 to 13.3×10^6 N/m^2, with mean junction temperatures up to 195°C for a selected range of surface conditions. The results of this investigation will be useful in the analysis of energy transfer problems involving dissimilar metal junctions.

Acknowledgment

This research was supported in part by NSF Grant ENG-24684.

References

[1]Sommers, R. R. II, Miller, J. W., and Fletcher, L. S., "Thermal Contact Conductance of Dissimilar Metals," Thermophysics and Thermal Control: AIAA Progress In Astronautics and Aeronautics, Vol. 65, edited by R. Viskanta, AIAA, New York, 1979, pp. 149-175.

[2]Padgett, D. L. and Fletcher, L. S., "An Apparatus to Determine the Thermal Contact Conductance between Dissimilar Metals," University of Virginia, Charlottesville, Va., RLES Report UVA/526268/MAE80/101, Sept. 1980.

[3]Clausing, A. M. and Chao, B. T., "Thermal Contact Resistance in a Vacuum Environment," University of Illinois, Urbana, Ill., Experiment Station Report ME-TN-242-1, Aug. 1963.

[4]Fletcher, L. S. and Gyorog, D. A., "Prediction of Thermal Contact Conductance between Similar Metal Surfaces," Heat Transfer and Spacecraft Thermal Control: AIAA Progress in Astronautics and Aeronautics, Vol. 24, edited by J. W. Lucas, MIT Press, Cambridge, Mass. 1971, pp. 273-288.

[5]Fried, E., "Study of Interface Thermal Contact Conductance, Summary Report," General Electric Company, Document No. 66SD4471, 1966.

[6]Barzelay, M. E., Jong, K. N., and Holloway, G. F., "Effect of Pressure on Thermal Conductance of Contact Joints," NACA TN 3295, May 1955.

[7]Lewis, D.V. and Perkins, H. C., "Heat Transfer at the Interface of Stainless Steel and Aluminum - The Influence of Surface Conditions on the Directional Effect," International Journal of Heat and Mass Transfer, Vol. 11, 1968, pp. 1371-1381.

[8]Somers, R. R. II, Miller, J. W., and Fletcher L. S., "The Thermal Contact Conductance of Dissimilar Metals," AIAA Paper 78-873, 2nd AIAA/ASME Thermophysics and Heat Transfer Conference, Palo Alto, Calif., May 1978.

[9]Clausing, A. M., "Heat Transfer at the Interface of Dissimilar Metals - The Influence of Thermal Strain," International Journal of Heat and Mass Transfer, Vol. 9, 1966, pp. 791-801.

[10]Barber, J. R., "The Effect of Thermal Distortion on Constricting Resistance," International Journal of Heat and Mass Transfer, Vol. 14, 1971, pp. 751-766.

Chapter V. Finite-Element Analysis Techniques

Finite-Element Thermal Analysis of Structures with Reusable Surface Insulation

Earl A. Thornton* and Kumar K. Tamma†
Old Dominion University, Norfolk, Va.

Abstract

A finite-element approach for efficient thermal analysis of structures with thermal protection systems (TPS) is described. The elements are developed for applications to reusable surface insulation (RSI), but the approach is applicable to other TPS. New two-dimensional finite elements are developed to model conduction in the TPS and supporting structure. A TPS/structural element predicts transient nonlinear temperature variations through the TPS thickness and structural temperatures at the TPS/structure interface. The performance of the TPS/structural element is demonstrated for a Shuttle wing truss subjected to reentry heating. The results indicate that the approach predicts detailed structural temperature distributions with significant reductions in model size and computer time. The approach offers potential for extension to more complex three-dimensional structures.

Nomenclature

A_s = cross-sectional area of structural member

Presented as Paper 82-0835 at the 3rd AIAA/ASME Joint Thermophysics, Fluids, Plasma & Heat Transfer Conference, St. Louis, Mo., June 7-11, 1982.

*Associate Professor, Mechanical Engineering and Mechanics Department.

+Research Assistant, Mechanical Engineering and Mechanics Department. Currently Assistant Professor, Mechanical and Aerospace Engineering Department, West Virginia University, Morgantown, W. Va.

[B] = temperature gradient interpolation matrix
c = specific heat
[C] = finite-element capacitance matrix
E = modulus of elasticity
F_T = equivalent thermal force, see Eq. (9)
$k_{Tx,Ty}$ = thermal conductivities of thermal protection system (TPS)
k_s = thermal conductivity of structure
[k] = thermal conductivity matrix
[K] = finite-element conductance matrix
h = thickness of TPS element, see Fig. 5
I,J,K,L = finite-element node numbers
ℓ = length of TPS element, see Fig. 5
m = $k_{Ty} w / k_s A_s b$, see Eq. (6)
n = number of subelements per finite element
[N] = finite-element interpolation function matrix
q = aerodynamic heating rate
{Q} = finite-element heat load vector
S = aerodynamic heating surface
t = time
T = temperature
T_{ref} = reference temperature for zero thermal force
u,v = displacement components
V = subelement volume
w = width of TPS element, see Fig. 4
x,y,z = Cartesian coordinates
α = coefficient of thermal expansion
ξ,η = nondimensional finite-element coordinates, see Eq. (4)
ρ = density

Subscripts

e = finite-element matrix or vector
s = structural
T = thermal protection system (TPS)

Introduction

The recent flights of Columbia have focused attention on the importance of the reusable surface insulation (RSI) thermal protection system (TPS). The

RSI tile system received much preflight publicity owing to mechanical problems associated with installation on the Shuttle vehicle.[1] The excellent performance of the TPS in the Shuttle's flights has validated the fundamental effectiveness of the system. However, research continues on the development of a TPS alternative to the RSI system, and one such concept,[2] a multiwall metallic TPS, is currently under investigation at the NASA Langley Research Center.

Less attention has been focused on the thermal analysis problems associated with the development of TPS for current and future space transportation. Yet, the addition of a TPS to a complex aerospace structure creates an analysis problem which taxes the capabilities of modern computers and thermal analysis programs. Furthermore, accurate thermal analysis is critical in evaluating the performance of the TPS and assessing the thermal structural response of the metallic flight structure. Thermal analysis of the combined TPS/vehicle structural system is difficult because the response is inherently nonlinear and transient, and the models are complex owing to the three-dimensional TPS/structural geometry. In addition, the effectiveness of the TPS causes the thermal transient to have a quite long duration, making analyses very expensive. For example, Shuttle structural components do not reach maximum temperatures until shortly before touchdown, which requires a re-entry thermal analysis for over 50 min. Research programs are currently underway at the NASA Langley Research Center to improve both the efficiencies and capabilities of thermal analysis methods for the analysis of space transportation vehicles. A significant portion of the research is devoted to improving finite-element thermal analysis methods, since the finite-element method is capable of both thermal and structual analysis.[3,5]

The purpose of this paper is to describe a finite-element approach currently under development for efficient, effective thermal analysis of structures with thermal protection systems. Past modeling techniques and thermal analysis approaches for the Shuttle TPS/wing structure are first discussed to illustrate current capabilities and analysis difficulties. Basic requirements for an effective thermal/structural analysis method are then described. Next, features of the finite-element TPS/structural thermal analysis approach are presented. Two TPS finite elements under development for effective modeling of Shuttle-type

TPS/structures are then described. Finally, the performance of the finite-element TPS/structural thermal analysis approach is demonstrated by analyzing three two-dimensional TPS/structural models by 1) highly detailed models with conventional finite elements, and 2) models employing the TPS finite elements.

Thermal Modeling of Shuttle Wing Structure

The Shuttle wing structure[6] is shown in Fig. 1. The aluminum structure uses Pratt trusses for the wing ribs and corrugated webs for the spars. The upper and lower wing surface covers are reinforced by stringers parallel to the spars. The RSI thermal protection system is cemented to the coverplates using a strain isolator pad; further details of the TPS are given in Ref. 1. Because of the complexity of the three-dimensional TPS/structural heat transfer, in design studies a complete wing thermal analysis was not tractable. Instead, a large number of one-dimensional "plug" models were used in Shuttle design and development. The location and details of a typical "plug" model are shown in Fig. 2. For the associated thermal stress analysis, the temperatures were interpolated between plug models. Uncertainties in the accuracy of the interpolated temperatures have raised questions about the validity of the "plug" model approach.

A finite-element model of the complete Shuttle wing (Fig. 3) used in an evaluation study[7] of transient algorithms is also useful to illustrate analytical difficulties. The model shown was originally developed as a structural dynamics model, but was adapted for the thermal algorithm study by adding conduction elements for heat transfer through the TPS to the structural joints. The complete model contained 2508 nodes, 1400 one- and two-dimensional conduction elements in the structure, and 2700 solid conduction elements in the TPS. The aerodynamic heating on the wing was represented by a time-dependent temperature specified on the external surface of the insulation on the underside of the wing. The figure shows the aerodynamic surface temperature history and the predicted very slow response of the structure. The solution time of 8600 s for this relatively crude model is indicative of the challenge involved in a realistic analysis of the full three-dimensional structure.

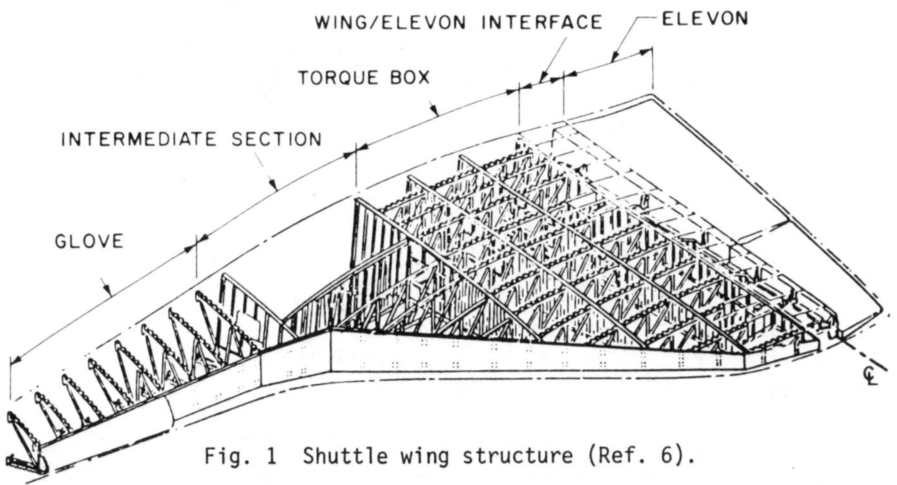

Fig. 1 Shuttle wing structure (Ref. 6).

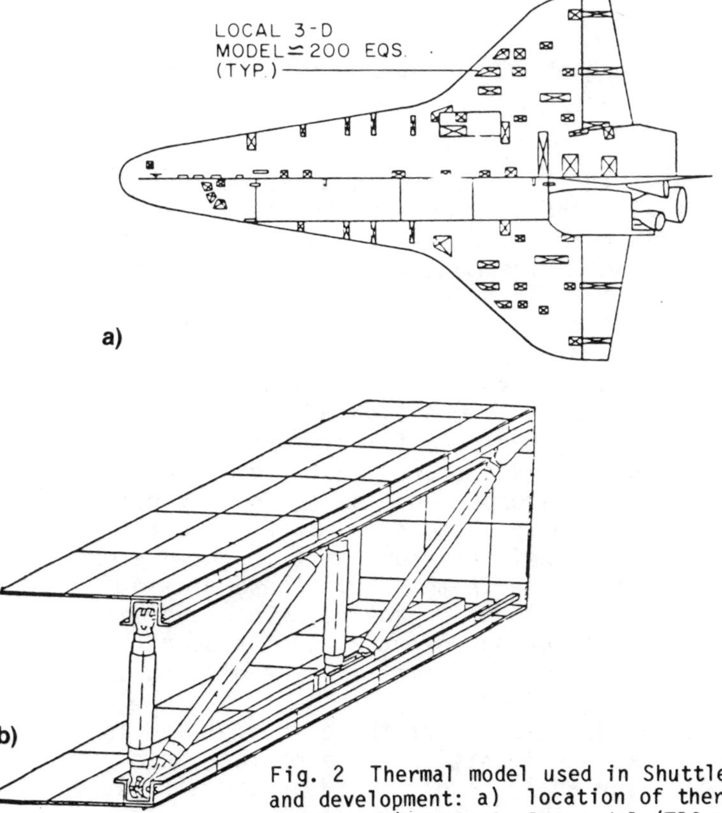

Fig. 2 Thermal model used in Shuttle design and development: a) location of thermal "plug" models; b) typical plug model (TPS not shown).

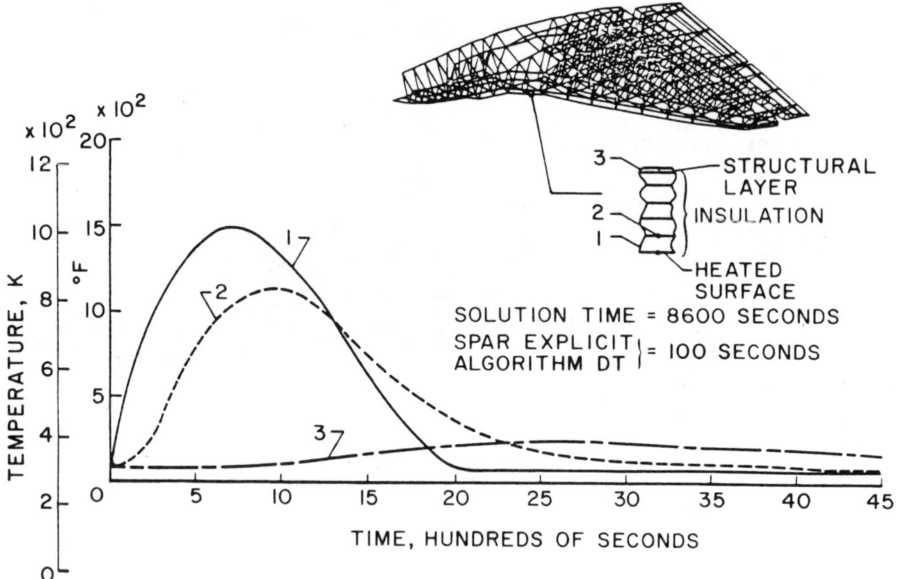

Fig. 3 Transient thermal response from NASA algorithm study (Ref. 7).

A model (Fig. 4) employed in a recent analysis[8] of the Shuttle wing thermal response for comparison with flight data is more indicative of the detail required in a realistic model. The model, which accounts for spanwise heat flow and effects of a rib truss in a wing segment, has 920 nodes. Conduction heat transfer in the structure was modeled with one- and two-dimensional elements and, in the TPS, with layers of three-dimensional elements. Comparison of predicted temperatures (not shown) and flight data showed good agreement. The analyses showed that radiation heat transfer between the upper and lower wing coverplates is significant. The models shown in Figs. 3 and 4 employed numerous three dimensional isoparametric solid elements to model heat transfer in the TPS. These elements caused the analyses to be expensive, since thermal properties varied during re-entry, and frequent recomputation of the isoparametric elements via Gauss integration was required.

The brief description of three thermal models of the Shuttle TPS has identified some of the basic requirements for analysis of a TPS/structural system during re-entry. The analysis must 1) compute

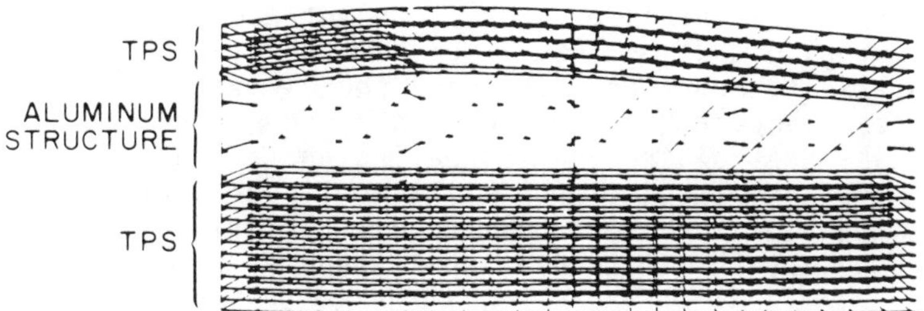

Fig. 4 Finite-element thermal model used in NASA Dryden study of Shuttle re-entry (Ref. 8).

temperatures and heat transfer for a thermal transient of quite long duration, 2) include variable thermal properties and radiation heat transfer, 3) model a complex three-dimensional TPS/structural geometry, 4) compute the heat transfer between the TPS and vehicle structure in sufficient detail to provide detailed structural temperature distributions for a thermal-stress analysis, and 5) efficiently interface with the structural analysis model. In order to meet these requirements using current computing technology, models of only TPS/structural segments are tractable and computational costs are high. High computer costs prohibit detailed three-dimensional thermal analysis for an entire vehicle component, such as a wing. If, however, a three-dimensional thermal analysis could be performed, the associated thermal-stress analysis would be tractable because 1) the structual model is relatively small compared to the thermal model (e.g., compare the number of TPS vs structural elements in Fig. 4), and 2) the thermal-stress analysis can be carried out as a succession of linear static analyses, i.e., quasistatic analyses, during the thermal transient.

TPS/Structure Finite Elements

The goal of the present research is to develop improved thermal models of the TPS/structural system which will eventually permit a complete three-dimensional analysis of a large structural component. The approach focuses on developing new finite-element capability for more effectively modeling heat transfer in the TPS. In this phase of the research, two-dimensional models are being employed to develop new TPS elements. Initial efforts have concentrated on developing conduction TPS elements which have the capability to model both the very sharply varying nonlinear temperature distributions through the TPS thickness due to aerodynamic heating and at the same time model the smoother but still nonlinear variation of the temperature at the TPS/structural interface which occurs because of structural heat transfer.

TPS Finite Elements

Two finite-element models for a TPS are shown in Fig. 5. Both of the elements have four nodes; nodes I and J connect the TPS elements to the structure, and

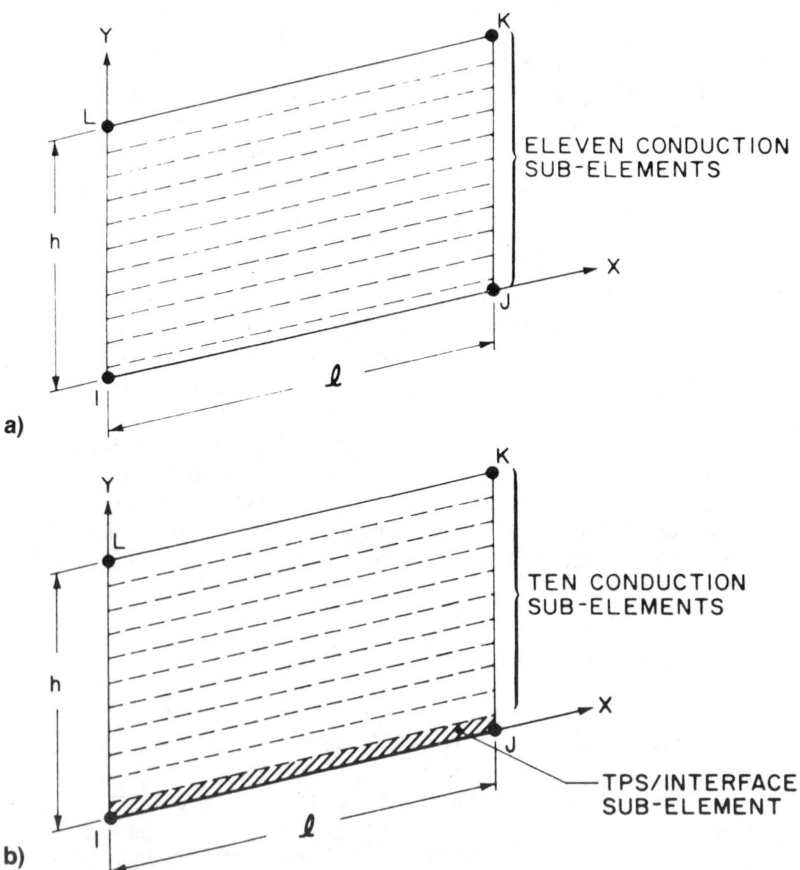

Fig. 5 TPS finite elements: a) TPS element with 11 subelements; b) TPS/structure element.

nodes L and K lie on the aerodynamic surface. The element length is ℓ , and the element height is h, where h represents the thickness of the TPS. The elements are made up of layers of subelements through the TPS thickness; the dotted lines in Fig. 5 represent subelement interfaces. The elements shown have 11 subelements, although the number of subelements required to represent temperature variations in the y direction accurately depends on the TPS thickness. Reference 8 showed for the Space Shuttle TPS that 11 element layers were adequate for the lower wing surface, but only four element layers were required for the upper wing surface.

The TPS element shown in Fig. 5a has four unknown temperatures at nodes I, J, K, L and 20 interior unknown temperatures at the corners of the subelements. Each of the 11 conduction subelements employs the bilinear temperature distribution used in conventional rectangular finite elements. Because of the large number of unknowns through the thickness h, a very sharply varying nonlinear temperature distribution can be represented, but the element permits only a linear temperature variation in the x direction. The linear temperature in the x direction was assumed to make the TPS element compatible with conventional conduction elements that are used to model structural heat transfer.

The TPS/structural element shown in Fig. 5b also has four unknown nodal temperatures at nodes I, J, K, L and 20 interior unknown temperatures. The element employs ten conduction subelements with bilinear temperature distributions and one TPS/interface subelement. The TPS/interface subelement represents heat transfer in the thermal "boundary layer" at the TPS/structure interface. It represents heat transfer in a thin layer of the TPS and, in addition, conduction in the supporting structure shown by the heavy line in Fig. 5b. Further details of the TPS/interface element will be given in the next section. The use of the TPS/interface subelement permits the TPS/structure element shown in Fig. 5b to represent a sharp variation of temperature through the TPS thickness and a nonlinear variation of temperature at the TPS/structural interface simultaneously.

The finite-element equations for the TPS finite elements can be derived from the governing heat conduction equations by the method of weighted residuals. For transient thermal analysis, the equations for a typical element are

$$[C]_e\{\dot{T}\}_e + [K]\{T\}_e = \{Q\}_e \tag{1}$$

where the element matrices are assembled from contributions of the subelements

$$[C]_e = \sum^n [C] \tag{2a}$$

$$[K]_e = \sum^n [K] \tag{2b}$$

$$\{Q\}_e = \sum^n \{Q\} \tag{2c}$$

where n denotes the number of subelements. The subelement equations are

$$[C] = \int_V \rho c\{N\}[N]dV \tag{3a}$$

$$[K] = \int_V [B]^T[k][B]dV \tag{3b}$$

$$\{Q\} = \int_S q\{N\}dS \tag{3c}$$

where integration takes place over the volume V of the subelement and surface S for the aerodynamic heating q. All thermal parameters may be temperature dependent in general but for simplicity are assumed constant herein.

Element matrices for the conduction subelements are based on the bilinear interpolation functions for a four-node plane element,

$$[N] = [(1-\xi)(1-\eta) \;\; \xi(1-\eta) \;\; \xi\eta \;\; (1-\xi)\eta] \tag{4}$$

where ξ and η denote nondimensional element local coordinates. The subelements may be either rectangular or quadrilateral in shape. Rectangular subelements have been selected because the element matrices, Eqs. (3a-3c), can be evaluated in closed form, avoiding the computational expense of Gauss integration customarily used in isoparametric quadrilateral elements.

TPS/Interface Elements

Figure 6 describes pictorially a typical TPS/interface subelement heat-transfer model. Aerodynamic heating is conducted normally through the thickness of the thermal protection system and is transferred to the subelement through nodes 3 and 4. The heat from the bottom of the thermal protection system is transferred nonuniformly to the structural member, where it is conducted along its length. For unsteady heat transfer, an energy balance on a small segment of the thermal protection system and attached structure gives the two-dimensional governing differential equations for the temperature distribution $T_T(x,y,t)$ and $T_s(x,t)$, where $T_T(x,y,t)$ and $T_s(x,t)$ are the temperature distributions in the TPS and structure, respectively,

TPS:

$$k_{Tx}\, w \frac{\partial^2 T_T}{\partial x^2} + k_{Ty}\, w \frac{\partial^2 T_T}{\partial y^2} = \rho_T c_T w \frac{\partial T_T}{\partial t} \qquad (5a)$$

structure:

$$k_s\, A_s \frac{\partial^2 T_s}{\partial x^2} + k_{Ty}\, w \frac{\partial T_T}{\partial y}(x,0) = \rho_s c_s A_s \frac{\partial T_s}{\partial t} \qquad (5b)$$

Heat transfer in the x direction in the TPS is small in comparison to heat transfer in the y direction and is neglected. This assumption permits an analytical solution to the steady-state form of the Eqs. (5a) and (5b):

$$T_T(x,y) = [1-\frac{y}{b} \quad \frac{y}{b}\frac{x}{\ell} \quad \frac{y}{b}(1-\frac{x}{\ell})] \begin{Bmatrix} T_s \\ T_3 \\ T_4 \end{Bmatrix} \qquad (6a)$$

and

$$T_s(x) = [(c_1-\frac{c_2 s_1}{s_2}) \quad (\frac{s_1}{s_2}) \quad (\frac{x}{\ell} - \frac{s_1}{s_2}) \qquad (6b)$$

$$(1 - \frac{x}{\ell} + \frac{c_2 s_1}{s_2} - c_1)] \begin{Bmatrix} T_1 \\ T_2 \\ T_3 \\ T_4 \end{Bmatrix}$$

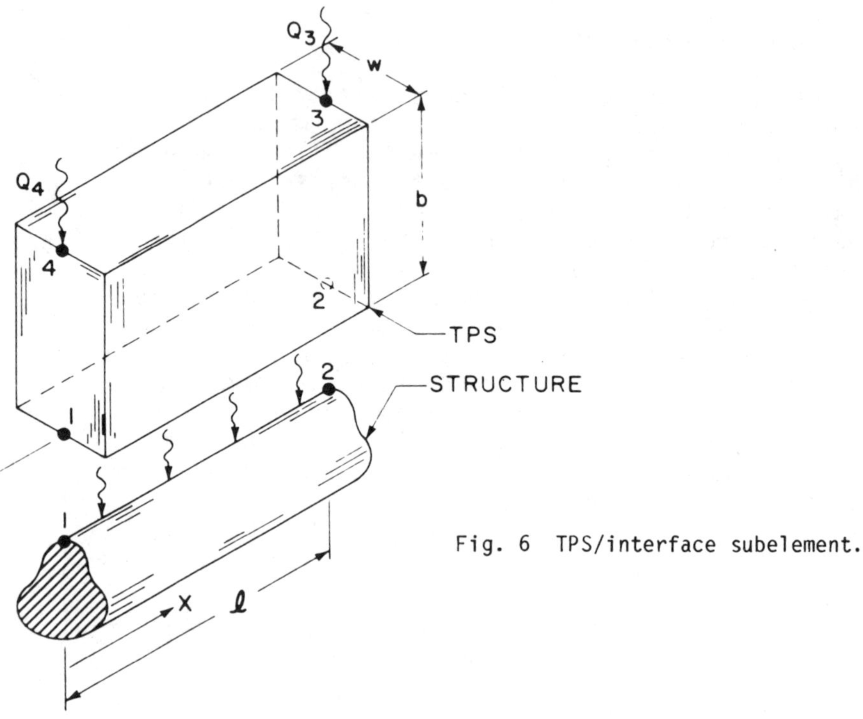

Fig. 6 TPS/interface subelement.

where $c_1 = \cosh\sqrt{m}x$; $c_2 = \cosh\sqrt{m}\ell$; $s_1 = \sinh\sqrt{m}x$; $s_2 = \sinh\sqrt{m}\ell$; and

$$m = k_{Ty}w/k_sA_sb$$

From the above analytical solution, the TPS and structure finite-element interpolation functions $[N_T]$ and $[N_s]$ can be expressed in terms of temperatures at nodes 1-4. Following the method of weighted residuals, the finite-element matrices for the coupled Eqs. (5a) and 5b) are derived as

TPS:

$$[C_T]\{\dot{T}_T\} + [K_T]\{T_T\} + [K_{TB}]\{T_T\} = \begin{Bmatrix} Q_1 \\ Q_2 \\ Q_3 \\ Q_4 \end{Bmatrix} \tag{7a}$$

$$[C_T] = \int_0^b\int_0^\ell \rho_T c_T w\{N_T\}[N_T]dxdy \tag{7b}$$

$$[K_T] = \int_0^b\int_0^\ell k_{Ty}w\{\frac{\partial N_T}{\partial y}\}[\frac{\partial N_T}{\partial y}]dxdy \tag{7c}$$

$$[K_{TB}] = \int_0^{x} k_{Ty} w \{N_T\}[N_T] dxdy \qquad (7d)$$

structure:

$$[C_s]\{\dot{T}_s\} + [K_s]\{T_s\} + [K_{sB}]\{T_T\} = \begin{Bmatrix} Q_1 \\ Q_2 \end{Bmatrix} \qquad (8a)$$

$$[C_s] = \int_0^{\ell} \rho_s c_s \{N_s\}[N_s] dx \qquad (8b)$$

$$[K_s] = \int_0^{\ell} k_s A_s \, [B_s]^T \, [B_s] dx \qquad (8c)$$

$$[K_{sB}] = \int_0^{\ell} k_{Ty} w \{N_s\}\left[\frac{\partial N_T}{\partial y}\right]_{y=o} dx \qquad (8d)$$

The interpolation functions based on Eq. (6) are used to evaluate the TPS/interface subelement matrices given in Eqs. (7-8). Subelement matrices were obtained in closed form by analytically evaluating the integrals.

The TPS/interface subelement when used for steady-state heat transfer with appropriate boundary conditions predicts exact nodal temperatures and the exact temperature distribution within the TPS and structure. For unsteady heat transfer, the subelement predicts approximate nodal and element temperature distributions, but with accuracy corresponding to higher-order polynomial functions. An alternate approach could be to use a quadratic or cubic variation of temperature along the structural member but additional interior nodes would be required on the x axis in Fig. 6. An advantage of the formulation above is that the TPS/structural thermal element is directly compatible with structural elements used in thermal-stress analysis. To facilitate the thermal-stress analysis, the structural equivalent thermal force can be calculated from

$$F_T(t) = \frac{E\alpha}{\ell} \int_0^{\ell} [T_s(x,t) - T_{ref}] dx \qquad (9)$$

where $T_s(x,t)$ is given by Eq. 6b with time-dependent nodal temperatures, and T_{ref} is the reference temperature for zero thermal force.

In results to be presented in the applications section that follows, only the TPS/structural element is used. Comparative results to be presented show this element to be very effective for analysis neglecting radiation heat transfer. Radiation heat transfer is neglected for simplicity in these analyses that demonstrate the concepts of the finite-element approach.

Applications

The performance of the finite-element TPS/structural thermal analysis approach is demonstrated by analyzing three two-dimensional TPS/structural models. Where possible, the effectiveness of the approach is evaluated by comparison with solutions based on a refined mesh of conventional finite elements. The models analyzed are based on dimensions, properties, and heating histories representative of a Space Shuttle wing truss. The four-bay wing truss analyzed is shown in Fig. 7. The models employed are for 1) a semi-bay, 2) a complete bay, and 3) the complete truss. A thermal model of conventional two-dimensional conduction elements is shown for one semi-bay on Fig. 7a. One-dimensional two-node rod elements represent conduction in the structure. The corresponding thermal model with

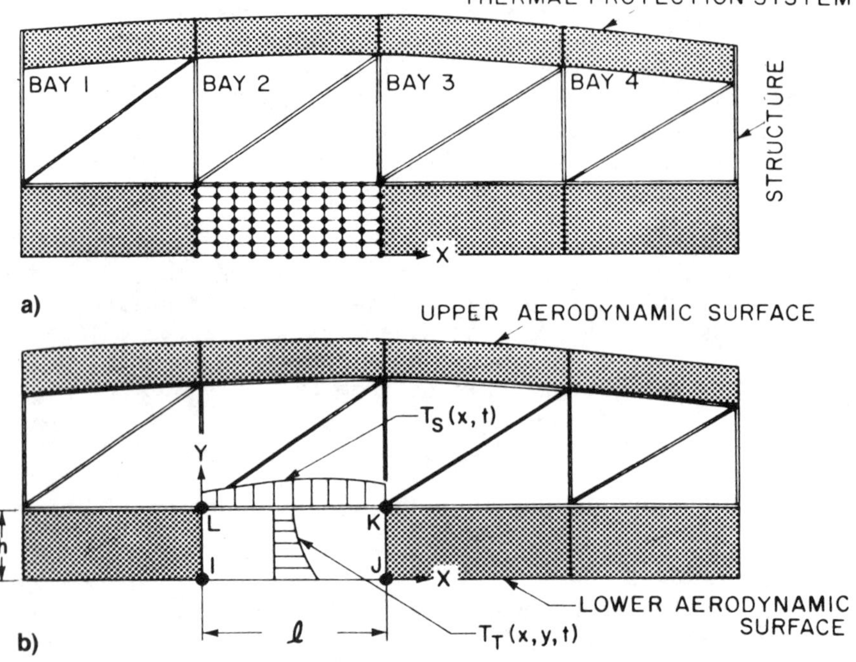

Fig. 7 Comparison of conventional and TPS element models of wing section: a) typical detailed conventional finite-element thermal model; b) thermal model with typical TPS element.

in the structure. The corresponding thermal model with one TPS/structural element (Fig. 5b) for a semi-bay is shown in Fig. 7b. Re-entry heating in all examples is represented by specified temperature histories (Fig. 8) on the aerodynamic surfaces.

Radiation heat exchanges are neglected, and the linear transient equations are solved either by an implicit Crank-Nicholson algorithm or an explicit Euler algorithm. The implicit algorithm uses consistent capacitance matrices defined by Eq. (3a). The explicit algorithm uses lumped capacitance matrices computed by "lumping" the consistent matrix along the diagonal and setting the off-diagonal terms equal to zero. For all computations, the implicit algorithm used a time step of 50 s, and the explicit algorithm used a time step of 4 s. The explicit algorithm is used in a special code developed for TPS elements that allows the user to input only TPS element nodal data, thereby greatly reducing input data preparation.

Semi-Bay Model

A semi-bay consisting of four truss members and a section of the TPS is analyzed using conventional and TPS/structural element models similar to those shown in Figs. 7a and 7b, respectively. The thermal response of the semi-bay model was first computed using a refined mesh of 110 conventional quadrilateral elements with 11 element layers through the TPS thickness, and 13 rod elements using the implicit transient algorithm. Then the response was computed with one TPS/structural element and three rod elements using both implicit and explicit algorithms. The refined mesh of conventional elements required solving 132 equations, and the TPS/structural element model required solving 24 equations. The semi-bay transient thermal response is presented in Figs. 9a-9c.

Transient temperature distributions through the TPS thickness (Fig. 9a) show excellent agreement. The results demonstrate that the 11-layer TPS element is capable of representing thickness temperature variations accurately, and there is no significant degradation of accuracy introduced by the lumped capacitance matrices employed in the explicit algorithm. The temperature distributions in the horizontal truss member supporting the TPS are shown in Fig. 9b at t = 3000 s. Structural temperatures are a maximum at this time, which is close to the Shuttle touchdown. Temperatures computed by the

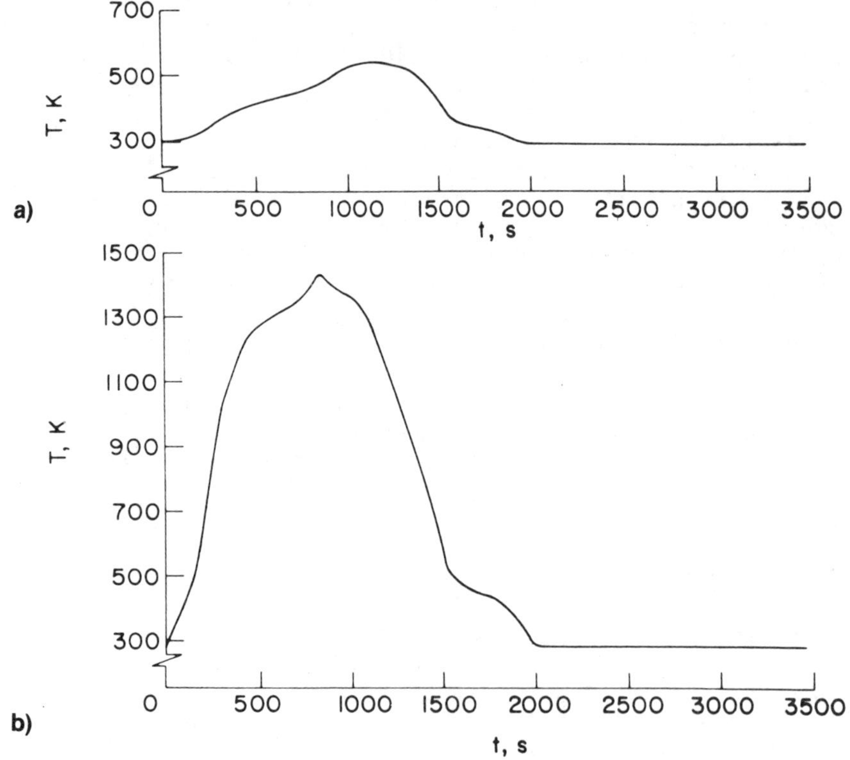

Fig. 8 Prescribed aerodynamic temperature histories: a) upper surface; b) lower surface.

TPS/structural element with either the implicit or explicit algorithmn show good agreement with the conventional refined mesh solution. The member center temperature is underestimated slightly, and the member end temperatures are overestimated. Member temperatures predicted by the explicit algorithm show the largest differences. For thermal-stress analysis, these differences are not significant, since structural member thermal forces depend on the average member temperature as indicated by the temperature integral that appears in the thermal force, Eq. (9). The truss member thermal force histories shown in Fig. 9c show excellent agreement throughout the response, verifying that the average member temperatures are predicted satisfactorily.

The semi-bay model validated the basic capability of the TPS/structural element and established that the explicit algorithm can be used with only an insignificant loss in accuracy. The explicit algorithm offers significant savings in computer time that will be demonstrated in the full-bay model.

Full-Bay Model

A full-bay model consisting of five truss members and the TPS on the upper and lower wing surfaces was analyzed using discretizations similar to those used in the semi-bay model. The conventional element model used 220 quadrilateral elements and 23 rod elements; the TPS element model used two TPS/structural elements and three rod elements. The conventional model required solving 264 equations, and the TPS element model required solving 48 equations. Aerodynamic heating on the upper

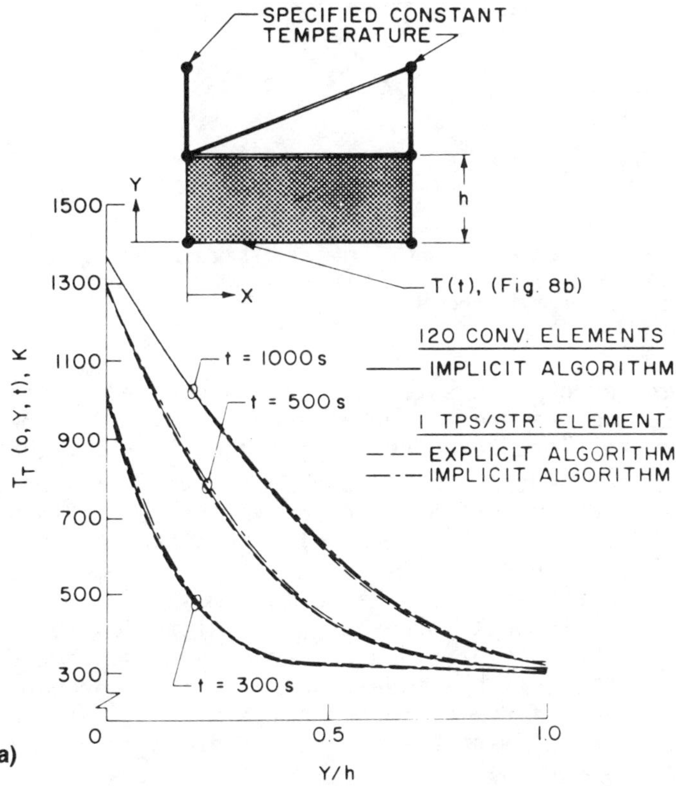

Fig. 9 Thermal response of a semi-bay model: a) TPS thermal response; b) structural temperature distribution, t = 3000 s; c) thermal force history.

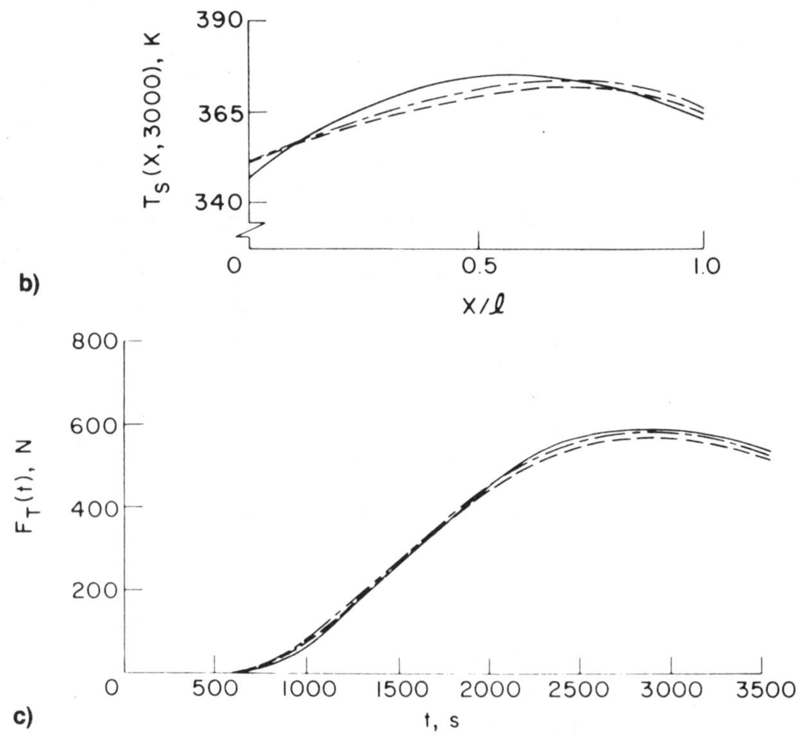

Fig. 9 (cont.) Thermal response of a semi-bay model: a) TPS thermal response; b) structural temperature distribution, t = 3000 s; c) thermal force history.

and lower aerodynamic surfaces was represented by the specified temperature histories shown in Fig. 8. The full-bay transient response is presented in Figs. 10a-10c.

The transient thermal response at two points on the upper and lower wing TPS/structure interfaces is shown in Fig. 10a. Temperature histories predicted by the simpler TPS element model show excellent agreement with the response predicted by the refined mesh of conventional elements. Thermal force histories for the horizontal truss members are compared in Figs. 9b and 9c. As in the semi-bay model, the excellent agreement for the thermal force histories demonstrates that the TPS/structural element predicts average member temperatures very well.

The transient solution of 264 equations in the conventional model by the implicit algorithm required 260 CPU seconds (CPUs); the transient solution of 48 equations in the TPS model by the explicit algorithm

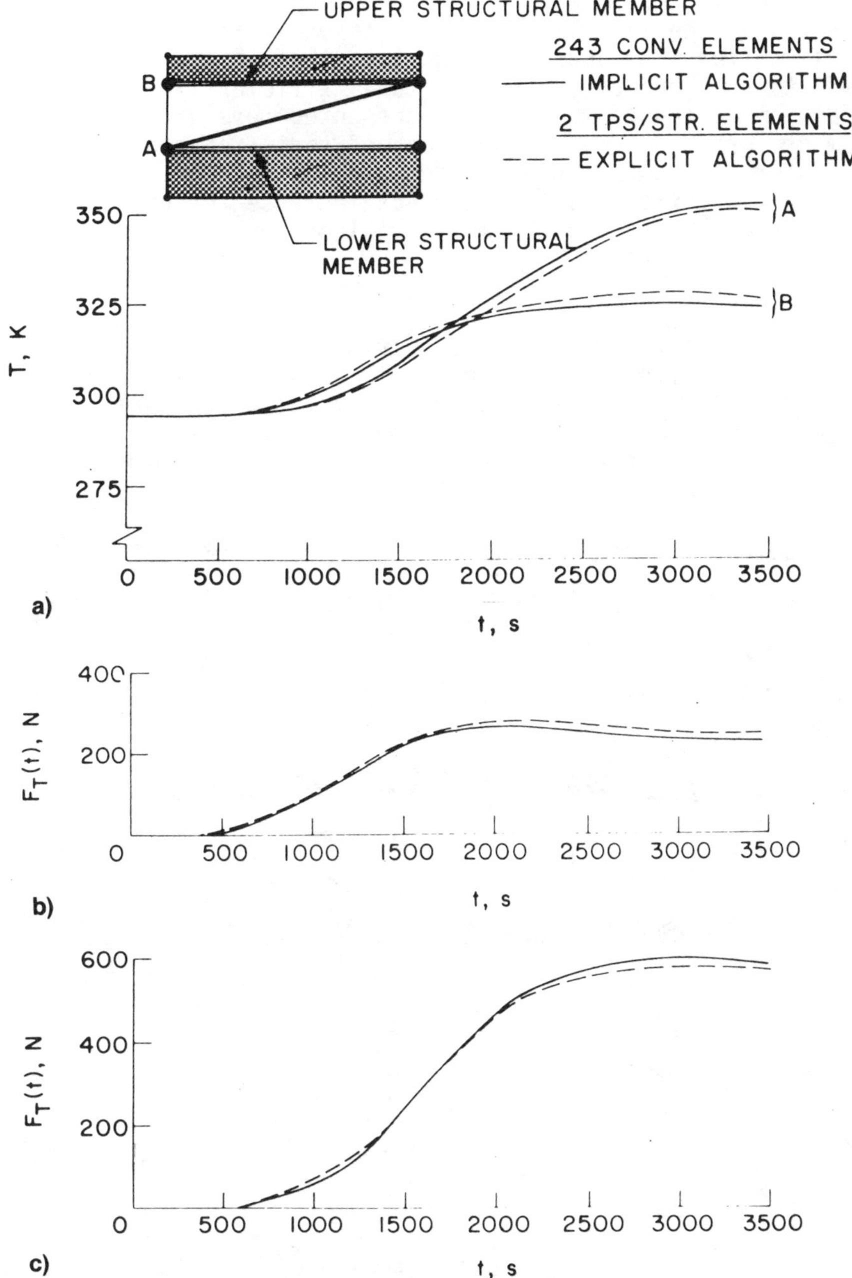

Fig. 10 Thermal-structural response of full-bay model: a) temperature histories; b) upper structural member thermal force history; c) lower structural member thermal force history.

required 105 CPUs. The reduction in computer time offered by the TPS elements with the explicit algorithm is significant because it means that the approach can be used to analyze multi-bay structures efficiently. In addition, although the results presented herein neglect temperature-dependent properties and radiation heat transfer, these nonlinear effects can be included in the explicit algorithm without a great increase in computer time in comparison to an implicit algorithm.

Complete Truss Model

To demonstrate the efficiency of the approach for a multi-bay analysis, the complete truss was thermally and structurally analyzed. Based on previous experience, a solution with a refined conventional finite-element model (1056 nodes and 969 elements) is estimated to

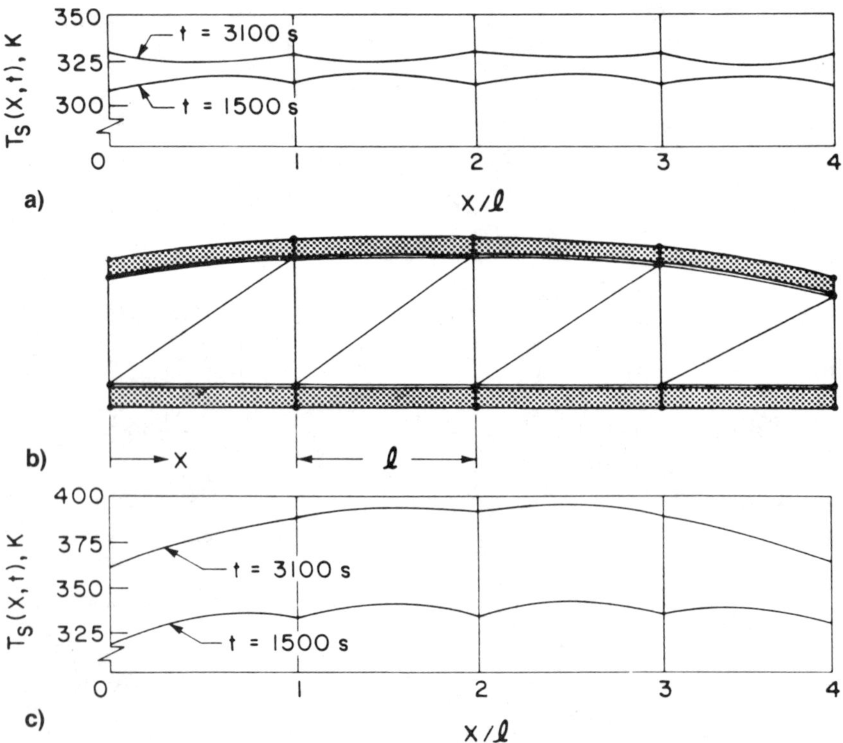

Fig. 11 Thermal-structural response of wing truss: a) upper structural member's temperature distribution; b) truss with TPS; c) lower structural member's temperature distribution.

require about 750 CPUs. Because of the large estimated cost of the conventional analysis and the good agreement demonstrated previously, the complete truss was analyzed using only the TPS element approach. The complete truss thermal model has 20 nodes, 8 TPS/ structural elements, and 9 rod elements. In the transient thermal analysis, 120 equations were solved by the explicit algorithm. The structural model represented the truss with 17 two-node rod elements. The structural response was computed every 100 s in the transient reponse. The thermal analysis required 155 CPUs of computer time, and the quasistatic structural analysis with 20 unknown displacements required 5 CPUs. The relatively high computer time required for the thermal analysis compared to the structural analysis illustrates the importance of an efficient thermal analysis.

Temperature distributions along the upper and lower horizontal truss members are shown in Fig. 11 at two times in the transient response. The "scalloped" shape of these distributions occurs because of conduction heat transfer through the interior truss members. Early in the response, the horizontal truss members have higher temperatures near their centers because of aerodynamic heating conducted through the TPS, but the interior truss members act as heat sinks, causing horizontal

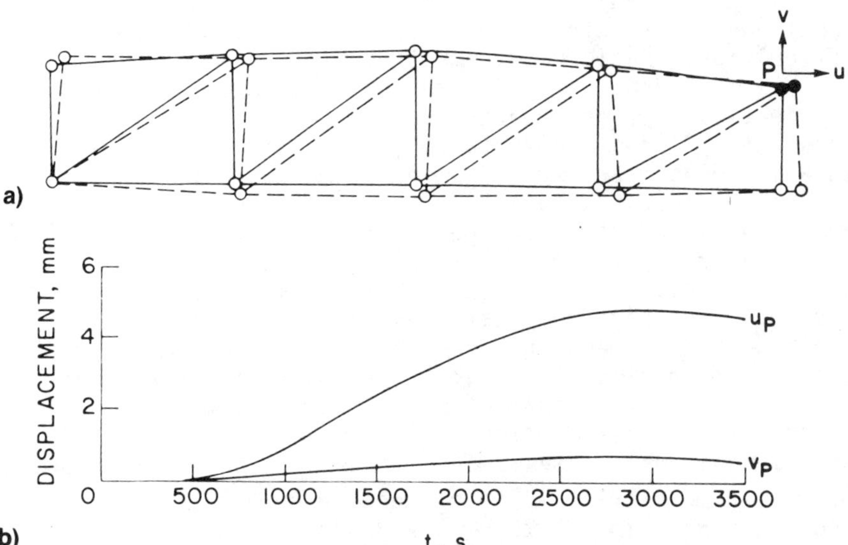

Fig. 12 Structural response of wing truss: a) deformed shape, t = 3100 s; b) typical displacement histories.

member temperatures to drop at joints. Later in the response, after aerodynamic heating has diminished, conduction heat transfer through the interior members causes the joint temperatures on the upper structural members to rise above the member center temperatures, yielding the inverted scalloped distribution. The scalloped temperature distributions at the early times in the response were predicted in Ref. 8 and agree well with flight data. However, the inverted scalloped distribution at later times was not predicted because of radiation heating of the upper surface by the hotter lower surface. The inverted scalloped temperature distribution is thus a deficiency in the present analysis, indicating the need to include radiation heat transfer.

The thermal-structural response of the truss is presented in Fig. 12. The truss is free to expand, and the results show that maximum displacements occur near Shuttle touchdown at 3000 s. Maximum elongations occur on the lower horizontal members of the truss, which experience the highest temperatures.

Concluding Remarks

A finite-element approach for efficient thermal analysis of structures with thermal protection systems (TPS) is described. A review of past thermal modeling techniques illustrates current capabilities and analysis difficulties. Basic requirements for an effective thermal-structural analysis are the capability to 1) efficiently compute temperatures in a long duration transient, 2) include variable thermal properties and radiation heat transfer, 3) model complex three-dimensional TPS/structural geometry, 4) compute temperatures in sufficient detail for the thermal-structural analysis, and 5) efficiently interface with the structural analysis model. Current modeling techniques and high computational costs prohibit detailed three-dimensional thermal analysis of large vehicle components. The goal of the present research is to develop improved finite-element thermal models of a TPS/structural system which will lead to a complete three-dimensional analysis of a large structural component.

Two-dimensional finite elements are developed for modeling conduction in the TPS and supporting structure. The elements are based on using layers of subelements through the TPS thickness to represent highly nonlinear thickness temperature gradients. A TPS/structural element uses an interface subelement to represent heat transfer between the TPS and the supporting structure. The interface subelement employs interpolation functions based on an analytical solution for steady-state heat transfer. To gain efficiency in computations, rectangular subelements are used and subelement matrices are evaluated in closed form.

The performance of the finite-element TPS thermal analysis approach is demonstrated for three two-dimensional models of a Space Shuttle wing truss subjected to re-entry heating. Two segments of the truss are analyzed by models with 1) TPS/structural elements, and 2) detailed models of conventional elements. Results show that the TPS finite-element approach predicts TPS and structural temperature distributions and member thermal forces accurately. Comparisons of TPS finite-element predicted temperatures by implicit and explicit algorithms showed that the faster explicit algorithm can be used with no significant loss in accuracy. Transient thermal analysis of the complete truss with the TPS element approach demonstrates the capability to efficiently model and analyze a complete structure. Temperature distributions for the complete truss analysis are in qualitative agreement with other Shuttle analyses for early times in the response, but comparisons of temperatures at later response times suggest the need to include internal radiation heat transfer.

The applications demonstrate that the TPS finite-element approach can simplify thermal models and reduce computational time significantly. The results suggest that the approach extended to three-dimensional elements offers the opportunity for long duration thermal-structural transient analysis of major vehicle components with TPS at acceptable costs.

References

[1]Cooper, P. A. and Holloway, P. F., "The Shuttle Tile Story," Astronautics and Aeronautics, Vol. 19, Jan. 1981, pp. 24-34.

[2]Shideler, J. L, Kelley, H. N., Avery, D. E., Blosser, M. L., and Adelman, H. M., "Multiwall TPS--An Emerging Concept," AIAA Paper 81-0586, 22nd Structures, Structural Dynamics and Material Conference, Atlanta, Ga., April 1981, AIAA CP811, pp. 349-356.

[3]Thornton, E. A., Dechaumphai, P., and Wieting, A. R., "Integrated Thermal-Structural Finite Element Analysis," AIAA Paper 80-0717, 21st Structures, Structural Dynamics and Materials Conference, Seattle, Wash., May 1980, AIAA CP804, Pt. 2, pp. 957-969.

[4]Thornton, E. A., Dechaumphai, P., Wieting, A. R., and Tamma, K. K., "Integrated Transient Thermal-Structural Finite Element Analysis," AIAA Paper 81-0480, 22nd Structures, Structural Dynamics and Materials Conference, Atlanta, Ga., April 1981, AIAA CP811, pp. 16-32.

[5]Mahaney, J., Thornton, E.A., and Dechaumphai, P., "Integrated Thermal-Structural Analysis of Large Space Structures," NASA CP-2216, Nov. 1981, pp. 179-198.

[6]Naps, N. and Zacher, M. J., "Design Development of the Space Shuttle Wing," AIAA Paper 80-0727, 21st Structures, Structural Dynamics and Materials Conference, Seattle, Wash., May 1980, AIAA CP804, Pt. 1, pp. 371-376.

[7]Adelman, H. M. and Haftka, R. T., "On the Performance of Explicit and Implicit Algorithms for Transient Thermal Analysis of Structures," NASA TM-81880, Sept. 1980.

[8]Ko, W. L., Quinn, R. L., Gong, L., Schuster, L. S., and Gonzales, D., "Reentry Heat Transfer Analysis of the Space Shuttle Orbiter," NASA CP-2216, Nov. 1981, pp. 295-325.

[9]Huebner, K. H. and Thornton, E. A., The Finite Element Method for Engineers, 2nd ed., Wiley, New York, 1982.

Control-Volume-Based Finite-Element Formulation of the Heat Conduction Equation

G.E. Schneider* and M. Zedan†
University of Waterloo, Waterloo, Canada

Abstract

A control-volume-based finite-element formulation is presented for application to conduction heat-transfer problems. The specific case of a linear, quadrilateral, isoparametric finite element is considered. The control-volume formulation offers the advantage over the conventional formulations in that it permits a direct, physical interpretation of the governing algebraic equations through enforcement of the conservation constraint for finite control volumes distributed throughout the domain. In addition, it is observed that reduced computational times and improved accuracy may also result through application of the procedure. The control-volume-based formulation is demonstrated by application to several test problems and the results compared with those from a Galerkin, weighted residual formulation.

Nomenclature

c	=	specific heat
C	=	boundary condition constant
CV	=	control volume
E	=	total energy within CV
h	=	heat-transfer coefficient
$\hat{i},\hat{j}$	=	unit vectors in the x,y directions

Presented as Paper 82-0909 at 3rd AIAA/ASME Joint Thermophysics, Fluids, Plasma & Heat Transfer Conference, St. Louis, Mo., June 7-11, 1982.

*Associate Professor, Department of Mechanical Engineering.
†Graduate Student, Department of Mechanical Engineering.

J	=	Jacobian of transformation matrix
$k_{i,j}$	=	coefficient of stiffness matrix
k_x, k_y, k_n	=	conductivity in the x,y,n directions
n	=	normal
N_i	=	shape function
p	=	heat generation rate per unit volume
P	=	total CV heat generation rate
q	=	heat flux
Q	=	heat flow rate
R_i	=	right-hand side of algebraic equation
s	=	local coordinate
S	=	surface
t	=	local coordinate, or time
T	=	temperature
V	=	volume
x,y	=	Cartesian coordinates
ρ	=	density

Introduction

The complexity of modern engineering systems, the growing utilization of materials having directionally dependent thermal properties, and the increasing demand placed on thermal analysts for precision in their prediction of the operation of engineering systems are among the factors which have led to the widespread use of discrete methods in thermal analysis. Of the available discrete methods, the finite-difference and finite-element methods are the most commonly employed discrete methods in conduction heat-transfer analysis.

The emergence of discrete methods for solving partial differential equations effectively began when Richardson[1] presented his paper to the Royal Society. To obtain an approximate procedure, he directed his efforts at determining an approximate representation of the equation, rather than at obtaining an approximation to the energy balance represented by the equation. Approximations for derivatives were obtained from truncated Taylor series expansions of the temperature field in the vicinity of the point of interest. Consequently, this form of the method of finite differences has become known as a Taylor series formulation. While this form of the finite-difference method has historically been restricted to rectangular domains and meshes, Schneider et al.[2] extended the method to include applicability to general, orthogonal, curvilinear coordinate systems. More recently, Robertson[3] applied the Taylor series approach to problems involving general, nonorthogonal, curvilinear coordinate systems.

In recent years, however, the control-volume approach has gained considerable popularity among heat-transfer analysts and researchers.[4] In this approach, an energy balance is applied to discrete control volumes and only where surface fluxes require approximation are the Taylor series representations employed. It is a property of the control-volume approach that the resulting finite-difference equations are conservative; that is, the discrete equations maintain an accurate accounting of the energy flows through the domain by ensuring that the approximation for surface fluxes is independent of the side from which the surface of the flux is viewed. This approach has been extended to general, orthogonal, curvilinear coordinate systems by Schneider et al.[5] and, more recently, to general, nonorthogonal, curvilinear coordinate systems by Zedan and Schneider.[6]

In the application of the above procedures, the coordinate system, whether orthogonal or nonorthogonal, must be known a priori over the solution domain. The coordinate systems employed include the conventional Cartesian, circular cylinder, and spherical coordinate systems; the more exotic orthogonal systems[2,5] analytically defined nonorthogonal systems;[3,6] and systems obtained as the result of a numerical coordinate transformation.[7] One disadvantage inherent in the above finite-difference approaches is that the coordinate system must be defined over the entire solution domain prior to effecting the discrete method, while the control-volume formulations possess the advantage of being conservative in nature. The finite-element methods, however, provide a mechanism for removing the above disadvantage by utilizing a coordinate system which is local to each individual element.

The development of finite-element methods, however, has followed a philosophically different approach from that followed by finite-difference methods. In stress analysis, for example, there exists a variational extremum principle such that the minimization of the strain energy within a continuum domain leads, naturally, to the formulation of a discrete model. In conduction heat transfer, however, such a natural formulation with a clear physical interpretation does not exist. The search for a variational formulation to the heat conduction problem has, however, led to several quasi- or pseudovariational formulations.[8-10] The Galerkin method of weighted residuals is more commonly employed, in recent years, to formulate the finite-element model for heat conduction.[11] Generally, however, the finite-element equations for heat conduction are nonconservative in nature. This nonconservative nature can be viewed as a disadvantage of the method in that direct control of the energy flow is not available to the user. In particular, this is viewed as a disadvantage of the method by control-volume finite-difference practitioners.

It is the purpose of this paper, therefore, to provide a control-volume-based finite-element procedure for application to conduction heat-transfer analysis. Baliga and Patankar,[12] through the use of specialized interpolation functions, have presented a control-volume-based finite-element procedure for convection-diffusion problems. Their procedure, however, is applicable only to linear triangular elements. The present formulation, while applied to linear, quadrilateral, isoparametric finite elements, is directly extendable to other element configurations. In the remainder of the paper, the procedure is

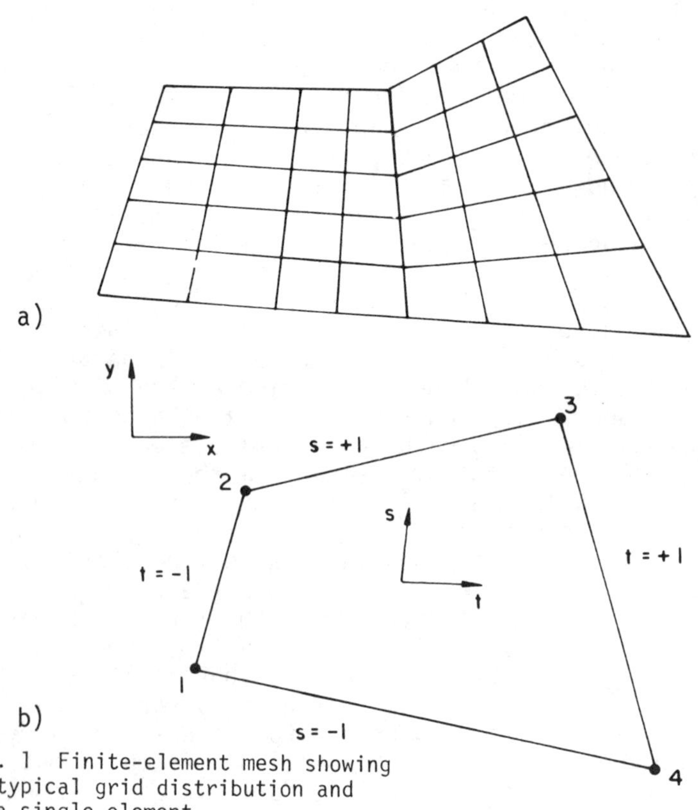

Fig. 1 Finite-element mesh showing a) typical grid distribution and b) a single element.

developed and application is made to several test problems to demonstrate the procedure.

Geometric Preliminaries

The domain illustrated in Fig. 1a is used to typify a solution domain which is subdivided into linear, quadrilateral, finite elements. A typical element is presented in Fig. 1b, where the local (s,t) coordinate system is shown. For the individual element, the local node numbers range from 1 through 4, as shown, and the temperature field T and the global coordinates (x,y) are expressed in the form[13]

$$T = \sum_{i=1}^{4} N_i T_i \tag{1}$$

$$x = \sum_{i=1}^{4} N_i x_i \tag{2}$$

$$y = \sum_{i=1}^{4} N_i y_i \tag{3}$$

where the shape functions N_i are given by

$$\begin{aligned} N_1 &= \tfrac{1}{4} (1-s)(1-t) \\ N_2 &= \tfrac{1}{4} (1+s)(1-t) \\ N_3 &= \tfrac{1}{4} (1+s)(1+t) \\ N_4 &= \tfrac{1}{4} (1-s)(1+t) \end{aligned} \tag{4}$$

From the expression for temperature, Eq. (1), the x and y derivatives of temperature can be determined as

$$\frac{\partial T}{\partial x} = \sum_{i=1}^{4} \frac{\partial N_i}{\partial x} T_i \tag{5}$$

and

$$\frac{\partial T}{\partial y} = \sum_{i=1}^{4} \frac{\partial N_i}{\partial y} T_i \tag{6}$$

Further, the x and y derivatives of the shape functions appearing in Eqs. (5) and (6) are determined in the usual fashion by

$$\begin{Bmatrix} \dfrac{\partial N_i}{\partial x} \\ \\ \dfrac{\partial N_i}{\partial y} \end{Bmatrix} = \frac{1}{\text{Det }[J]} \begin{bmatrix} \dfrac{\partial y}{\partial t} & \dfrac{-\partial y}{\partial s} \\ \\ \dfrac{-\partial x}{\partial t} & \dfrac{\partial x}{\partial s} \end{bmatrix} \begin{Bmatrix} \dfrac{\partial N_i}{\partial s} \\ \\ \dfrac{\partial N_i}{\partial t} \end{Bmatrix} \tag{7}$$

where

$$\text{Det }[J] = \frac{\partial x}{\partial s}\frac{\partial y}{\partial t} - \frac{\partial y}{\partial s}\frac{\partial x}{\partial t} \tag{8}$$

Finally, considering the general line segment of Fig. 2, being traversed in the direction from point 1 to point 2 in that figure, the outward normal $d\vec{S}$ as shown in the figure, can be expressed in the form

$$d\vec{S} = dy\ \hat{i} - dx\ \hat{j} \tag{9}$$

which will be valid irrespective of the orientation of the line segment provided that the signs of dx and dy are correct and correspond to the direction of traverse of the line segment.

The above relations are adequate to permit the derivation of the finite-element equations from a control-volume point of view. This will be performed in the following section.

Element Property Equations

Following a procedure analogous to that followed by the conventional formulation of the finite-element equations, a single, isolated, finite element will be first considered. This will enable element "property" relations to be established in the form of matrix equations expressing nodal temperature vs nodal heat flow dependencies. Appropriate assembly rules will then be employed to construct the global equation system from the elemental equations.

To derive the elemental equations, the isolated finite element illustrated in Fig. 3 is considered. In this figure, the element is subdivided into four internal control volumes, each of which can be associated with a corresponding node of the element definition vector. In the linear quadrilateral element shown in the figure, the control-volume boundaries are chosen to be coincident with

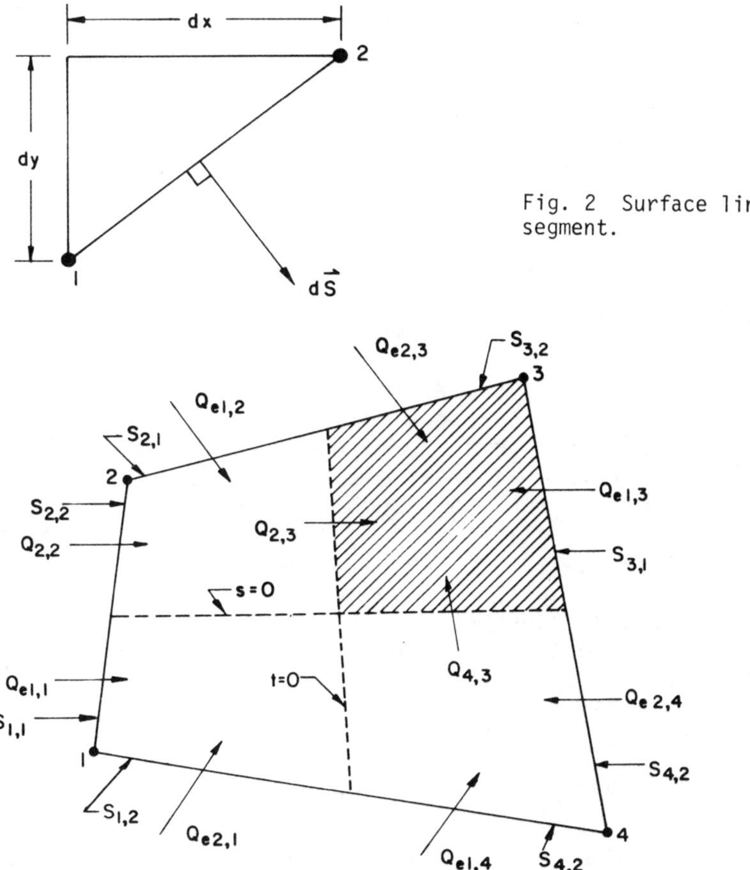

Fig. 2 Surface line segment.

Fig. 3 Single element with control-volume subdivision and heat flows.

the element exterior boundaries and with the local coordinate surfaces defined by s=0 and t=0. This latter choice is somewhat arbitrary, but, as will become apparent later in the assembly process, the choice must be consistent from element to element within the global structure. The particular selection of s=0 and t=0 serves this purpose well.

In words, an energy balance applied to a single control volume can be expressed as

[Net rate of conduction into control volume]

\+ [Rate of generation within control volume]

= Rate of change of energy within the control volume (10)

With reference to the control volume associated with node 3 of Fig. 3, this equation can be expressed mathematically in the form

$$Q_{2,3} + Q_{4,3} + Q_{e1,3} + Q_{e2,3} + \iint_{CV} p dV$$

$$= \frac{\partial}{\partial t} \iint_{CV} \rho\, c\, T\, d\, V \tag{11}$$

where the integration limits are those corresponding to the control volume associated with node 3. The subscripts el and e2 in Eq. (11) refer to energy flows into control volume 3 through surfaces which are on the exterior of the element and which emanate either from the physical domain boundary or from adjacent elements. These heat flows will be considered when the assembly procedure is effected or when the application of boundary conditions is considered. The remaining terms of Eq. (11) are considered, individually, in the discussion below.

The heat flows within the element are considered first. In general, the heat flow through a surface can be expressed in the form

$$Q = - \int_S \vec{q} \cdot \vec{dS} \tag{12}$$

where the heat flux vector $\vec{q}$ is given as

$$\vec{q} = -k_x \frac{\partial T}{\partial x} \hat{i} - k_y \frac{\partial T}{\partial y} \hat{j} \tag{13}$$

Using the expression for an element of surface from Eq. (9), and noting that the surface is traversed in a counter-clockwise sense, the evaluation of $Q_{2,3}$, for example, becomes

$$Q_{2,3} = \int_{s=1}^{s=0} \left\{ -k_y \frac{\partial T}{\partial x} dy + k_y \frac{\partial T}{\partial y} dx \right\} \Bigg|_{t=0} \tag{14}$$

where the path of integration has been indicated. Along this path, since t=0, the differentials dx and dy become functions of the single coordinate s. Thus the integral expressed by Eq. (14) becomes simply expressible as

$$Q_{2,3} = -\int_{s=0}^{1} \left[k_x \frac{\partial T}{\partial x}\frac{\partial y}{\partial s} - k_y \frac{\partial T}{\partial y}\frac{\partial x}{\partial s} \right] \Bigg|_{t=0} ds \tag{15}$$

Utilizing the relations for the derivatives of temperature, as given by Eqs. (5) and (6), Eq. (15) can be written as

$$Q_{2,3} = -\sum_{i=1}^{4} \left\{ \int_{0}^{1} \left[k_x \frac{\partial N_i}{\partial x}\frac{\partial y}{\partial s} - k_y \frac{\partial N_i}{\partial y}\frac{\partial x}{\partial s} \right] \Bigg|_{t=0} ds \right\} T_i \tag{16}$$

where Eqs. (7) and (8) can be used to evaluate the derivatives $\partial N_i/\partial x$ and $\partial N_i/\partial y$. In a similar manner, the heat flow $Q_{4,3}$ can be determined to be

$$Q_{4,3} = +\sum_{i=1}^{4} \left\{ \int_{0}^{1} \left[k_x \frac{\partial N_i}{\partial x}\frac{\partial y}{\partial t} - k_y \frac{\partial N_i}{\partial y}\frac{\partial x}{\partial t} \right] \Bigg|_{s=0} dt \right\} T_i \tag{17}$$

For completeness, the heat flows $Q_{2,1}$ (i.e., from volume 2 into volume 1) and $Q_{4,1}$ are given here as

$$Q_{2,1} = -\sum_{i=1}^{4} \left\{ \int_{-1}^{0} \left[k_x \frac{\partial N_i}{\partial x}\frac{\partial y}{\partial t} - k_y \frac{\partial N_i}{\partial y}\frac{\partial x}{\partial t} \right] \Bigg|_{s=0} dt \right\} T_i \tag{18}$$

and

$$Q_{4,1} = +\sum_{i=1}^{4} \left\{ \int_{-1}^{0} \left[k_x \frac{\partial N_i}{\partial x}\frac{\partial y}{\partial s} - k_y \frac{\partial N_i}{\partial y}\frac{\partial x}{\partial s} \right] \Bigg|_{t=0} ds \right\} T_i \tag{19}$$

From the above four heat flows, the remaining heat flows required for conservation can be determined, since

$$Q_{3,2} = -Q_{2,3} \tag{20}$$

or, in general,

$$Q_{i,j} = -Q_{j,i} \qquad i,j = 1,4 \quad i \neq j \tag{21}$$

The heat generation terms are considered next. These terms are relatively straightforward to evaluate, since, if the generation rate per unit volume is known, the control-volume heat generation rate is given simply by the evaluation, for control volume 3, for example, of the integral

$$P_3 = \int_0^1 \int_0^1 p \; |\mathrm{Det}\;[J]| \; ds \; dt \tag{22}$$

Finally, the energy storage terms are considered. Employing the approximation for temperature as given by Eq. (1), the rate of energy storage is given by, for volume 3,

$$\frac{\partial E_3}{\partial t} = \frac{\partial}{\partial t} \sum_{i=1}^{4} \left\{ \int_0^1 \int_0^1 \rho \; c \; N_i \; |\mathrm{Det}\;[J]| \; ds \; dt \right\} T_i \tag{23}$$

For temporally invariant properties, this expression can be rearranged to the form

$$\frac{\partial E_3}{\partial t} = \sum_{i=1}^{4} \left\{ \int_0^1 \int_0^1 \rho \; c \; N_i \; |\mathrm{Det}\;[J]| \; ds \; dt \right\} \dot{T}_i \tag{24}$$

with a suitable difference expression used to approximate $\dot{T}_i$. For a fully implicit approach, Eq. (24) would be approximated by the expression

$$\frac{\partial E_3}{\partial t} = \sum_{i=1}^{4} \left\{ \frac{1}{\Delta t} \int_0^1 \int_0^1 \rho \; c \; N_i \; |\mathrm{Det}\;[J]| \; ds \; dt \right\} \times (T_i^n - T_i^o) \tag{25}$$

where the superscripts n and o denote the current and previous time planes, respectively.

For control volume 3, then, the algebraic equation can now be formed in the form

$$[k_{3,1}^{e,*} \quad k_{3,2}^{e,*} \quad k_{3,3}^{e,*} \quad k_{3,4}^{e,*}] \begin{Bmatrix} T_1 \\ T_2 \\ T_3 \\ T_4 \end{Bmatrix} = [R_3^{e,*}] \tag{26}$$

where, after division of all terms by -1, the terms are

$$k_{3,i}^{e,*} = + \int_0^1 \left[k_x \frac{\partial N_i}{\partial x} \frac{\partial y}{\partial s} - k_y \frac{\partial N_i}{\partial y} \frac{\partial x}{\partial s} \right] \Bigg|_{t=0} ds$$

$$- \int_0^1 \left[k_x \frac{\partial N_i}{\partial x} \frac{\partial y}{\partial t} - k_y \frac{\partial N_i}{\partial y} \frac{\partial x}{\partial t} \right] \Bigg|_{s=0} dt$$

$$+ \frac{1}{\Delta t} \int_0^1 \int_0^1 \rho \; c \; N_i \; |\text{Det} \; [J]| \; ds \; dt \qquad (27)$$

and where

$$R_3^{e,*} = + \int_0^1 \int_0^1 p \; |\text{Det}[J]| \; ds \; dt$$

$$+ \sum_{i=1}^{4} \left\{ \frac{1}{\Delta t} \int_0^1 \int_0^1 \rho \; c \; N_i \; |\text{Det} \; [J]| \; ds \; dt \right\} T_i^0$$

$$+ (Q_{e1,3} + Q_{e2,3}) \qquad (28)$$

In the above, the superscript e denotes that the indicated terms have been evaluated from elemental considerations. Equations similar to Eq. (26) can be written for control volumes 1, 2, and 4. While the details of these additional equations will not be provided here, the reader is reminded that computational savings can be accrued through the realization that $Q_{i,j} = -Q_{j,i}$, $i \neq j$. At this point, therefore, the element stiffness equations can be generated from a control-volume, conservation point of view. It remains to assemble the element stiffness matrices to form the global system of equations and to facilitate the implementation of boundary conditions.

Assembly Procedure and Boundary Condition Application

Considering first the interior of the global computational domain, it is noted that the term $(Q_{e1,3} + Q_{e2,3})$

from the right-hand side of matrix equations, as given by Eq. (28), represents externally applied heat flows into the particular control volume from regions outside of the element itself. In considering interior elements adjacent to the element under consideration, it is realized that similar terms will arise for these elements at their common boundaries. Since the sum of all such heat flows at common control volume boundaries must be supplied externally, the assembly process will be defined to equate their sum to the externally applied heat flows. This interpretation is entirely consistent with that given the surface integrals resulting from a Galerkin weighted residual formulation, and, in most situations of practical interest, results in a net sum of zero being applied. In practice, the implementation results that these terms need not be evaluated at all.

There is an additional implication of the above assembly procedure. With reference to Fig. 4, the implication is that the conservation principle is not applied over element quadrants as implied earlier, but, rather, that conservation is being applied over the "effective" control volume as shown in the figure. The apparent increased control-volume size, from the element quadrant originally examined, however, is analogous to the "domain of influence" of conventional finite-element formulations where contributions arise from all elements which share a common node. The remaining terms in Eqs. (26-28) are formed at global level, in the usual manner, through appropriate summation of elemental contributions.

In considering those elements which lie on a physical boundary of the computational domain, the cancellation of

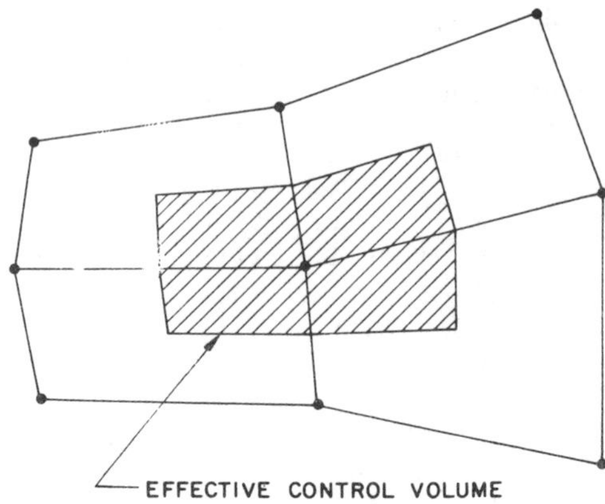

Fig. 4 Effective control volume resulting from assembly procedure.

externally applied heat flows does not occur. However, these terms now provide a mechanism for the implementation of boundary conditions. The boundary condition that will be considered is the convective condition given by

$$k_n \frac{\partial T}{\partial n} = -hT + C \tag{29}$$

applied on the exterior surface of the physical domain. Through appropriate choice of the constants h and C, the Neuman and Dirichlet conditions can be determined as special cases of the more general condition, Eq. (29). Integrating the left-hand side of Eq. (29) over a control-volume surface which lies on the physical boundary of the domain, using $S_{3,1}$ of Fig. 3 as typical, leads to the result that

$$\int_{S_{3,1}} k_n \frac{\partial T}{\partial n} \, dS_{3,1} = + Q_{e1,3} \tag{30}$$

Using this result, together with the integrated form of the right-hand side of Eq. (29), results in

$$Q_{e1,3} = \int_{S_{3,1}} (-hT + C) \, dS_{3,1} \tag{31}$$

Further, using the expression for the temperature distribution given by Eq. (1), Eq. (31) can be expressed in the form

$$Q_{e1,3} = \sum_{i=1}^{4} \left\{ -\int_{S_{3,1}} h \, N_i \, dS_{3,1} \right\} T_i + \int_{S_{3,1}} C \, dS_{3,1} \tag{32}$$

where the shape functions N are evaluated on the surface S_3. Extension of this result to include all boundaries which lie on a physical boundary is direct and will not be presented here. The final form of the equation representing conservation of energy for control volume 3 of Fig. 3 can now be written in the form

$$[k^e_{3,1} \;\; k^e_{3,2} \;\; k^e_{3,3} \;\; k^e_{3,4}] \begin{Bmatrix} T_1 \\ T_2 \\ T_3 \\ T_4 \end{Bmatrix} = [R^e_3] \tag{33}$$

where

$$k^e_{3,i} = + \int_0^1 \left[k_x \frac{\partial N_i}{\partial x} \frac{\partial y}{\partial s} - k_y \frac{\partial N_i}{\partial y} \frac{\partial x}{\partial s} \right] \Bigg|_{t=0} ds$$

$$- \int_0^1 \left[k_x \frac{\partial N_i}{\partial x} \frac{\partial y}{\partial t} - k_y \frac{\partial N_i}{\partial y} \frac{\partial x}{\partial t} \right] \Bigg|_{s=0} dt$$

$$+ \frac{1}{\Delta t} \int_0^1 \int_0^1 \rho \; c \; N_i \; |\text{Det } [J]| \, ds \; dt + \int_{S_3} h \; N_i \; dS_3 \tag{34}$$

$$R^e_3 = + \int_0^1 \int_0^1 p \; |\text{Det } [J]| \; ds \; dt$$

$$+ \sum_{i=1}^{4} \frac{1}{\Delta t} \int_0^1 \int_0^1 \rho \; c \; N_i \; |\text{Det } [J]| \; ds \; dt \; T^\circ_i + \int_{S_3} C \; dS_3 \tag{35}$$

where the surface integrals are evaluated only when the surface coincides with the physical domain boundary, and for all surfaces on the domain boundary.

Expressions analogous to those provided in Eqs. (33-35) can be readily determined for the remaining control-volume sections within the element. Upon forming the complete elemental property or stiffness matrix, the usual assembly rules apply. It is instructive, however, to examine the character of the assembled coefficient matrix and to compare this with that which results from the assembly of linear triangular and linear quadrilateral finite elements when adopting a Galerkin formulation of the equations. This will be done through the use of "numerical molecules" for a uniform mesh, as shown in Fig. 5.

Figure 5a presents the numerical molecule representation for triangular elements within a square network of nodal points. It is noted that the relative weighting, through the coefficients, of the nodal points involved corresponds, in the case of steady-state conduction considered here, to the familiar central difference finite-

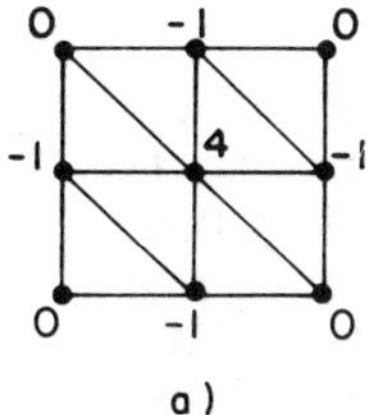

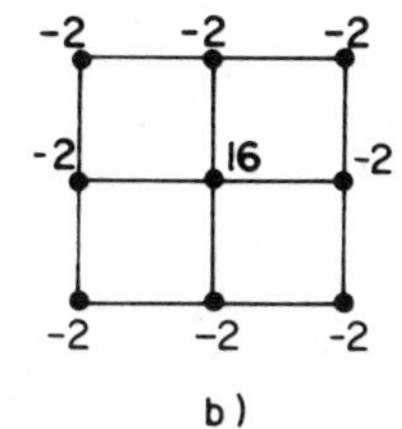

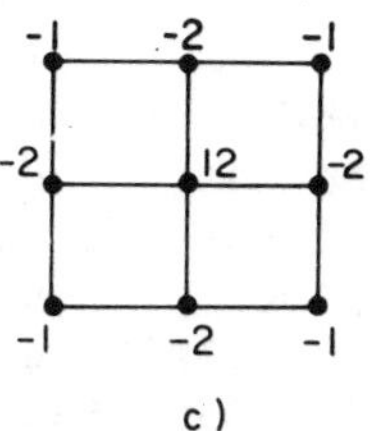

Fig. 5 Numerical molecule sturcture for a) triangular Galerkin elements, b) quadrilateral Galerkin elements, and c) quadrilateral control-volume formulation.

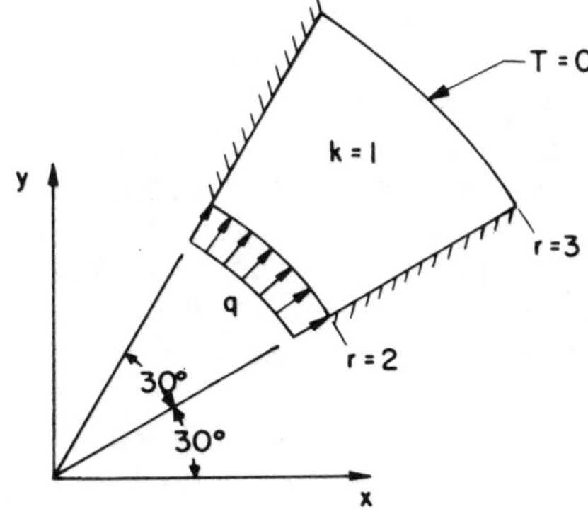

Fig. 6 First test problem configuration.

difference formulation. That is, the scheme is a five-point scheme with values of magnitude 1 and 4 for the coefficients. Conversely, the linear quadrilateral molecule of Fig. 5b results in a nine-point scheme where all nodal points surrounding the central node exert an equal influence on the value at the center of the molecule, irrespective of their differing distances from that point. Finally, the control-volume-based formulation molecule is shown in Fig. 5c. It is seen here that the corner nodal points exert, in this scheme, a reduced influence from the top, bottom, and side nodes in the figure. This is physically consistent with the realization that the corner nodes are located a greater distance from the central node than are the top, bottom, and side nodes. Although the actual magnitudes of these coefficients may not be optimal, the qualitatively correct relative magnitudes of these coefficients may be expected to result in greater accuracy for a specified grid network.

Application to Test Problems

The control-volume-based finite-element formulation for the heat conduction problem was coded and tested to validate the formulation. The initial testing reported in this paper is restricted to steady-state heat conduction within a continuum with zero source strength. The reporting of these test results is considered significant, since the treatment of the diffusion phenomenon is fundamentally different, in the control-volume formulation, from conventional formulations.

Following verification testing of the code through application to various one-dimensional Cartesian configurations, test problems of successively increasing complexity were examined. The first of these is shown in Fig. 6, with boundary conditions as shown in the figure. Although this problem is one-dimensional in the radial, circular cylinder coordinate, the temperature distribution resulting from a Cartesian formulation is fully two-dimensional. Further, while the mesh is not strongly nonorthogonal, the elements used in the discretization are not orthogonal. The results obtained for this problem showed rapid convergence to the exact, analytical solution, thus verifying the applicability and correct operation of the code in a two-dimensional environment.

The second problem examined is the constriction resistance problem shown, with boundary conditions illustrated, in Fig. 7. The analytical solution to this problem has been determined by Schneider et al.[14] and they provide expressions for both the temperature distribution and for the thermal constriction resistance. The numerical results obtained from the control-volume formulation, when applied to this problem, converged rapidly to the analytically determined value for the thermal constriction resistance. A comparison was also made of the temperature distributions

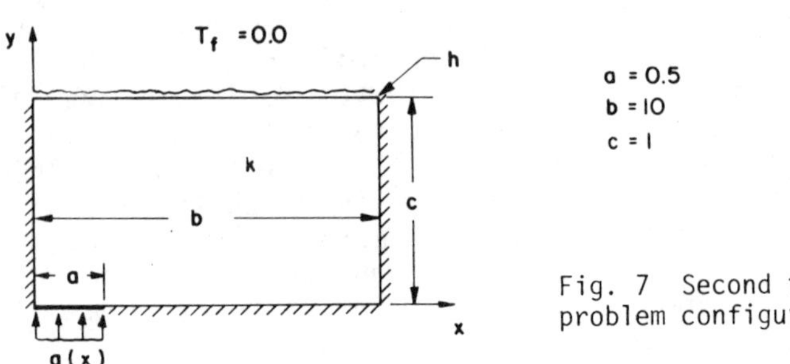

Fig. 7 Second test problem configuration.

obtained from the control-volume formulation and from the Galerkin formulation when compared to the analytically determined temperature field. The comparison was made for an 8x8 element distribution, with the results presented in Fig. 8 for three different, horizontal surfaces. The ordinate of the figure is the percent error of the finite-element solutions.

Figure 8a presents the results for the two formulations for the temperature distribution along the surface defined by y = 0.0. It is seen from the figure that, near the thermally communicating portion of this lower boundary, the errors arising from both methods of formulation are comparable. However, as the lower boundary is traversed towards the outer edge, the difference between the two formulations becomes appreciable. Indeed, at the outer edge, the control-volume formulation results in an error which is 40% less than that resulting from the Galerkin formulation.

Similar results are displayed by Figs. 8b and 8c, where it is again seen that near the leftmost boundary, the errors are comparable; while at the rightmost boundary, the control-volume formulation results, again, in errors which are 40% less than those resulting from the Galerkin formulation.

The final test problem is that shown in Fig. 9, where the boundary conditions are also indicated. The problem is again a thermal constriction problem, but one for which the predominant heat flow direction changes within the domain through 90 deg. The analytical solution for the temperature field for this problem has been determined by the authors in order to make a comparison of formulations. The comparison is presented in Fig. 10 for 12 elements in the x direction and 11 elements in the y direction. The comparison is presented for the three surfaces defined by y = 0, 5, and 10. For both the y = 0 and the y = 5 surfaces, the error of the control-volume formulation is considerably lower than that of the Galerkin formulation. For the y = 10 boundary, the control-volume formulation error is generally less than the Galerkin formulation error except in the vicinity of the flux prescribed portion of the surface. In all cases presented, however, the errors, for this mesh subdivision, are quite small and quite acceptable irrespective of the formulation method employed.

Notwithstanding the above observations, the motivation for development of a control-volume-based finite-element formulation has not been to provide a finite-element formulation of enhanced accuracy, but, rather, to

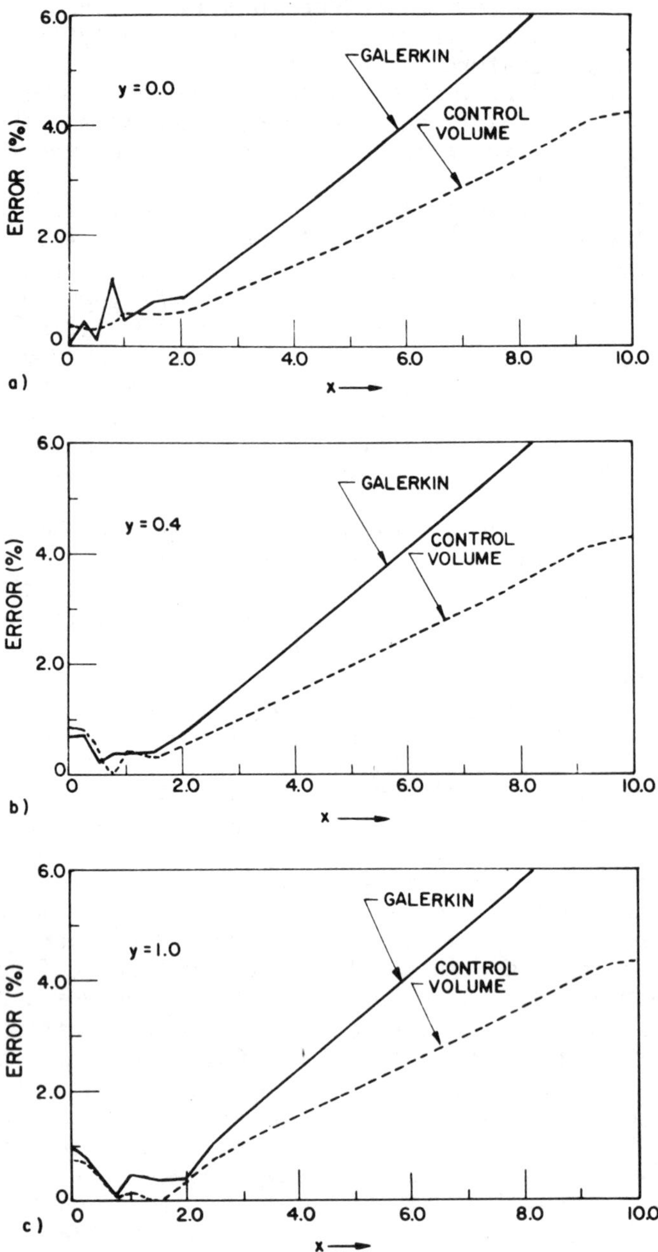

Fig. 8 Comparison of percent error for the Galerkin vs control-volume formulation for second test problem for a) y = 0.0, b) y = 0.4, and c) y = 1.0.

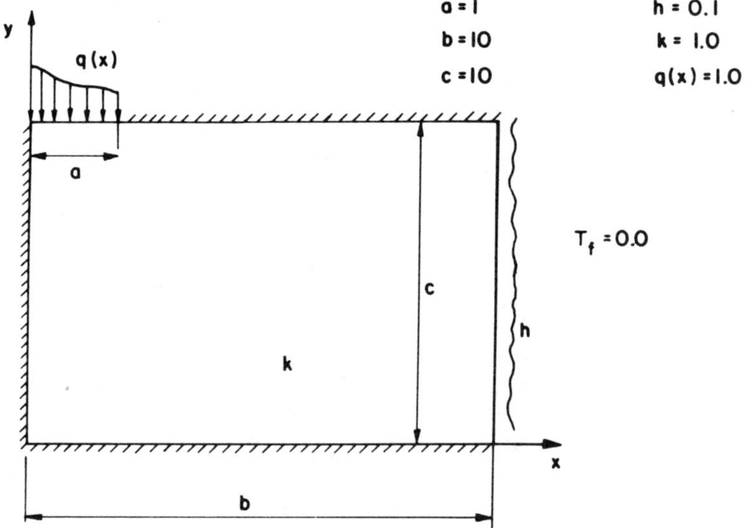

Fig. 9 Third test problem configuration.

provide a mechanism for deriving the finite-element equations through direct application of a conservation principle and to remove some of the obscurity that accompanies the application of error distribution principles or variational principles, where these exist. Indeed, the control-volume approach can be immediately extended to include the influence of fluid flow even though variational principles do not exist for the general flow situation. Thus any benefit derived from the method in the form of enhanced accuracy is an added benefit in addition to accomplishing the primary objective as described above.

Discussion and Conclusions

A control-volume-based finite-element method has been presented for application to the solution of conduction heat-transfer problems. The motivation for developing the method has been to provide a framework for deriving the finite-element equations through stringent application of a conservation principle to control volumes dispersed throughout the computational domain. In this way, the resulting algebraic equations possess a more direct physical interpretation than do those which result from the conventional methods of formulation. This result has been achieved.

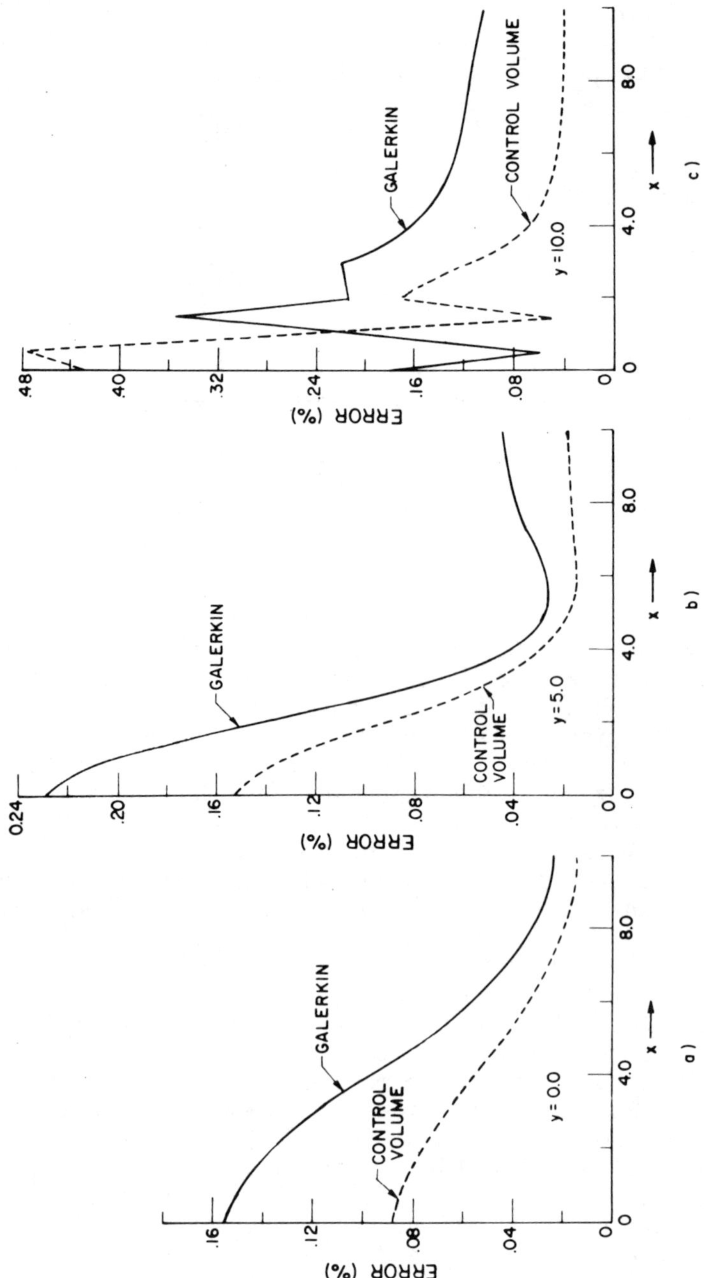

Fig. 10 Comparison of percent error for the Galerkin vs control-volume formulation for third test problem for a) y = 0.0, b) y = 5.0, and c) y = 10.0.

The formulation was presented by applying the conservation balance to control volumes located within individual elements. The method of choosing the boundaries for the control volumes within the element as presented in this paper is not a unique method, but was chosen to ensure compatibility with other elements when the assembly process is implemented. For the specific case considered here of the linear quadrilateral element, the internal coordinate surfaces given by $s = 0$ and $t = 0$ serve this purpose well. It was observed, consistent with conventional formulations, that the actual control volumes for which the conservation principle is applied comprise portions of all elements which share the associated node in their element definition vectors.

The conservation equation, in discrete form, has been derived in this paper for one specific control volume. The procedure was presented in detail for this control, with the remaining control-volume equations being readily obtainable by direct extension of the procedure detailed herein. A method of implementing boundary condition specifications was presented, and it has been shown that the usual assembly procedures apply to the control-volume formulation.

The method was demonstrated by application to several test problems. The initial testing performed in this paper has been restricted to the case of steady, zero source strength, conduction heat transfer. The method performed well on all problems examined and demonstrated convergence to the correct solution. For the final two test problems, where detailed comparisons have been made, the control-volume-based formulation yielded improved accuracy in the temperature field, over the Galerkin formulation, throughout almost the entire solution domain. While the improvement in solution accuracy was not the primary goal of the procedure presented in this paper, it does provide an additional benefit to be gained through adoption of the control-volume approach. More extensive testing, however, is required to determine the extent of the universality of this benefit.

An additional benefit, although not explicitly addressed in this paper, is the potential for reduced computational requirements in the element stiffness determination/assembly phase of code operation. The potential benefit arises particularly for the steady-state case where numerical integration is required only over the control-volume surfaces rather than over the element volume. The extent of this saving depends, to a large extent, on the quadrature order required by the particular problem. It is the authors' experience, however, that computational savings in the element stiffness generation phase of operation are in excess of 25%.

In summary, then, a control-volume-based formulational framework has been provided in this paper to permit direct application of governing conservation principles in the determination of the algebraic finite-element equations. The procedure presented herein can readily be extended to other, higher-order finite-element configurations. Aside from any computational advantages which accrue from this procedure, the procedure possesses the benefit of permitting a more direct physical interpretation of the resulting, algebraic, finite-element equation system.

Acknowledgments

The authors express their sincere thanks to the Natural Sciences and Engineering Research Council of Canada for their financial support of this project in the form of an operating grant to G. E. Schneider.

References

[1]Richardson, L. F., "The Approximate Arithmetical Solution by Finite Differences of Physical Problems Involving Differential Equations, with an Application to the Stresses in a Masonry Dam," Transactions of the Royal Society of London, Series A, Vol. 120, 1910, pp. 307-357.

[2]Schneider, G. E., Strong, A. B., and Yovanovich, M. M., "Finite Difference Modelling of the Heat Conduction Equation in General Orthogonal Curvilinear Coordinates Using Taylor Series Expansion," Proceedings of the AICA International Symposium on Computer Methods for Partial Differential Equations, edited by R. Vichnevetsky, AICA,1975, pp. 312-317.

[3]Robertson, S. R., "A Finite Difference Formulation of the Equation of Heat Conduction in Generalized Coordinates," Numerical Heat Transfer, Vol. 2, No. 1, 1979, pp. 61-80.

[4]Patankar, S. V., Numerical Heat Transfer and Fluid Flow, McGraw-Hill Book Co., New York, 1980.

[5]Schneider, G. E., Strong, A. B., and Yovanovich, M. M., "A Physical Approach to the Finite Difference Solution of the Conduction Equation in Orthogonal Curvilinear Coordinates," ASME Paper 75-WA/HT-94, ASME Winter Annual Meeting, Houston, Texas, Nov. 1975.

[6]Zedan, M. and Schneider, G. E., "A Physical Approach to the Finite Difference Solution of the Conduction Equation in Generalized Coordinates," Numerical Heat Transfer, Vol. 5, No. 1, 1982, pp. 1-20.

[7]Thompson, J. F., Thames, F. C., and Mastin, C. W., "Automatic Numerical Generation of Body-Fitted Curvilinear Coordinate System for Field Containing Any Number of Arbitrary Two-Dimensional Bodies," Journal of Computational Physics, Vol. 15, July 1974, pp. 299-319.

[8]Zrenkicwicz, O. C. and Parekh, C. J., "Transient Field Problems: Two-Dimensional and Three-Dimensional Analysis by Isoparametric Finite Elements," International Journal for Numerical Methods in Engineering, Vol. 2, No. 1, 1970, pp. 61-71.

[9]Zienkiewicz, O. C., Finite Elements in Engineering Science, McGraw-Hill Book Co., New York, 1971.

[10]Brot, M. A., "Variational Principles in Irreversible Thermodynamics with Application to Viscoelasticity," Physical Review, Vol. 97, No. 6, 1955, pp. 1463-1469.

[11]Finlayson, B. A. and Scriven, L. E., "On the Search for Variational Principles," International Journal for Heat and Mass Transfer, Vol. 10, No. 5, 1967, pp. 799-821.

[12]Baliga, B. R. and Patankar, S. V., "A New Finite Element Formulation for Convection-Diffusion Problems," Numerical Heat Transfer, Vol. 3, No. 4, 1980, pp. 393-409.

[13]Huebner, K. H., The Finite Element Method for Engineers, Wiley, New York, 1975.

[14]Schneider, G. E., Yovanovich, M. M., and Cane, R. L. D., "Thermal Constriction Resistance of a Convectively Cooled Plate with Non-Uniform Flux over its Opposite Face," Journal of Spacecraft and Rockets, Vol. 17, No. 4, 1980, pp. 372-376.

Finite-Element Analysis of Planar Conductive and Radiative Transfer with Flux Boundary

R. Fernandes* and J. Francis†
University of Oklahoma, Norman, Okla.

Abstract

The analysis of transient combined conduction and radiation in a planar medium with temperature boundary conditions has been solved by numerous investigators. In this study we address the problem of a heat flux imposed on one boundary and a temperature prescribed on the other. The Galerkin finite-element method represents a unique procedure for solving this problem. The medium is considered to be an absorbing, emitting, and isotropically scattering infinite slab with diffuse boundaries. Results are presented for temperature and radiative flux profiles, and compared to others wherever possible. The agreement is excellent.

Nomenclature

$E_n(\tau)$ = $\int_0^1 \mu^{n-2} \exp(-\tau/\mu)\, d\mu$

G = incident radiation or integrated intensity
$I^+(\tau,\mu)$ = radiant intensity in the positive μ direction
$I^-(\tau,\mu)$ = radiant intensity in the negative μ direction
$I_b(T)$ = blackbody intensity
J = radiosity
k = thermal conductivity
N = dimensionless parameter = $k\beta/4n^2\sigma T_{ref}^3$

Presented as Paper 82-0910 at the 3rd AIAA/ASME Joint Thermophysics, Fluids, Plasma & Heat Transfer Conference, St. Louis, Mo., June 7-11, 1982.

*Presently Professor, Mechanical Engineering Department, Guru Nanak Dev Engineering College, Bidar, Karnataka, India.

†Professor, School of Aerospace, Mechanical and Nuclear Engineering.

n = index of refraction
Q = dimensionless radiative heat flux = $q/k\beta T_{ref}$
q = radiative heat flux in the x direction
q'' = total (conductive and radiative) heat flux
T = absolute temperature
β = extinction coefficient = $\kappa + \sigma$
ε = emissivity of a surface
η = dimensionless incident radiation = $G/n^2\sigma T_{ref}{}^4$
θ = dimensionless temperature = T/T_{ref}
κ = absorption coefficient
μ = cosine of polar angle
σ = scattering coefficient, also Stefan-Boltzmann constant
τ = optical depth of radiating material = $\int_0^x \beta dx$
τ' = dummy integration variable
τ_o = optical thickness of radiating material
ϕ = azimuthal angle
χ = dimensionless radiosity = $J/n^2\sigma T_{ref}{}^4$
Ψ = dimensionless heat flux $q''/k\beta T_{ref}$
ω_o = albedo for single scattering = σ/β

Subscripts

1,2 = walls 1 and 2, respectively

Introduction

The solution to combined conductive and radiative heat-transfer problems of planar geometries has been the subject of numerous investigations. Various approaches have been followed which usually involve a simplification of the radiative contribution. These simplifications take various forms, with one of the most common being the Rosseland model. The solution of the transient problem with an isotropically scattering medium and prescribed temperature boundary conditions using finite elements has been presented previously by the authors.[1] The finite-element methodology as applied to combined conduction and radiation problems has also been presented by Wu et al.[11] Other authors[2-10] have presented solutions for the planar problem with temperature boundary conditions. However, the case with a heat flux prescribed on one boundary has not been presented. While the finite-element procedure remains essentially the same, the formulation is quite different. This paper presents the transient solution for combined conductive and radiative heat transfer in an absorbing, emitting, and scattering planar medium bounded by diffuse surfaces and subject to a prescribed heat flux on one boundary and a prescribed temperature on the other.

Analysis

For plane parallel geometry, the governing energy equation is given by

$$\rho c_\rho \frac{\partial T}{\partial t} + \frac{\partial}{\partial x}\left[-k\frac{\partial T}{\partial x} + q\right] = 0 \qquad (1)$$

where q is the radiative flux given by

$$q(\tau) = 2\pi\int_0^1 [I^+(\tau,\mu) - I^-(\tau,\mu)]\ \mu d\mu \qquad (2)$$

The intensities I^+ and I^- are in the positive and negative directions, respectively (see Fig. 1). These are found from the solution to the transient equations

$$\frac{dI^+}{d\tau}(\tau,\mu) + \frac{1}{\mu} I^+(\tau,\mu) = \frac{\kappa n^2}{\mu\beta} I_b(T) + \frac{\sigma}{4\pi\beta\mu} G(\tau) \qquad (3)$$

and

$$-\frac{dI^-}{d\tau}(\tau,\mu) + \frac{1}{\mu} I^-(\tau,\mu) = \frac{\kappa n^2}{\mu\beta} I_b(T) + \frac{\sigma}{4\pi\beta\mu} G(\tau) \qquad (4)$$

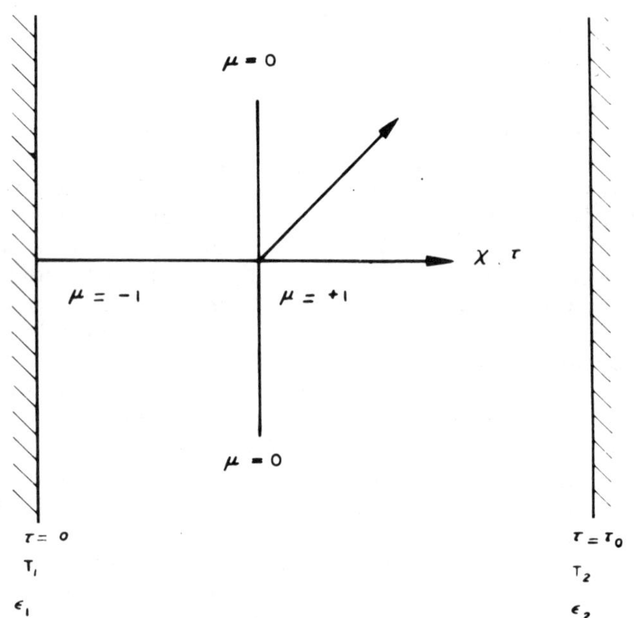

Fig. 1 Coordinate geometry.

where the irradiation $G(\tau)$ is given by

$$G(\tau) = 2\pi\int_0^1 [I^-(\tau,\mu') + I^+(\tau,\mu')]\, d\mu' \tag{5}$$

In this formulation the properties are considered to be constant and the scattering is treated as isotropic. The radiative boundary conditions are for diffusely reflecting surfaces, and the total heat flux is prescribed on one surface while the temperature is known on the other surface.

After nondimensionalizing, the following equations are obtained:

$$\frac{\partial\theta}{\partial t^*} - \frac{\partial^2\theta}{\partial\tau^2} + \frac{\partial Q}{\partial\tau} = 0 \tag{6}$$

where $t^* = \dfrac{\kappa\beta^2 t}{\rho c_\rho}$

$$\eta(\tau) = 2\Big[\chi_1 E_2(\tau) + \chi_2 E_2(\tau_0-\tau) + \int_0^{\tau_0} \Big\{(1-\omega_0)\,\theta^4(\tau') + \frac{\omega_0}{4}\,\eta(\tau')\Big\}E_1(\tau-\tau'1)d\tau'\Big] \tag{7}$$

$$4NQ(\tau) = 2\Big[\chi_1 E_3(\tau) - \chi_2 E_3(\tau_0-\tau) + \int_0^{\tau} \Big\{(1-\omega_0)\,\theta^4(\tau') + \frac{\omega_0}{4}\,\eta(\tau')\Big\}E_2(\tau-\tau')d\tau' - \int_\tau^{\tau_0}[(1-\omega_0)\,\theta^4(\tau') + \frac{\omega_0}{4}\,\eta(\tau')]\,E_2(\tau'-\tau)d\tau'\Big] \tag{8}$$

where

$$\chi_1=\varepsilon_1\theta_1{}^4+2(1-\varepsilon_1)\chi_2E_3(\tau_0) + 2(1-\varepsilon_1)\Big[\int_0^{\tau_0}\Big\{(1-\omega_0)\theta^4(\tau') + \frac{\omega_0}{4}\,\eta(\tau')\Big\}E_2(\tau')d\tau'\Big] \tag{9}$$

and

$$\chi_2=\varepsilon_2\theta_2{}^4 + 2(1-\varepsilon_2)\chi_1E_3(\tau_0) + 2(1-\varepsilon_2)\Big[\int_0^{\tau_0}\Big\{(1-\omega_0)\theta^4(\tau')+ \frac{\omega_0}{4}\,\eta(\tau')\Big\}E_2(\tau_0-\tau')d\tau'\Big] \tag{10}$$

The total heat transfer is given by $q'' = -k\, dT/dx + q$, which is nondimensionalized as

$$\Psi = -\frac{d\theta}{d\tau} + Q \tag{11}$$

The finite-element procedure is to divide the domain 0 to τ_0 into a finite number of elements. The unknown functions $\theta(\tau)$, $\eta(\tau)$, $Q(\tau)$ are approximated over each element

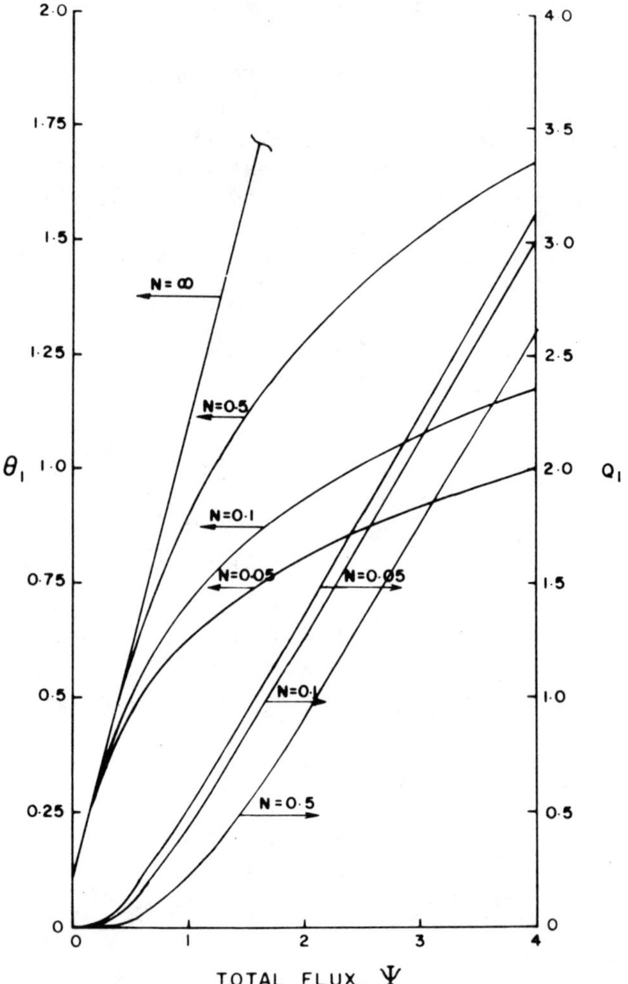

Fig. 2 Steady state for $\omega_o = 0$, $\varepsilon_1 = \varepsilon_2 = 1$, $\tau_o = 1$, and $\theta_2 = 0.1$.

in the form

$$F_e(\tau) = \sum_{i=1}^{2} \psi_e^{(i)}(\tau)\, F_e^{(i)} \tag{12}$$

where $F(\tau)$ stands for any one of $\theta(\tau)$, $\eta(\tau)$, or $Q(\tau)$. The functions $\psi_e^{(i)}(\tau)$ are the local interpolating functions given by

$$\psi_e^{i}(\tau) = \frac{\tau_e^{i+1} - \tau}{\tau_e^{i+1} - \tau_e^{i}} \tag{13a}$$

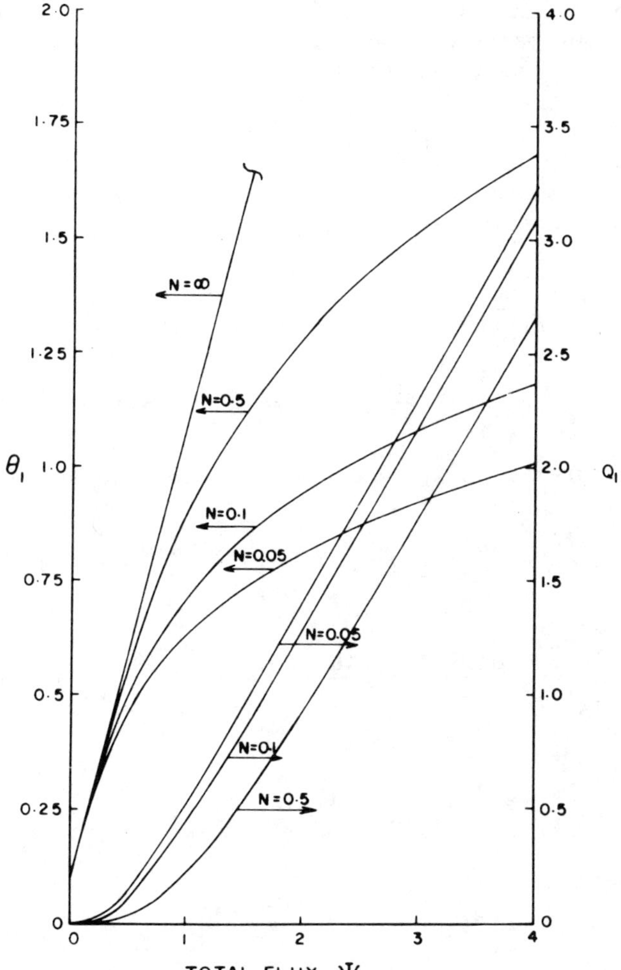

Fig. 3 Steady state for $\omega_o = 0.5$, $\varepsilon_1 = \varepsilon_2 = 1$, $\tau_o = 1$, and $\theta_2 = 0.1$.

$$\psi_e^{i+1}(\tau) = \frac{\tau - \tau_e^{i}}{\tau_e^{i+1} - \tau_e^{i}} \tag{13b}$$

where the superscripts are node indices and the subscripts are element indices.

The integral terms are represented by a summation of integrals over the individual elements in the domain. Substitution of the assumed form of the unknown functions into the governing equations, multiplication by the shape function, and integrating over the individual elements constitute the Galerkin formulation. This procedure involves integration by parts of the terms involving second derivatives in the energy equation as shown below:

$$\frac{\partial\theta}{\partial t^*} - \frac{\partial^2\theta}{\partial\tau^2} + \frac{\partial Q}{\partial\tau} = 0$$

multiplying by the interpolation function and integrating over an element we have

$$\int_{\tau_e^{(1)}}^{\tau_e^{(2)}} \frac{\partial\theta}{\partial t^*}\,\psi_e^{(j)}\,d\tau - \int_{\tau_e^{(1)}}^{\tau_e^{(2)}} \frac{\partial^2\theta}{\partial\tau^2}\,\psi_e^{(j)} d\tau + \int_{\tau_e^{(1)}}^{\tau_e^{(2)}} \frac{\partial Q}{\partial\tau}\,\psi_e^{(j)}\,d\tau = 0 \quad j=1,2 \qquad (14)$$

Using integration by parts on the second and third terms we obtain

$$\int_{\tau_e^{(1)}}^{\tau_e^{(2)}} \frac{\partial\theta}{\partial t^*}\,\psi_e^{(j)}\,d\tau + \int_{\tau_e^{(1)}}^{\tau_e^{(2)}} \frac{\partial\theta}{\partial\tau}\,\frac{d\psi_e^{(j)}}{d\tau}\,d\tau - \int_{\tau_e^{(1)}}^{\tau_e^{(2)}} Q\,\frac{d\psi_e^{(j)}}{d\tau} = \left(\frac{\partial\theta}{\partial\tau} - Q\right)\psi_e^{(j)}\Big|_{\tau_e^{(1)}}^{\tau_e^{(2)}} \quad j = 1,2 \qquad (15)$$

From Eq. (13a),

$$\psi_e^{(1)}\Big|_{\tau_e^{(1)}} = 1, \quad \psi_e^{(1)}\Big|_{\tau_e^{(2)}} = 0$$

and from Eq. (13b),

$$\psi_e^{(2)}\Big|_{\tau_e^{(1)}} = 0, \quad \psi_e^{(2)}\Big|_{\tau_e^{(2)}} = 1$$

For j=1, the right-hand side of Eq. (15) becomes $\left(-\,\partial\theta/\partial\tau + Q\right)\Big|_{\tau_e^{(1)}}$. This expression denotes the combined flux (radiative and conductive) at node 1 of element e. For the first element it represents the flux boundary condition. Upon adding the equations obtained from Eq. (15) for j = 1 and 2 at each element in the manner described by Eq. (1), the combined flux at the interior nodes cancels out. This leaves only total flux at the first and last nodes along with net radiative flux and temperatures at all nodes as unknowns. The total flux at the first node and the temperature at the last node are given as boundary conditions. The transient term was handled using the Crank-Nicolson scheme.

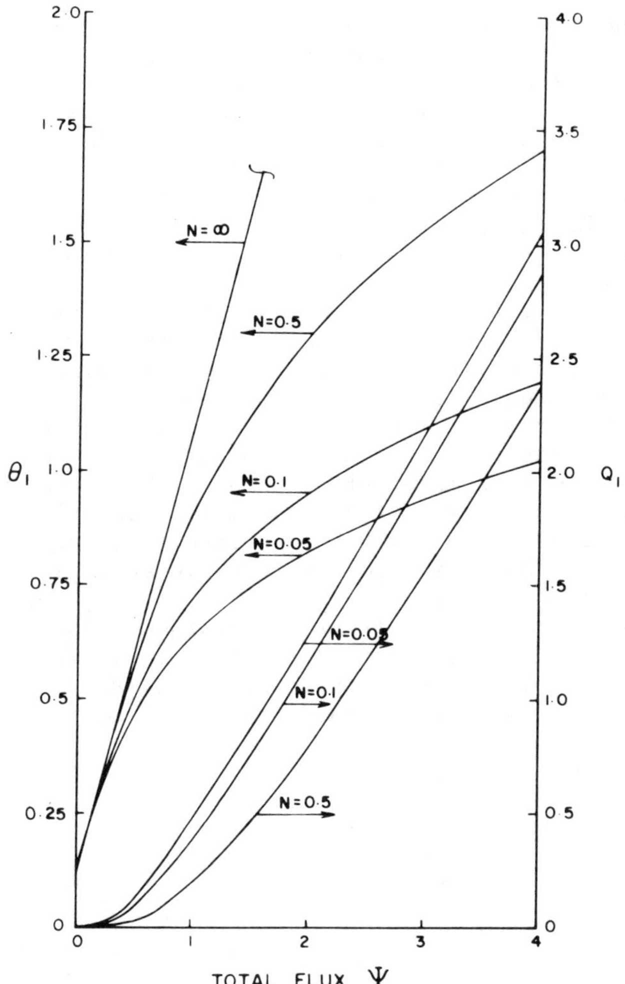

Fig. 4 Steady state for $\omega_o = 1$, $\varepsilon_1 = \varepsilon_2 = 1$, $\tau_o = 1$, and $\theta_2 = 0.1$.

A four-point quadrature was employed to determine the integrals. In all the results presented 22 nodes were used. The first three nodes were an optical distance of 0.025 apart, while the remaining nodes were 0.05 apart. The time step was 0.005.

Results

The steady-state results are presented in both tabular and graphical form. Wherever possible our results were compared with those of others through the inverse problem.

By the inverse problem we mean a similar problem except the formulation is for prescribed temperatures at the boundaries and the flux is calculated from the solution. One of the results of the prescribed temperature boundary problem, i.e., the inverse problem, is the total (conductive and radiative) flux at the first node (i.e., $\tau = 0$). In this paper we impose the total flux at the first node and prescribe a temperature at the last node, i.e., $\tau = \tau_o$. Thus we can compare the results of this paper with those of the inverse problem.

For black surfaces our results were compared with Lii and Ozisik,[9] Crosbie and Viskanta,[5] and Viskanta.[2] The agreement was excellent. However, for low-emissivity bounding surfaces, small albedo ω_o, and small N, the results of this paper did not compare favorably with those of Viskanta.[2] Differences of up to 15% occurred for the temperature slopes. It is interesting to note that these differences were also observed for the combination of low emissivity, low albedo, and low N in the results of Ref. 1, where θ^4 was treated as linear across each element. In Ref. 1, for the case $\omega_o = 0$, $\varepsilon_1 = \varepsilon_2 = 0.1$, $\tau_o = 1$, $\theta_1 = 1$, and $\theta_2 = 0.5$, the following results were obtained: for $d\theta/d\tau|_{\tau=0}$, -1.0252 (Ref. 2), -0.873 (Ref. 1); Q_1, 0.1247 (Ref. 2), 0.13 (Ref. 1); ψ_1, 1.1499 (Ref. 2), 1.003 (Ref. 1). However, when setting the combined flux equal to 1.15 and θ_2 =

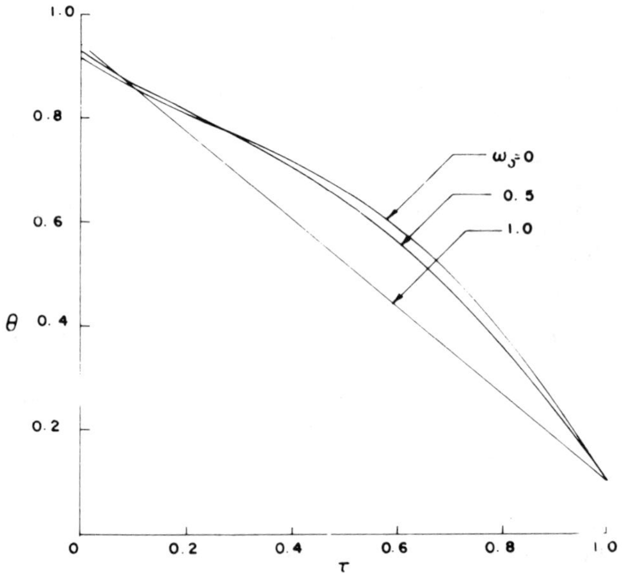

Fig. 5 Effect of scattering on the steady-state temperature profile for $\varepsilon_1 = \varepsilon_2 = 1$, N = 0.05, flux = 3, $\theta_2 = 0.1$, and $\tau_o = 1$.

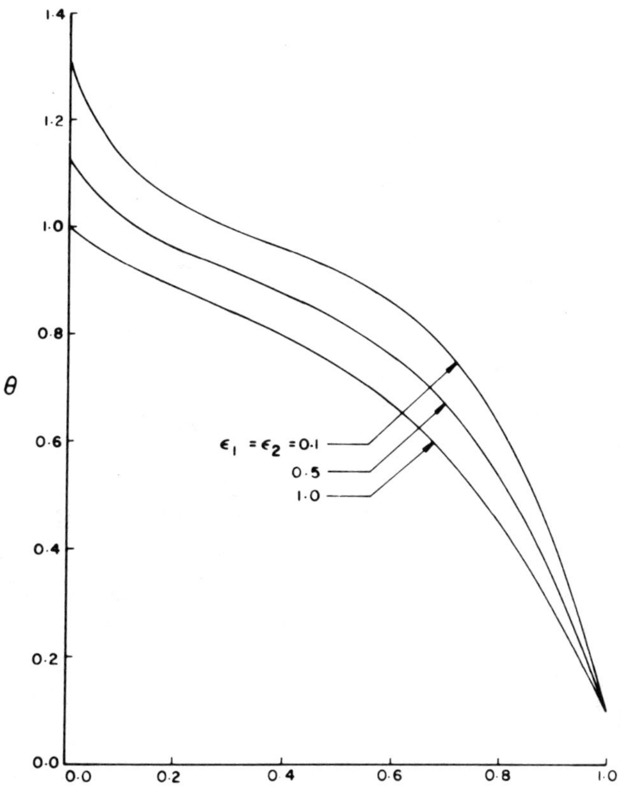

Fig. 6 Effect of surface emissivity on the steady-state temperature distribution for $\omega_o = 0$, $N = 0.05$, $\tau_o = 1$, $\theta_2 = 0.1$, and flux = 4.

0.5 in this study, we obtained $\theta_1 = 1.045$ and $Q_1 = 0.1616$, and when the combined flux was set equal to 1.003, we obtained $\theta_1 = 0.9987$ and $Q_1 = 0.1318$. This demonstrates the consistency for the two problems solved by the finite-element method even though they are formulated in a different manner. This gives us confidence in our solution and we would recommend using it. In Tables 1-3, results are presented for albedo ω_o of 0 and 0.5 with various combinations of total (applied) flux, boundary surface emissivity, and conduction radiation parameter N.

These results are for steady-state conditions, with the solutions presented for various locations within the material. While completing these various cases it was determined that the larger the N or the lower the incident flux, the longer it took to achieve steady state. These

Table 1 Steady-state results: $\omega_o = 0$, $\theta_2 = 0.1$, $\tau_o = 1$

Total flux		Temperature θ at				Radiative flux Q at		
at $\tau = 0$	$\varepsilon_1 = \varepsilon_2$	$\tau = 0$	$\tau = 0.25$	$\tau = 0.5$	$\tau = 0.75$	$\tau = 0$	$\tau = 0.5$	$\tau = 1$
				(a) N = 0.5				
3.0	0.5	1.665	1.344	1.050	0.650	1.269	1.666	0.496
3.0	0.1	1.860	1.446	1.137	0.715	0.414	1.623	0.111
4.0	0.5	1.860	1.516	1.221	0.777	1.949	2.614	0.800
4.0	0.1	2.105	1.637	1.333	0.873	0.678	2.611	0.187
				(b) N = 0.1				
3.0	0.5	1.195	0.986	0.825	0.560	1.618	2.228	0.720
3.0	0.1	1.366	1.067	0.907	0.638	0.595	2.258	0.176
4.0	0.5	1.315	1.091	0.938	0.659	2.344	3.249	1.096
4.0	0.1	1.520	1.184	1.038	0.764	0.916	3.319	0.279
				(c) N = 0.05				
3.0	0.5	1.027	0.853	0.734	0.521	1.737	2.418	0.820
3.0	0.1	1.185	0.926	0.811	0.600	0.671	2.467	0.208
4.0	0.5	1.126	0.939	0.827	0.608	2.480	3.448	1.227
4.0	0.1	1.314	1.022	0.920	0.713	1.018	3.528	0.325

Table 2 Steady-state results: $\omega_o = 0.5$, $\theta_2 = 0.1$, $\tau_o = 1$, $\varepsilon_1 = \varepsilon_2 = 0.5$

Total flux at $\tau = 0$	Temperature θ at				Radiative flux Q at		
	$\tau = 0$	$\tau = 0.25$	$\tau = 0.5$	$\tau = 0.75$	$\tau = 0$	$\tau = 0.5$	$\tau = 1$
(a) N = 0.5							
1.0	0.961	0.753	0.546	0.329	0.143	0.153	0.064
2.0	1.425	1.129	0.835	0.495	0.687	0.747	0.314
3.0	1.706	1.371	1.043	0.629	1.394	1.552	0.661
4.0	1.909	1.551	1.211	0.745	2.158	2.452	1.060
(b) N = 0.1							
1.0	0.786	0.630	0.475	0.300	0.314	0.349	0.149
2.0	1.059	0.863	0.677	0.431	1.019	1.164	0.507
3.0	1.226	1.012	0.821	0.536	1.799	2.106	0.939
4.0	1.350	1.123	0.936	0.627	2.604	3.097	1.414
(c) N = 0.05							
1.0	0.701	0.568	0.439	0.286	0.393	0.445	0.193
2.0	0.919	0.757	0.610	0.403	1.135	1.326	0.590
3.0	1.054	0.878	0.732	0.497	1.932	2.303	1.056
4.0	1.155	0.968	0.828	0.578	2.748	3.313	1.563

Table 3 Steady-state results: $\omega_o = 0.5$, $\theta_2 = 0.1$, $\tau_o = 1$, $\varepsilon_1 = 0.1$, $\varepsilon_2 = 0.5$

Total flux	Temperature θ at				Radiative flux Q at		
at $\tau = 0$	$\tau = 0$	$\tau = 0.25$	$\tau = 0.5$	$\tau = 0.75$	$\tau = 0$	$\tau = 0.5$	$\tau = 1$
(a) N = 0.5							
1.0	1.019	0.792	0.570	0.341	0.038	0.103	0.011
2.0	1.607	1.244	0.918	0.545	0.237	0.631	0.071
3.0	1.978	1.537	1.178	0.721	0.543	1.439	0.166
4.0	2.248	1.751	1.388	0.879	0.901	2.380	0.282
(b) N = 0.1							
1.0	0.877	0.688	0.517	0.324	0.104	0.286	0.033
2.0	1.238	0.971	0.768	0.497	0.411	1.110	0.132
3.0	1.461	1.146	0.950	0.644	0.794	2.104	0.265
4.0	1.628	1.276	1.094	0.774	1.218	3.153	0.420
(c) N = 0.05							
1.0	0.797	0.629	0.485	0.314	0.140	0.390	0.046
2.0	1.086	0.855	0.700	0.474	0.484	1.304	0.162
3.0	1.268	0.997	0.854	0.608	0.894	2.338	0.312
4.0	1.406	1.103	0.974	0.725	1.345	3.401	0.486

results are consistent, since radiation heat transfer is an instantaneous phenomenon and large N implies conduction dominated. For lower fluxes the radiation fluxes are a smaller fraction of the total.

Figures 2-4 present the effect of albedo and N on the steady-state results for black bounding surfaces. Both the temperature and net radiative flux at the first node are plotted for various total fluxes imposed at the boundary. It is seen that the effect of scattering on temperature and radiative flux at the first node is negligible. However, the effect of varying the parameter N is substantial, ranging from the linear solution if pure conduction ($N = \infty$) to the radiation dominated case of $N = 0.05$.

Figure 5 shows the effect of scattering on the temperature profile throughout the medium of a radiation dominated case. These results are for a dimensionless flux equal to 3 and an imposed temperature of 0.1 at the other surface.

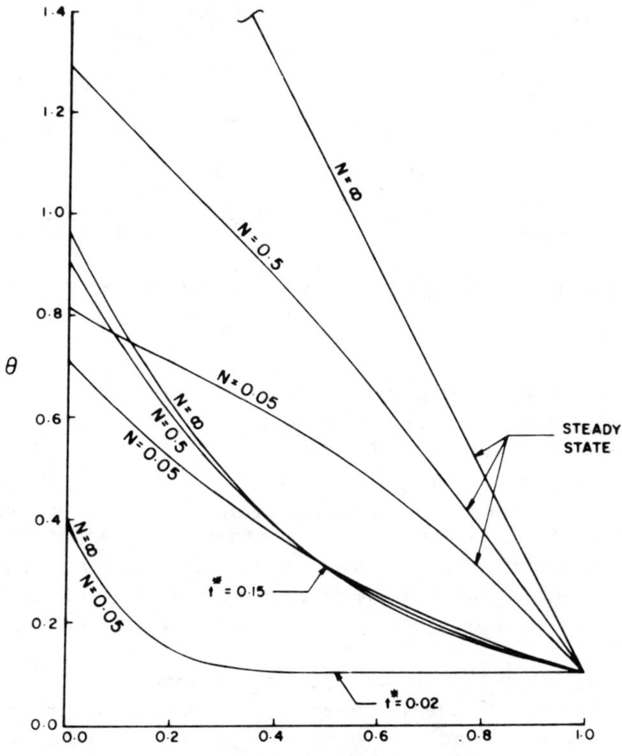

Fig. 7 Effect of N on the transient temperature distribution for $\omega_o = 0.5$, $\varepsilon_1 = \varepsilon_2 = 1$, $\theta_{initial} = \theta_2 = 0.1$, $\tau_o = 1$, and flux = 2.

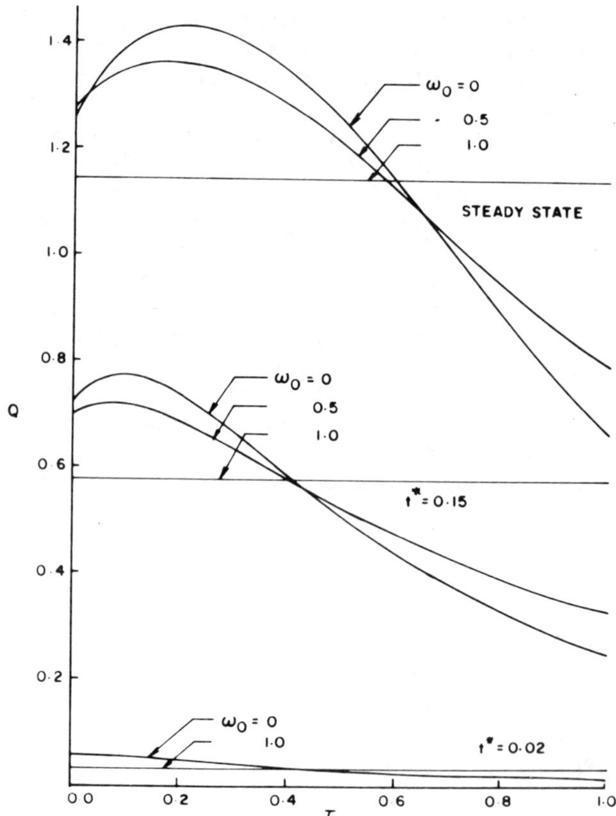

Fig. 8 Effect of scattering on the transient net radiative flux for $\varepsilon_1 = \varepsilon_2 = 1$, $\tau_o = 1$, flux = 2, and N = 0.1.

The boundaries are again black and the parameter N = 0.05. As expected, the results for albedo $\omega_o = 1$ or pure scattering are linear.

Figure 6 shows the effect of surface emissivity on the steady-state temperature distribution. It is seen that reducing the surface emissivity increases the temperature and slope at the first node. This is because the net radiative flux is reduced at the surface and thus more energy is transferred by conduction, thereby increasing the temperature.

Figures 7 and 8 are transient solutions. Figure 7 shows the effect of the parameter N on temperature distribution, and Fig. 8 shows the effect scattering has on the radiative flux distribution.

Conclusions

The finite-element solution is shown to give excellent results and is particularly well suited to problems of this nature because the formulation yields the flux and temperature directly as unknowns in the formulation.

The effect of conduction-radiation parameter N, surface emissivity, albedo ω_o, and incident flux are all considered. Solutions were compared with existing solutions with excellent agreement.

The problem addressed has numerous applications including the guarded hot plate thermal conductivity apparatus. However, the most important aspect of the solution is as a means of addressing other problems. It is a viable approach to numerous problems encountered in the field, particularly in the analysis of insulating materials.

References

[1]Fernandes, R., Francis, J., and Reddy, J. N., "A Finite Element Approach to Combined Conductive and Radiative Heat Transfer in a Planar Medium," Heat Transfer and Thermal Control: AIAA Progress in Astronautics and Aeronautics, Vol. 78, edited by A. L. Crosbie, AIAA, New York, 1981, pp. 92-109.

[2]Viskanta, R., "Heat Transfer by Conduction and Radiation in Absorbing and Scattering Materials," Journal of Heat Transfer, Vol. 87, Feb. 1965, pp. 143-150.

[3]Amlin, D. W. and Korpela, S. A., "Influence of Thermal Radiation on the Temperature Distribution in a Semi-Transparent Solid," Journal of Heat Transfer, Vol. 101, Feb. 1979, pp. 76-80.

[4]Viskanta, R. and Grosh, R. J., "Heat Transfer by Simultaneous Conduction and Radiation in an Absorbing Medium," Journal of Heat Transfer, Vol. 84, Feb. 1982, pp. 63-72.-72.

[5]Crosbie, A. L. and Viskanta, R., "Interaction of Heat Transfer by Conduction and Radiation in a Nongray Planar Medium," Warme-und Stoffubertragung, Vol. 4, 1971, pp. 205-212.

[6]Weston, K. C. and Hauth, J. L., "Unsteady, Combined Radiation and Conduction in an Absorbing, Scattering, and Emitting Medium," Journal of Heat Transfer, Vol. 95, Aug. 1973, pp. 357-364.

[7]Viskanta, R. and Grosh, R. J., "Heat Transfer by Simultaneous Conduction and Radiation," International Journal of Heat Mass Transfer, Vol. 5, Aug. 1962, pp. 729-734.

[8]Berganam, J. B. and Seban, R. A., "Heat Transfer by Conduction and Radiation in Absorbing and Scattering Materials," Journal of Heat Transfer, Vol. 93, May 1971, pp. 236-238.

[9]Lii, C. C. and Ozisik, M. N., "Transient Radiation and Conduction in an Absorbing, Emitting, Scattering Slab with Reflective Boundaries," International Journal of Heat Mass Transfer, Vol. 15, No. 5, May 1972, pp. 1175-1179.

[10]Viskanta, R. and Grosh, R. J., "Heat Transfer in a Thermal Radiation Absorbing and Scattering Medium," International Developments in Heat Transfer, Part IV, ASME, New York, 1961, pp. 820-828.

[11]Wu, S. T., Fergerson, R. E., and Altgilbers, L. L., "Application of Finite-Element Techniques to the Interaction of Conduction and Radiation in a Participating Medium," Heat Transfer and Thermal Control: AIAA Progress in Astronautics and Aeronautics, Vol. 78, edited by A. L. Crosbie, AIAA, New York, 1981, pp. 61-91.

Author Index for Volume 86